Cavitation and Bubble Dynamics

Cavitation and Bubble Dynamics

Guest Editor

Rui Han

MDPI

Basel • Beijing • Wuhan • Barcelona • Belgrade • Novi Sad • Cluj • Manchester

Guest Editor
Rui Han
Heilongjiang Province Key
Laboratory of Nuclear Power
System & Equipment
Harbin Engineering University
Harbin
China

Editorial Office
MDPI AG
Grossspeteranlage 5
4052 Basel, Switzerland

This is a reprint of the Special Issue, published open access by the journal *Fluids* (ISSN 2311-5521), freely accessible at: https://www.mdpi.com/journal/fluids/special_issues/ETQLS218U1.

For citation purposes, cite each article independently as indicated on the article page online and as indicated below:

Lastname, A.A.; Lastname, B.B. Article Title. *Journal Name* **Year**, *Volume Number*, Page Range.

ISBN 978-3-7258-8017-1 (Hbk)
ISBN 978-3-7258-8018-8 (PDF)
https://doi.org/10.3390/books978-3-7258-8018-8

Contents

fluids

Review

Non-Spherical Cavitation Bubbles: A Review

Boxin Jia * and Hitoshi Soyama

Department of Finemechanics, Tohoku University, Sendai 980-8579, Japan; soyama@mm.mech.tohoku.ac.jp
* Correspondence: jia.boxin.q3@dc.tohoku.ac.jp; Tel.: +81-22-795-6899; Fax: +81-22-795-3758

Abstract: Cavitation is a phase-change phenomenon from the liquid to the gas phase due to an increased flow velocity. As it causes severe erosion and noise, it is harmful to hydraulic machinery such as pumps, valves, and screw propellers. However, it can be utilized for water treatment, in chemical reactors, and as a mechanical surface treatment, as radicals and impacts at the point of cavitation bubble collapse can be utilized. Mechanical surface treatment using cavitation impacts is called "cavitation peening". Cavitation peening causes less pollution because it uses water to treat the mechanical surface. In addition, cavitation peening improves on traditional methods in terms of fatigue strength and the working life of parts in the automobile, aerospace, and medical fields. As cavitation bubbles are utilized in cavitation peening, the study of cavitation bubbles has significant value in improving this new technique. To achieve this, many numerical analyses combined with field experiments have been carried out to measure the stress caused by bubble collapse and rebound, especially when collapse occurs near a solid boundary. Understanding the mechanics of bubble collapse can help to avoid unnecessary surface damage, enabling more accurate surface preparation, and improving the stability of cavitation peening. The present study introduces three cavitation bubble types: single, cloud, and vortex cavitation bubbles. In addition, the critical parameters, governing equations, and high-speed camera images of these three cavitation bubble types are introduced to support a broader understanding of the collapse mechanism and characteristics of cavitation bubbles. Then, the results of the numerical and experimental analyses of non-spherical cavitation bubbles are summarized.

Keywords: cavitation; cavitation bubble; cloud cavitation; vortex cavitation; numerical analysis; mechanical treatment

1. Introduction

Bubbles are a normal phenomenon that can be observed in the fluid phase. According to Bernoulli's equation, decreasing static pressure until flow velocity reaches a specific value can create bubbles. This phenomenon of a phase change from the liquid phase to the gas phase is called "cavitation". Cavitation can even occur in animals in the natural world [1–9]. This phenomenon produces microjets that create plastic damage to a solid boundary [10,11]. Accordingly, this process causes erosion to hydraulic machinery, such as impellers, pumps, and pipes [12], but it can, nonetheless, be utilized in many fields. Cavitation can be used, for instance, in chemical reactors to promote oxidization in industrial processes. For example, the high pressures and temperatures caused by cavitation are proven to promote chemical reactions, and in this way, it can be applied in wastewater treatment as a type of advanced oxidation process [13,14], which is widely used today. In addition, cavitation can save energy and reduce costs compared to traditional processes [15–17]. Moreover, cavitation generated by ultrasound and hydrodynamics has also been used in microbial disinfection, biological wastewater treatment, and sonochemical reactors in biochemical engineering [18]. Elsewhere, in the case of biodiesel production, a hydrodynamic cavitation reactor was shown to be more effective and less costly than a traditional reactor [19–22]. Furthermore, cavitation reactors are highly effective at degrading specific contaminants in wastewater; for example, such harmful substances include industrial raw materials from manufacturing plants, such as bisphenol [23]; a kind of antibiotic from stock raising, sulfamerazine [24]; a typical pollutant in wastewater, ammonia nitrogen [25]; fatty acids [26]; and algae (microcystis aeruginosa) [27]. Venturi-type [28], jet-type [29], swirling vortex-type [30], and rotor-type [31] reactors are widely used to create cavitation. When surface treatments with a cavitation impact are carried out by these devices, this is called "cavitation peening". This peening method uses cavitation impacts in the same way as shot peening, but a shot is not used in this method. Experiments have shown that it has advantages compared to traditional methods, as shown in Figure 1. One of its advantages is that the treated surface is smoother than that of traditional methods, such as shot peening [32]. Cavitation peening can be used, for instance, in automobile parts to extend their service life until failure [33–35]. However, as cavitation bubbles' collapse and rebound cause plastic deformation, it is essential that we develop an understanding of these bubbles' characteristics and properties if we are to avoid erosion.

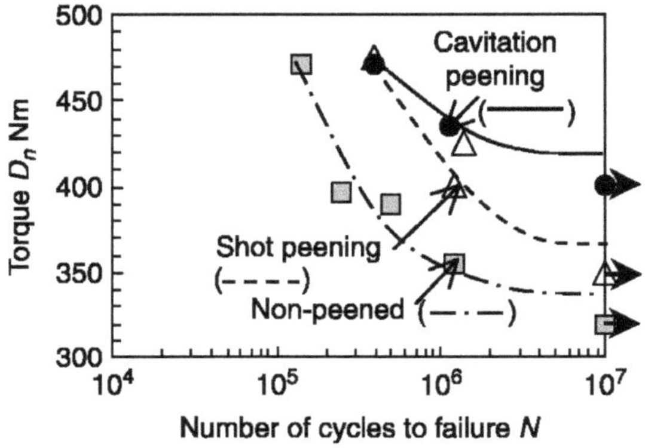

Figure 1. Improvement in the fatigue strength of the test gear via cavitation peening. Reprint from [35], with permission from Taylor and Francis, 2009, License Number 5803511487624.

In previous cavitation bubble research, Zhang et al. [36] comprehensively reviewed the interactions between cavitation bubbles and particles in silt-laden flows. Their work addressed the critical role of cavitation in hydraulic machinery and sediment transport, emphasizing the microscopic-scale interactions that contribute to material erosion and hydraulic efficiency. Such insights are invaluable for designing more durable hydraulic systems and understanding natural erosion processes. Further delving into the mechanics of cavitation, Wang and Chen [37] explored the thermodynamic behavior and energy transformation mechanisms during cavitation bubbles' collapse near rigid walls. Their numerical study elucidated how shock waves and jetting contribute to wall erosion and temperature changes, pinpointing the complex multi-phase flow dynamics. This work is pivotal for industries concerned with preventing cavitation-induced damage in fluid machinery and understanding the energy aspects of bubble collapse. Yu et al. [38] presented an innovative application of cavitation bubbles in fluorescence enhancement for bio-imaging. Through femtosecond laser-induced cavitation, they successfully amplified fluorescence signals in dye solutions without the drawbacks of nanoparticle use, such as biotoxicity and thermal damage. This advancement opens up new avenues for high-resolution imaging in biological research, where safe and effective fluorescence enhancement is crucial. On a practical front, Subramanian and Coutier-Delgosha [39] conducted intrusive experiments to better understand extreme conditions within collapsing cavitation bubbles. Reuter et al. [40] detected cavitation bubbles' dynamic impacts on a solid boundary by applying chronoamperometry. They found that the vortex ring created by collapsing bubbles has essential applications in deposition, mixing, and cleaning. Yasui [41] shed light on generating oxygen radicals from cavitation bubbles under ultrasound treatment, which expands our understanding of the chemical processes that occur within and around a collapsing bubble. The researchers identified significant amounts of O radicals, alongside traditionally studied OH radicals, which opens up new avenues for exploring sonochemical reaction mechanisms and their potential applications. Ghasemian et al. [42] presented high-speed ultrasound imaging as a potent tool for non-invasively capturing the fast dynamics of single cavitation bubbles in tissue-mimicking material, highlighting the importance of advanced imaging techniques in observing these rapid phenomena. The review by Krell [43] on the degradation and protection of materials from cavitation erosion synthesized findings from various studies to assess the effectiveness of numerous protective treatments. Their review is critically important for understanding the relationships between material hardness, different treatment methods (such as plasma nitriding, shot peening, and coating deposition), and the improvement in or deterioration of resistance against cavitation erosion. Moreover, the dynamics of cavitation bubbles and their interaction with nearby solid boundaries were scrutinized in the research by Reuter et al. [44], which highlights the intricate flow fields and vortex dynamics associated with collapsing bubbles. The present investigation complements the detailed imaging studies already performed and provides a more robust theoretical framework for predicting and analyzing the impact forces and erosion mechanisms of cavitation on solid surfaces. Above all, we have taken an integrated approach to studying cavitation, which enriches the current scientific discourse and paves the way for future research and technological innovations around one of nature's most powerful and destructive phenomena.

Three types of cavitation bubbles will be introduced: single cavitation bubbles, cloud cavitation bubbles, and vortex cavitation bubbles. The stand-off distance (dimensionless distance) $\gamma = L/R_{max}$, where L is the distance between the center of the bubble and the wall surface, and R_{max} is the maximum bubble radius, is used to clarify the types of single bubbles [11,45,46]. Previous studies initiated research into cavitation bubble collapse, with the cavitation bubbles treated as spherical [47–49]. However, non-spherical bubbles occur in cavitation, especially when bubbles are in the final stages of collapsing [50–52]. With the development of visual techniques, it has been proven that these non-spherical bubbles are created in cavitation, as has been observed clearly in cavitation peening [28,29,53–55]. For example, bubbles are non-spherical in a type of cavitation peening termed laser peening (Figure 2) [56]. Furthermore, the X-ray images of the bubbles in vortex cavitation reveal that they are angulated [55], not spherical or spherical-like, as previous numerical analyses indicated. Today, to close the gap between numerical analyses and field experiments, it is of significant value to research non-spherical bubbles in cavitation peening.

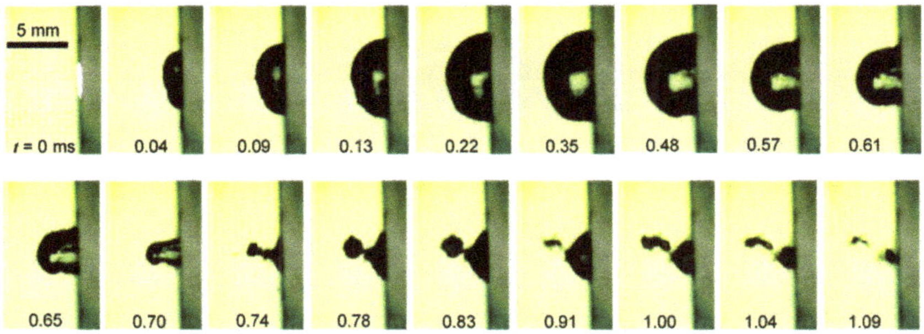

Figure 2. Images of laser cavitation on the wall near the glass chamber [56].

Returning to the initial study of bubbles, Figure 3 depicts the collapse of a solid cavitation bubble near a boundary in water [49]. As this shows, the bubble becomes non-spherical as it collapses and finally disappears. Roles are played in this process by stress and pressure. In this review, based on numerical analyses and field experiments, we explain the change in bubble shape and introduce the volume of pressure released near the solid boundary during that process, applying these ideas to scenarios from a single non-spherical cavitation bubble to complicated gathering bubbles. For complicated gathering bubbles, such as in cloud cavitation and vortex cavitation, more accurate simulations are presented than those in previous work, wherein we consider that these two cavitation processes involve non-spherical bubbles. Above all, this review emphasizes that the impact of bubbles collapsing near a solid wall is deeply related to the shapes of the bubbles in question.

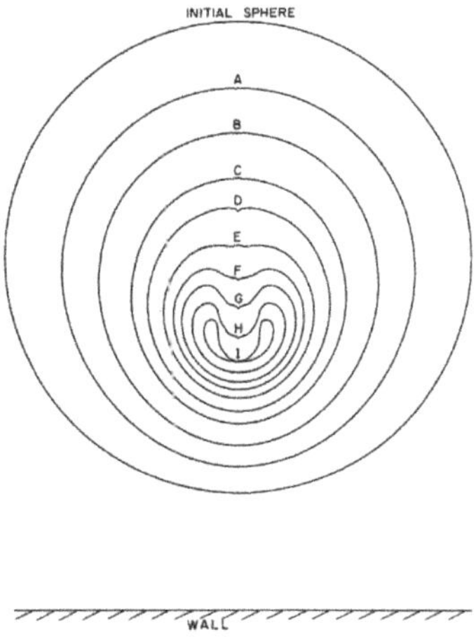

Figure 3. Collapse of a solid cavitation bubble near a boundary in water, computed by Plesset and Chapman. A–I represent different shapes as time goes on. Reprint from [49], with permission from Cambridge University Press, 1971, License Number 5724030138904.

In sum, this review describes and discusses non-spherical cavitation bubbles to help readers understand cavitation and its applications [57–59]. This review also discusses the relationship between single cavitation bubbles and complicated cavitation, such as the cases of cloud cavitation and vortex cavitation.

2. Single Cavitation Bubbles

This section will review research on single spherical and non-spherical cavitation bubbles. Figures 4 and 5 clearly show a cavitation bubble near a solid boundary, as published in early research [11,48]. When this research field first formed, investigations sought to find numerical solutions for the collapse and rebound of a spherical bubble in water [12]. They focused on the effect of liquid compressibility, and a numerical solution was obtained through a simple gas-like model. The calculations of the motion of the bubble wall and the gross effect of the gas content of the bubble are shown in Figure 6 (based on the irrotational assumption because the motion is spherically symmetric) [47]. In this figure, U is the velocity of the bubble's center, and c is the velocity of the sound in liquid. γ explains the gross behavior of the gas, while $\gamma > 1$ indicates that some of the heat of compression is retained in the gas.

Although this numerical analysis focused on spherical bubbles, it established crucial groundwork for us to gain an understanding of bubble collapse and rebound mechanics. Chiefly, it proved that the formation of cavitation bubbles damages a solid surface, which arises from pressure waves that originate from the bubbles. This understanding can be extended to the analysis of non-spherical cases.

Continuing this trajectory, a new wave of numerical analysis of spherical bubble collapse near a solid boundary has been introduced [49]. Here, the simulation is based on small steps used to calculate the velocity of the movement of the bubble surface. Since flow occurs via irritation, and viscosity and compressibility are neglected, the calculation is based on cylindrical coordinates. The finite difference method is applied to assess the interaction of a collapsing bubble with a solid surface, which is also the process through which spherical bubbles become non-spherical, as shown in Figure 3. This field of study critically discusses non-spherical bubbles and explains the "water hammer", highlighting the complexities introduced by the boundary condition, which may affect the symmetry of collapse and potentially lead to asymmetric or non-spherical bubble shapes. This sets the stage for the numerical analyses of bubbles collapsing that lack spherical symmetry.

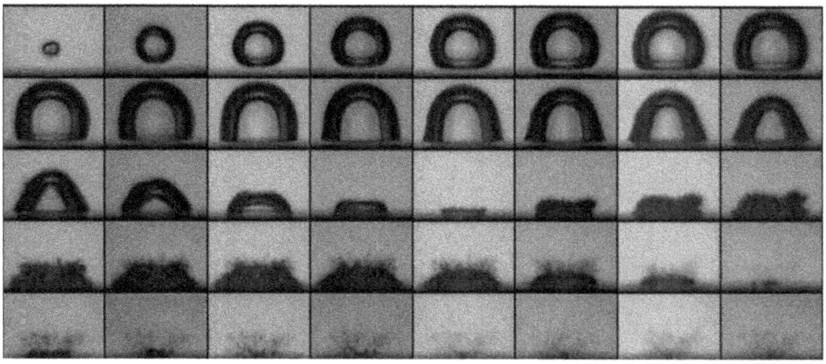

Figure 4. Cavitation bubble dynamics in the vicinity of a solid boundary. Reprint from [11], with permission from Cambridge University Press, 1998, License Number 5831440015629.

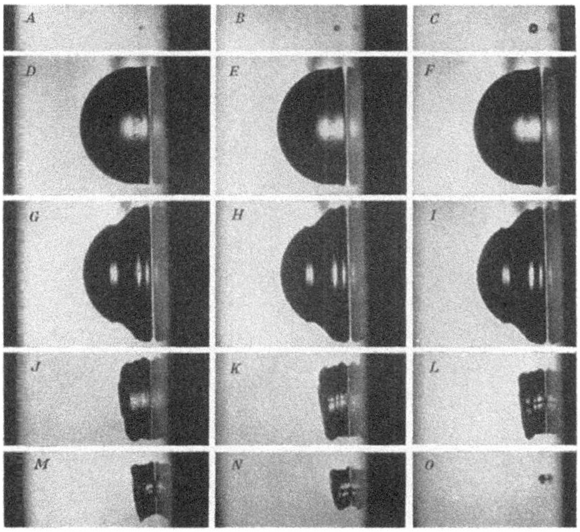

Figure 5. Growth of cavity from nucleus very close to a solid wall, and subsequent collapse. Timing: (**A–C**) at 0, 0.2, and 0.4; (**D–F**) at 5.8, 6.0, and 6.2; (**G–I**) at 11.4, 11.6, and 11.8; (**J–O**) at 16.8 to 17.8 ms. Reprint from [48], with permission from Royal Society (Great Britain), 1966, License Number 1492359-1.

To further research the bubble-collapsing effect, the dynamics of the shock-induced collapse and Rayleigh collapse of a bubble near a planar rigid surface and in a free field have been numerically analyzed [52]. This research has charted the whole process of bubble collapse, from the unsteady Euler equation used for calculations to the bubble's properties. Bubble collapse has been divided into two parts: free-field shock-induced collapse and Rayleigh collapse. Importantly, Rayleigh collapse is not spherically symmetric, meaning it applies to non-spherical bubbles. Shockwave collapse, including the precursor, water hammer, and pressure released, is demonstrated in Figure 7. The wall pressures of these two types of collapse have also been calculated. Through these numerical analyses, researchers have gained a greater understanding of bubble dynamics for both theoretical frameworks and practical applications, showing bubble behavior influences the system performance and material integrity.

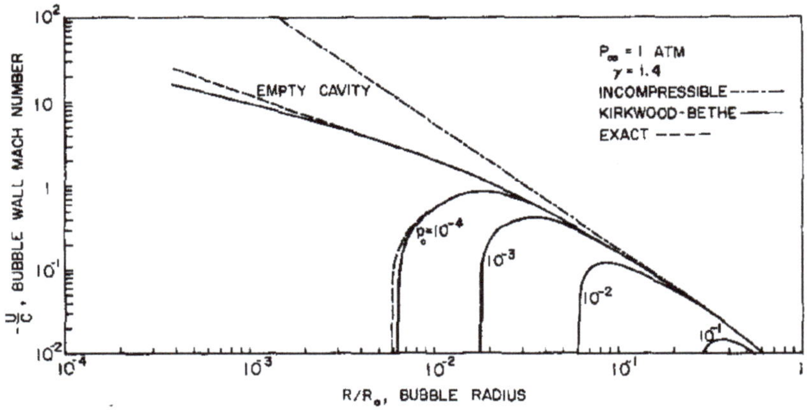

Figure 6. The bubble–wall Mach number as a function of the bubble radius with a decreasing gas content. The gas content is determined by its initial pressure p_0 in atmospheres. The index γ has the value 1.4, and the ambient pressure p_∞ is 1 atm. Reprint from [47], with permission from AIP Publishing, 1964, License Number 5806350103845.

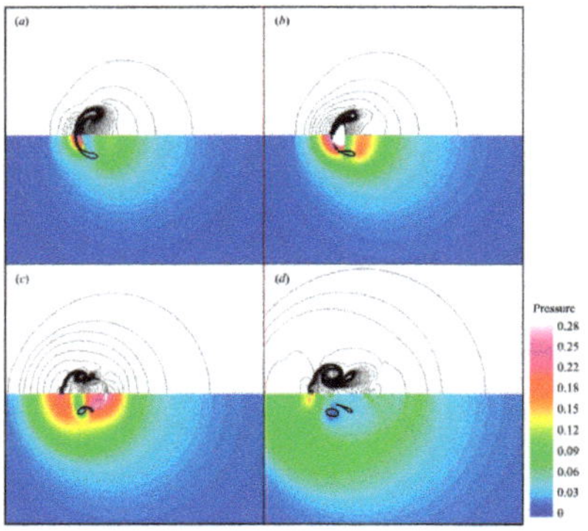

Figure 7. Water hammer shock in free-field shock-induced collapse (p_s/p_o = 353) at times (**a**) $t/(R_o/c_L)$ = 8.49, (**b**) $t/(R_o/c_L)$ == 8.68, (**c**) $t/(R_o/c_L)$ == 9.05, (**d**) $t/(R_o/c_L)$ == 9.52. Top: density lines; bottom: pressure contours non-dimensionalized by $\rho_L c_L^2$ (approximate interface location: γ = 1.42 contour). p_s: The shock pressure; p_o: The relevant pressure ahead of the compression; R_o: The radius of bubble; ρ_L: The ambient density of the water; c_L: The sound speed of the water. Reprint from [52], with permission from Cambridge University Press, 2009, License Number 5830760763951.

Numerical analyses and experimental observations have since been introduced that focus on the impact between the front of the re-entrant jet and the opposite side of the bubble, with the jet penetrating the slower-moving liquid close to the rigid boundary, which is called the "final stage of the bubble collapsing to a rigid boundary" [50]. The experimental research on this topic uses high-speed photography, with up to 5 million images per second. An incompressible impact model has enabled the comprehensive numerical analyses of bubble–boundary interaction mechanisms. The photographs of bubbles collapsing and the corresponding numerical analyses are shown in Figure 8. The evolution of the velocity field and pressure contours in the toroidal stage of bubble collapse for 1.1 is shown in Figure 9. This research has observed a transformation of kinetic energy from the radial flow into the main bubble to a rotational flow around the cavity next to the wall during the final stage of a bubble collapsing, and this transformation has ensured a "boundary" $\gamma = 1$. When $\gamma < 1$, the shock wave emission has weakened, which causes a reduction in the damage to the wall. This field of study is pivotal in identifying bubble interaction during cavitation events.

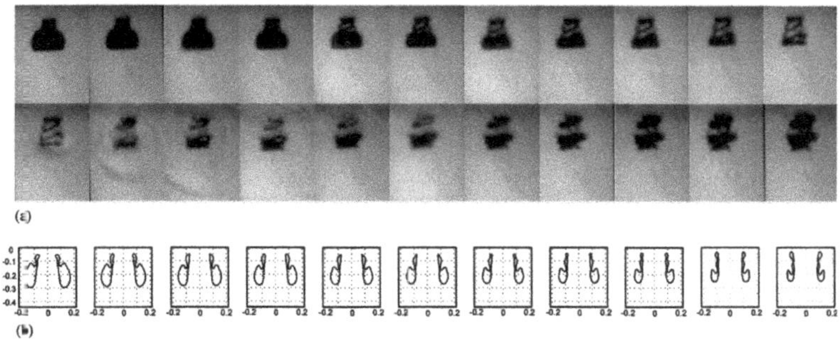

Figure 8. The final stage of the collapse of a cavitation bubble for $\gamma = 1.1$. (**a**) High-speed photographic record of bubble motion; frame interval 200 ns, frame width 1.4 mm. The location of the rigid boundary is at the upper limit of the frames. (**b**) Computed bubble profiles covering the time interval correspond to the photographic record's first row. The nondimensional times are t = 2.0084, 2.0097, 2.011, 2.0123, 2.0136, 2.0149, 2.0161, 2.0174, 2.0187, 2.02, and 2.0213. Reprint from [50], with permission from AIP Publishing, 2002, License Number 5806350497331.

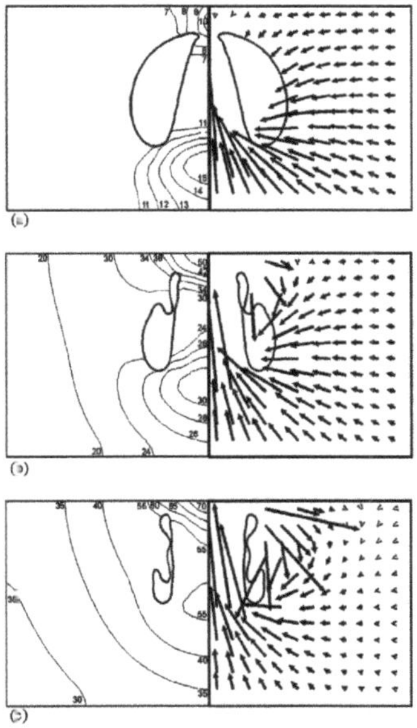

Figure 9. Evolution of the velocity field (**right**) and pressure contours (**left**) for the toroidal stage of bubble collapse for γ = 1.1. The letters correspond to the following nondimensional times: (**a**) 1.9981, (**b**) 2.011, and (**c**) 2.0213. Reprint from [50], with permission from AIP Publishing, 2002, License Number 5806350497331.

Research on single cavitation bubbles has practical implications for reducing material erosion caused by cavitation. The effect of liquid droplet impingement (LDI) on a material surface has been thoroughly researched [60]. Using the model of droplet capture at a gas–liquid interface, the fluid pressure on the material surface and the normal stress in the vertical direction of that material surface have been calculated. A simulation of the pressure in a fluid and the equivalent stress in a material on a wet surface is shown in Figure 10. Compared to the same situation on a dry surface, the simulations reveal a significant finding—the impinged droplet pressure decreases, which can diminish the equivalent stress inside the material. This damping effect can be attributed to the liquid film's ability to absorb and disperse the energy imparted by the impinging droplet, culminating in a reduced impact force. Furthermore, the research has led to the development of an empirical formula that quantifies the damping effect, relying on the critical fluid factors of LDI. This research provides an insightful analysis of the damping effects of a liquid film on a material when subjected to high-speed LDI. Its model allows for predicting and optimizing the responses of various materials to LDI scenarios, potentially leading to enhanced erosion resistance in fluid systems. It is also essential in such work to note the effect caused by bubbles collapsing, which should inspire further research and development in this field.

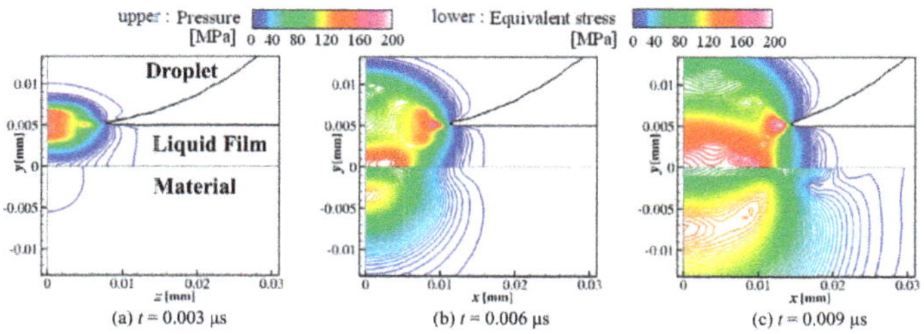

Figure 10. Time variation in pressure in fluid and equivalent stress in material on wet surface. Reprint from [60], with permission from J-stage, 2016.

The relationship between a non-spherical bubble collapse and the stress state of the nearby boundary/wall [45] has been examined to ascertain how varying distances between the bubble and the wall, called the stand-off distance (as mentioned before, γ), impact material stress. The mathematical model is shown in Figure 11, which corresponds to one-quarter of a three-dimensional space. Here, R_0 is the bubble radius and L is the distance between the bubble center and the wall. The governing equation used is similar to [60]. By conducting coupled fluid–material numerical simulations, studies have revealed that different collapse modes—pancake-shaped, hemispherical, microjet, and detaching modes—affect the stress distribution within adjacent materials. Some of the calculated results are shown in Figure 12. Furthermore, the relationship between the stand-off distance and the maximum equivalent stress is shown in Figure 13. The result is calculated by a supercomputer, and the CFD (computational fluid dynamics) code is Fortran in the Linux system. The calculation time for one stand-off distance is about 60 h. These figures provide a complete numerical analysis of the first stage of non-spherical bubble collapse. The peak of equivalent stress on the material surface can be observed when the stand-off distance is 0.3–1.0. This area of study underscores the complexity of bubble–material interactions and the potential for significant material damage, highlighting the importance of understanding these dynamics for protective material design and fluid system maintenance.

Further expanding on bubble interactions, Aganin et al. [61] introduces a particle model to simulate the behavior of weakly non-spherical bubbles. The model provided in that research focuses on the bubble's interaction. Unlike typical research, compressibility effects and the liquid viscosity effect are considered. Applying this model, a bubble collapsing near a solid boundary was simulated, along with the simultaneous interaction of three bubbles, as shown in Figures 14 and 15. This research has been instrumental in progressing the field of study from individual bubble dynamics to the interactions between multiple bubbles, which is crucial in many practical scenarios where bubble populations rather than isolated bubbles influence the system's overall behavior. Furthermore, the model accurately shows the relationship between the stand-off distance and stress caused by a bubble collapsing. One finding is that if the distance between the bubbles is smaller, this will produce a better result in the simulation. Above all, the particle model approach allows for more sophisticated simulations of bubble interactions, offering insights into the collective dynamics of bubble populations under various conditions, and the model has been proven to have a goodness of fit in the simulations of bubble collapse.

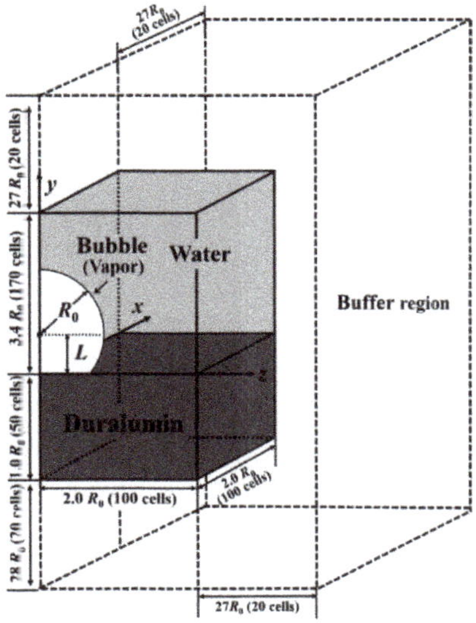

Figure 11. The mathematical model of the non-spherical bubble [45].

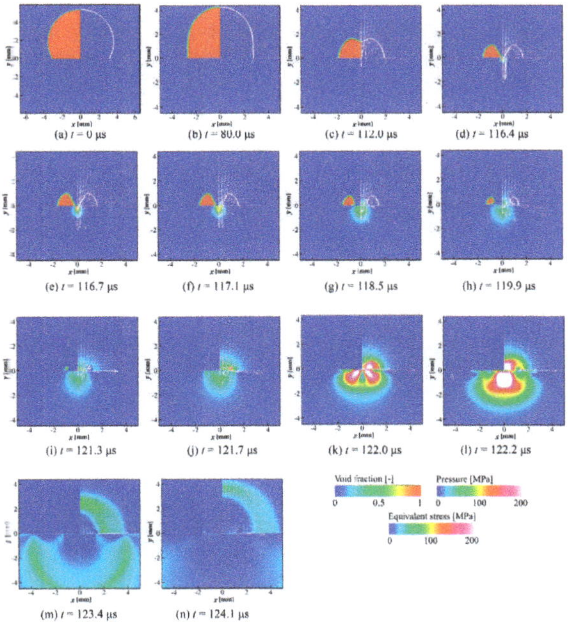

Figure 12. Time evolution of the void fraction (**upper left** of each sub-figure), pressure, and velocity vectors (**upper right**) in the fluid domain, and equivalent stress in the material domain (**lower**) ($\gamma = 0.5$) [45].

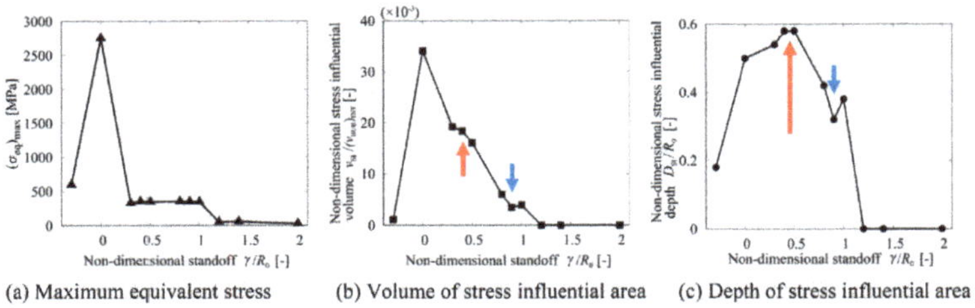

Figure 13. Variation in maximum equivalent stress and stress influence area with the stand-off distance between the bubble and the wall. The blue and red arrows point $\gamma/R_0 = 1$ and 0.5. [45].

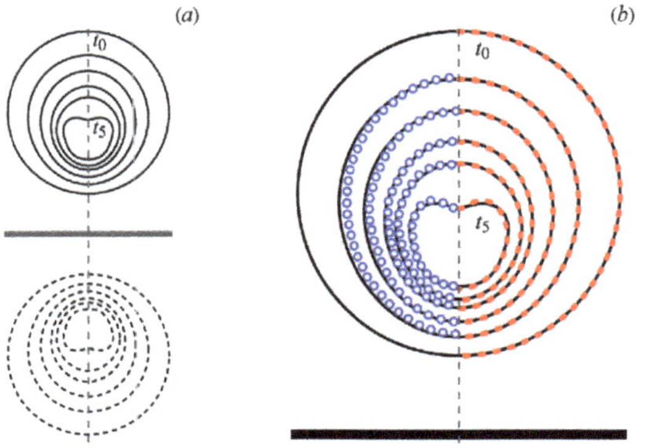

Figure 14. Evolution of the axial section of a bubble collapsing near a plane wall (t_0 corresponds to the beginning of collapse): (**a**) the results of a simulation of the interaction between two bubbles by the proposed model (solid lines correspond to the bubble under consideration; dashed lines present its mirror image), and (**b**) the same results for the bubble considered (solid lines), along with experimental data and the results obtained by the boundary element method (dashed lines). Reprint from [61], with permission from Elsevier, 2024, License Number 5830790844530.

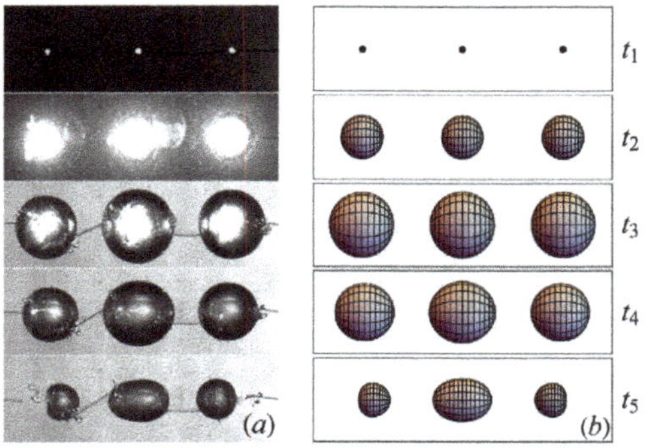

Figure 15. Interaction between three bubbles simultaneously produced collinearly at nearly equal inter-bubble distances: (**a**) in the experiments by Cui et al., and (**b**) in the simulation by the present model. Reprint from [61], with permission from Elsevier, 2024, License Number 5830790844530.

These studies have contributed significantly to broadening our understanding of cavitation bubbles, including spherical and non-spherical bubble dynamics in fluid systems. However, most research on non-spherical bubbles has a central issue—uncertainty around the erosion of the material surface caused by the bubble collapsing. To address that, many numerical analyses and experiments have been performed, seeking to explain the interaction and relationship between a bubble collapsing and a solid boundary. The progression from single-bubble dynamics to complex interactions between bubbles and with materials underlines the evolving landscape of bubble research, paving the way for further explorations and innovation in bubble behavior modeling and its application to material processing and fluid mechanics.

3. Cloud Cavitation

The exploration of bubble cloud dynamics, especially near rigid boundaries, in non-spherical bubble research has formed an active area of fluid dynamics study given the physical applications for the research findings, including those relating to cavitation peening. In this section, we introduce several studies on bubble clouds involving complex physical phenomena, such as shock wave emission, jet formation, and pressure fluctuation, which significantly impact the efficiency and effectiveness of cavitation bubble processes.

Building upon the foundation of single-bubble dynamics, Chahine and Duraiswami [62] expanded the investigation to multiple bubbles, presenting a study on the dynamic interactions in multi-bubble clouds. Their methodology combined analytical techniques and numerical simulations to capture the complex dynamics of bubble–bubble interactions within a cloud. Significantly, they found an agreement between the two approaches within their studied range, highlighting the role of bubble–bubble interactions in determining the collective behavior of bubble clouds. Their work is crucial for understanding the mesoscale dynamics within these clouds, which bridge the gap between individual bubble behavior and the macroscopic flow characteristics.

An early contribution to understanding the bubble cloud dynamics near a rigid boundary came in the form of experimental and numerical investigations of the final stage of the collapse of a cavitation bubble close to a rigid boundary [63]. This research offers invaluable insights into the behavior of liquid jets developed during bubble collapse and the shock wave emissions corresponding to the "splash" effect, as determined by employing high-speed photography alongside numerical simulations. Here, the wave emission caused by bubble clouds was researched to explain the mechanism of adverse effects. The study introduced the "cloud interaction" parameter to describe cloud bubbles in numerical analyses:

$$\beta = \alpha_0(1-\alpha_0)A_0^2/R_0^2 \tag{1}$$

where the initial radius and void fraction of the cloud are denoted by A_0 and α_0, respectively, and the initial radius of the bubbles within the cloud is denoted by R_0. The cloud bubbles can be divided into several steps: traveling bubble cavitation, maximum cavity length, re-entrant jet, bubbly mixture, local pulse phenomena, and global pulse phenomena. Figure 16 shows the global pulse phenomena created by the shock wave released. The findings shed light on cavitation erosion mechanisms, linking the bubble–boundary interaction with potential damage risks in hydraulic machinery and other fields.

Experimental measurements using a high-speed camera have shown that the bubble cloud damages the material surface, which was ascertained by detecting the occurrence of impulse pressure during the incubation of the bubble. The reason for the impulse pressure is that various distinct cavitation structures arise from the mixture of bubbles. However, the experiment highly depends on the beta parameter, which decides the size of a bubble cloud. Here, a large bubble cloud seems suitable for the numerical analysis model. This research has identified important features of fluid flows and explained the mechanism of impulsive pressure caused by bubble clouds.

The exploration of bubble cloud dynamics close to rigid boundaries was furthered in [51]. In this investigation, the researchers meticulously documented the final collapse stage of a bubble cloud near a rigid boundary, providing critical insights into the complex interactions that occur as a bubble cloud collapses. By using ultra-high-speed photography, which can produce precise images, the researchers observed two secondary shock wave emission types during cloud rebound, as shown in Figure 17. The first type, secondary shock wave emission, is a consequence of the free collapse of a bubble within the cloud as a result of the ambient pressure in the fluid. The second type of shock wave emission is due to the interaction of the cloud collapse-induced shock wave with microbubbles close to the cloud's collapse site. In the study, it seems that no jet formation was observed during the process of the cloud bubble collapsing. In this scenario, the single bubbles are probably too small to induce the formation of liquid jets. The study highlights the influence of rigid boundaries on the collapse process, presenting a detailed analysis of the pressure fields and mechanical effects involved, which can aid future research work to find a method for reducing collateral damage.

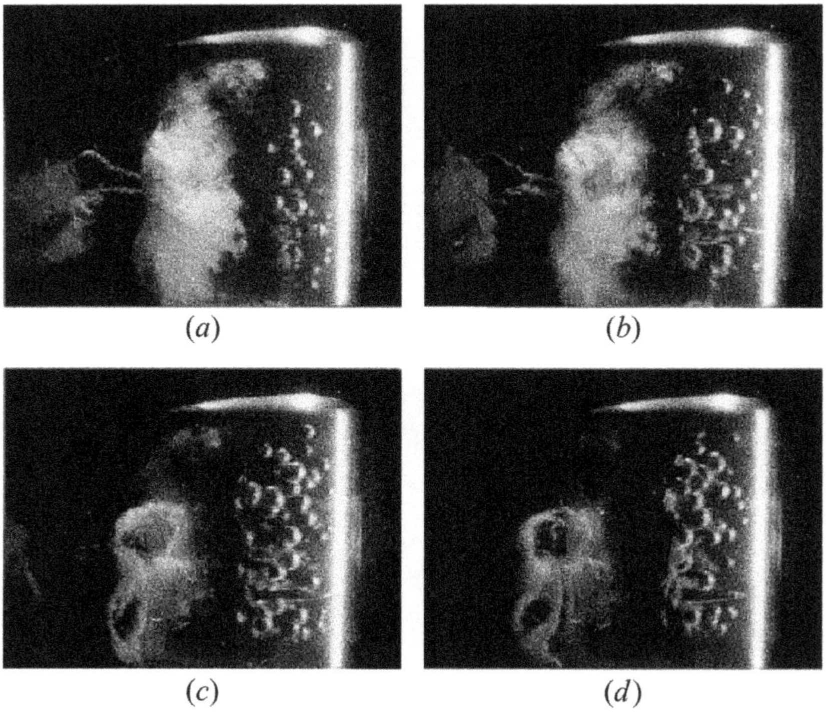

Figure 16. Consecutive high-speed movie frames (2 ms apart) showing global cloud collapse. (**a–d**) depict four consecutive frames from one such movie. Reprint from [63], with permission from Cambridge University Press, 1998, License Number 5830810037327.

Expanding the numerical methods available to explore bubble dynamics, a significant contribution arose with the development of a two-way coupled Eulerian–Lagrangian model that elucidates the microbubble cloud dynamics, accounting for incoming pressure waves and the resultant pressure loads on nearby rigid walls [64].

To simulate the two-phase medium as a continuum and solve the N-S equations, two models are introduced: a continuum homogeneous mixture flow model and a bubble motion and volume change model. According to the homogeneous mixture flow model, the mixture density is defined as follows:

$$\rho_m = (1 - \alpha)\rho_l + \alpha\rho_g \tag{2}$$

where the subscripts "l" and "g" represent liquid and gas properties, respectively, while is the gas volume fraction.

The governing equation for bubble motion and volume change takes the form of a modified Rayleigh–Plesset–Keller–Herring equation:

$$\left(1-\frac{\dot{R}}{c_m}\right)R\ddot{R}+\frac{3}{2}\left(1-\frac{\dot{R}}{3c_m}\right)\dot{R} = \frac{\left(u_{enc}-u_b\right)^2}{4}+\frac{1}{\rho_m}\left(1+\frac{\dot{R}}{c_m}+\frac{R}{c_m}\frac{d}{dt}\right)\left[p_v+p_g-p_{enc}-\frac{2\gamma}{R} -4\mu_m\frac{\dot{R}}{R}\right] \quad (3)$$

In the above, c_m is the local sound speed in the mixture where the bubble is located, μ_m is the medium viscosity, p_v is the liquid vapor pressure, p_g is the initial gas pressure in the bubble, k is the gas polytropic constant, γ is the surface tension, and u_b is the bubble center velocity. A pressure term is added, $(u_{enc} - u_b)^2/4$, due to the bubble slip velocity relative to the liquid, and the surface-averaged pressure (SAP) model is used to determine the encountered liquid quantities. p_{enc} and u_{enc} are the encountered pressures and velocities, respectively, in the liquid averaged over the bubble surface. This averaging procedure accounts for non-uniform pressures and velocities along the bubble surface. Using p_{enc} results in a significant improvement over classical models, which use the pressure at the bubble center. Calculation results for bubble locations and pressures can be obtained from this equation.

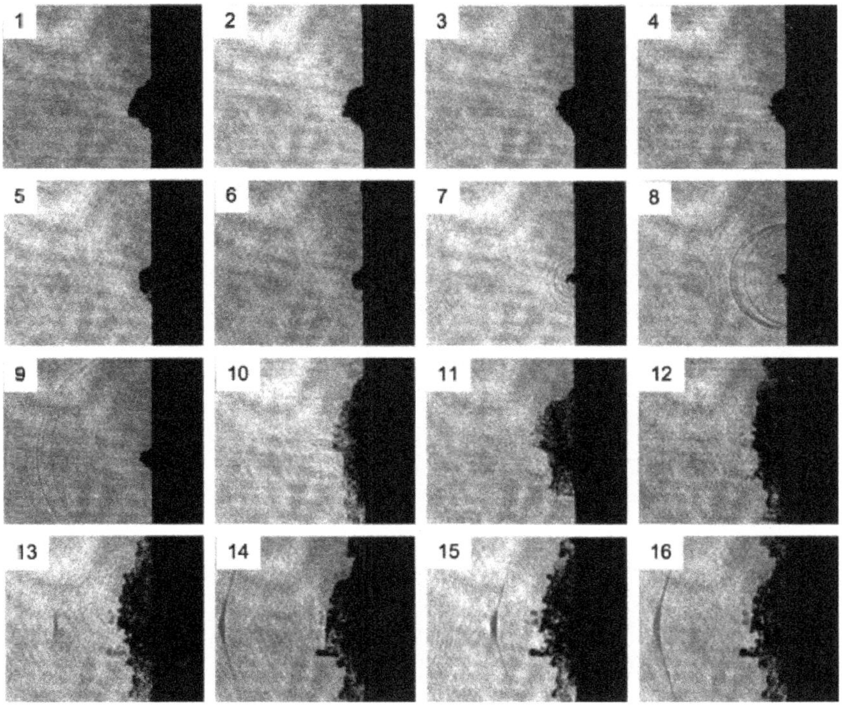

Figure 17. Collapse phase of a hemispherical bubble cloud. Frame interval 500 ns. The first frame was taken 159.55 μs after the application of ultrasound. Frame width 2.56 mm. The collapse point of the cloud occurs between frames 7 and 8. The main shock wave emission during cloud rebound can be seen in frames 8 and 9. Reprint from [51], with permission from Elsevier, 2011, License Number 5830811192279.

Moreover, the field of study of the dynamics of a bubble cloud near a rigid wall extends our comprehension of how external pressure frequencies and amplitudes influence bubble behavior and the subsequent pressures exerted on adjacent surfaces [65]. Furthermore, this research identifies resonance frequencies that significantly differ from natural frequencies, which exemplifies the complexity of bubble dynamics, particularly in terms of nonlinear effects and the generation of high pressures at the contacting walls. Figure 18 shows bubble radii versus time for different bubbles in the cloud and the corresponding pressure variations. As the resonance frequency, a pressure peak order of magnitudes higher than the excitation pressure occurs, which corresponds to a match between the driving pressure reaching its maximum and the collective individual bubbles collapsing.

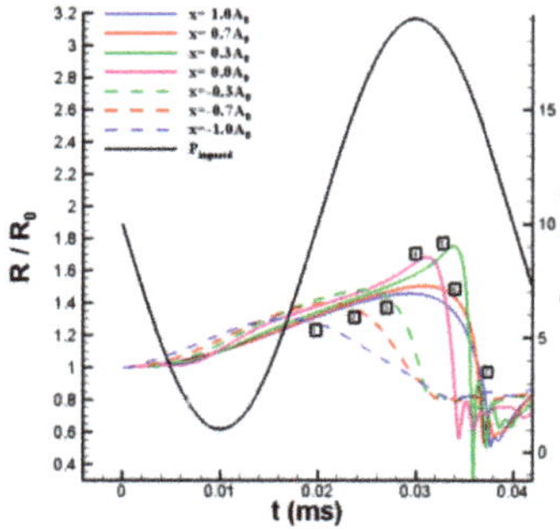

(**a**)

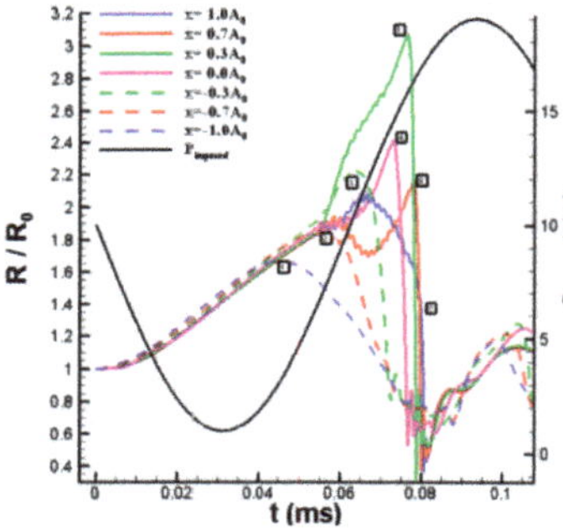

(**b**)

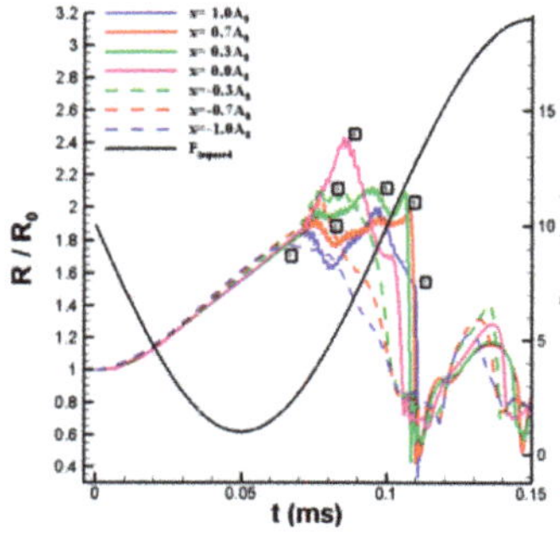

(**c**)

Figure 18. Comparison of bubble radii at different cloud locations for a bubble cloud excited at different pressure frequencies: (**a**) f = 25 kHz; (**b**) f = 8 kHz, and (**c**) f = 5 kHz [65].

Such modeling efforts highlight the critical role of the driving pressure's frequency in determining bubble cloud dynamics, which allows for distinguishing between collective cluster behavior and non-synchronized oscillations based on resonance conditions. Building upon this empirical groundwork, the authors of [66] present simulation-based insights into acoustically driven bubble cloud dynamics near a rigid wall. The numerical study provides a comprehensive view of how acoustic pressures influence bubble behavior, detailing bubble clouds' formation, growth, and collapse near rigid surfaces. In addition, the impacts of the amplitude and frequency of the excitation pressure, the cloud and bubble sizes, the void fraction, and the initial stand-off distance on the bubbles are all discussed and subject to numerical analysis. The researchers' mathematical model of cloud bubbles is shown in Figure 19, where R_0 is the radius of a single bubble, A_0 is the radius of the cloud, and X_0 is the distance between the solid boundary and the center of the cloud. Their analysis also produced important findings on cloud bubble distributions and internal pressures, as shown in Figure 20.

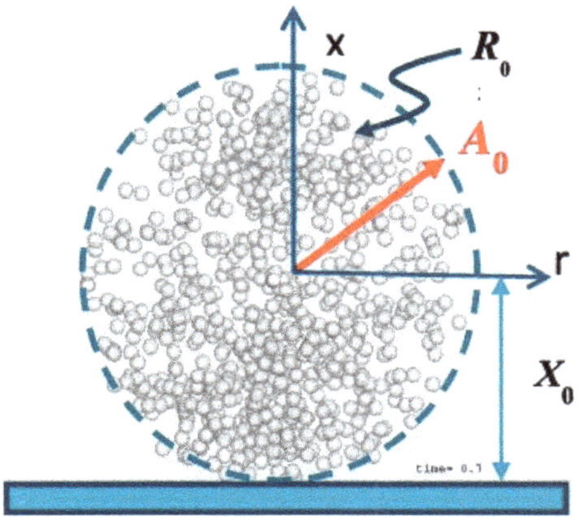

Figure 19. The mathematical model of cloud bubbles [66], with permission from Elsevier, 2018, License Number 5880601502087.

An experimental perspective was provided through a focused study on bubble–sphere interactions near a rigid wall [53], which delved into the varying degrees of interaction between oscillating bubbles and movable spheres. The research categorized the interactions into weak, intermediate, and firm based on distinct bubble shapes at the maximum volume, offering a classification that underlines the significance of bubble–sphere dynamics in the broader context of bubble behavior near rigid boundaries. The intense interaction between the bubble and sphere near a rigid boundary is shown in Figure 21.

Utilizing the Euler–Lagrange method we have mentioned, a new numerical simulation was performed of a solid's stress due to a bubble cloud's collapse [67]. The bubble radius and void fraction distribution for various stand-off distances are shown in Figure 22. The study discovered that the critical point of the stand-off distance $Y_C = 0.4R_c$. At this point, the von Mises stress has the highest value, and then it decreases. This research team's models may become more accurate in the future when considering mass transfer, but the published work to date gives a clear indication of the direction they are taking in developing their numerical analysis.

The early-stage dynamics have been investigated of hydrodynamic cloud cavitation, focusing on bubble size distribution (BSD) [68]. The novel application of a fiber optic probe to assess the BSD within a cavitation cloud sheds light on the significant role of turbulence in bubble fragmentation processes. The cloud cavitation in four successive shedding cycles is shown in Figure 23, taken from another literature work [69]. This study identifies two distinct power-law scales corresponding to different stages of bubble development, thereby improving our understanding of the mechanisms driving bubble formation and breakup in cavitation clouds. This research direction is instrumental in elucidating the microscale processes that precede the macroscopic phenomena observed in cavitation erosion and noise generation.

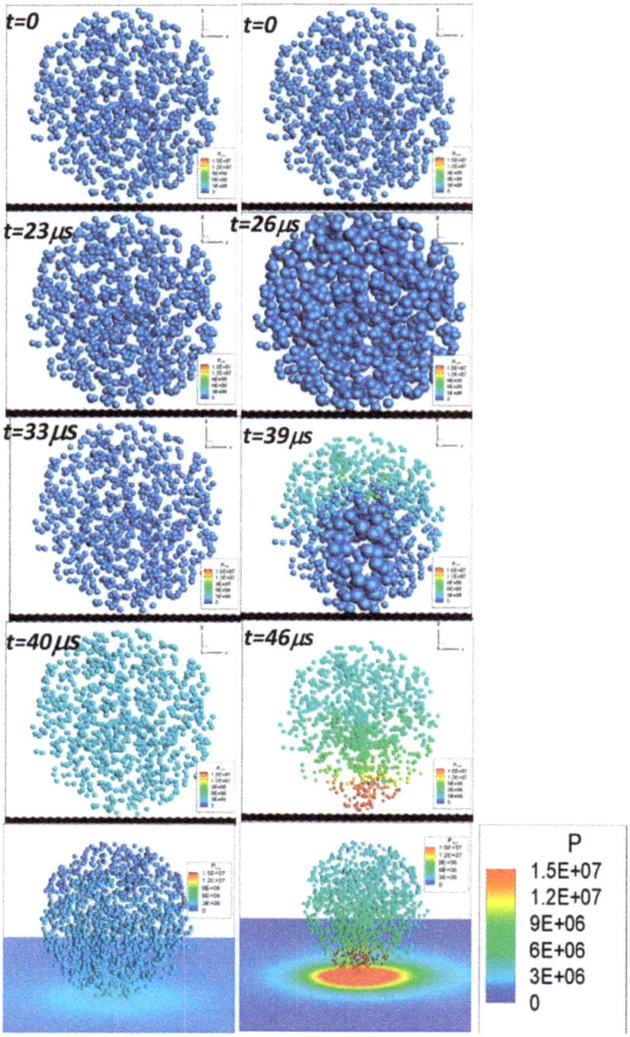

Figure 20. Time sequence of bubble distributions and internal pressures (indicated by the colored contours) in a bubble cloud driven by pressure excitation with two amplitudes. The last frame of each row is a view of the pressures at the rigid wall [66], with permission from Elsevier, 2018, License Number 5880601502087.

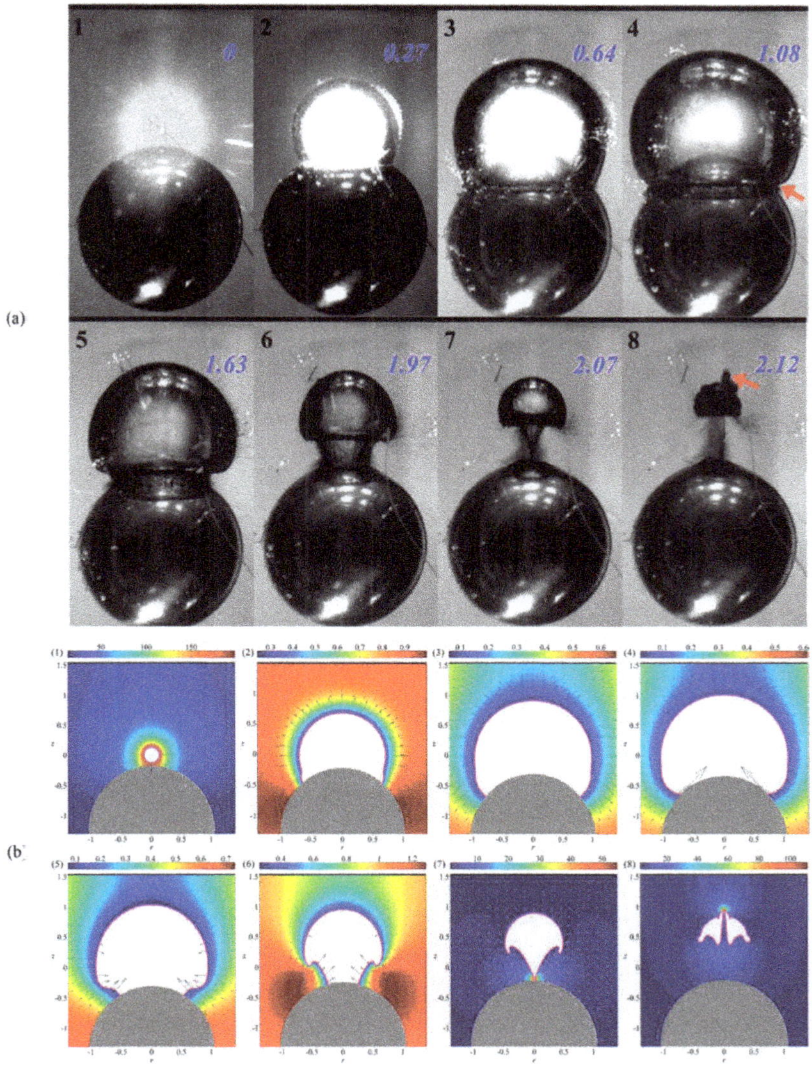

Figure 21. Strong interaction between the bubble and sphere near a rigid boundary. (**a**) Experimental images. Frames 1–9 depict the first cycle of the bubble and it is noted that the bubble bottom gains a higher local curvature (frames 4–6) due to the retardation effect from the sphere. An upward jet forms at the bubble bottom afterwards, and a strong spray at the jet tip can be observed in frame 7, leading to the unstable bubble surface after the jet impact. Frames 10–14 depict the second cycle of the bubble. The bubble migrates towards the wall during the rebounding phase and collapses onto the wall directly in the recollapse phase (frames 13 and 14). (**b**) Numerical results. The jet tip assumes a mushroom shape, which has favorable agreement with the experiment. Reprint from [53], with permission from AIP Publishing, 2017, License Number 5830821246785.

There has also been further exploration of the intricacies of bubble dynamics, clouds' internal structures, and their formation mechanism within cloud cavitating flows [70]. This study, centered on numerical investigations, elucidated the complex internal arrangements of bubbles within a cavitating cloud, mainly focusing on the mechanisms driving these formations. While the inquiry prioritized cloud cavitation, the insights gleaned have substantial relevance for understanding bubble dynamics near rigid boundaries since they shed light on the bubble cloud's structural evolution and interactions with surrounding flows, as shown in Figures 24 and 25. Such insights are valuable for refining bubble dynamics models, especially for use in predicting and mitigating the effects of cavitation in engineering systems and enhancing the efficacy of therapeutic applications involving ultrasound.

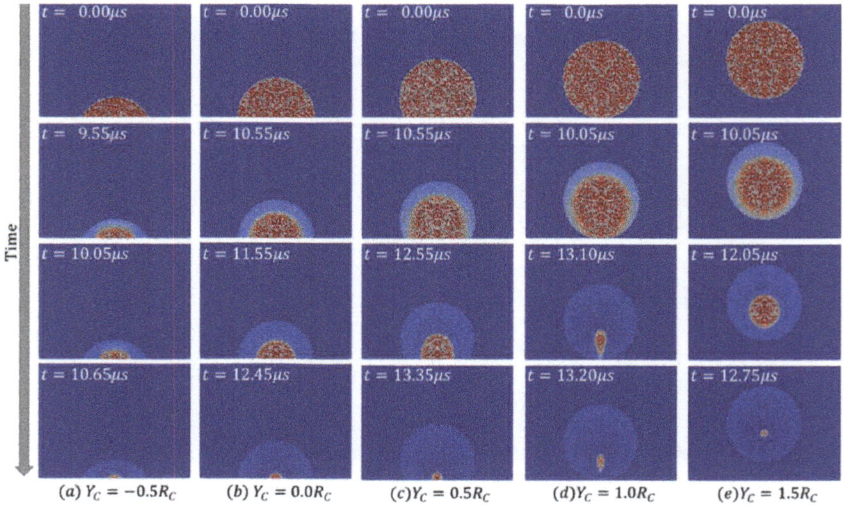

Figure 22. Time evolution of bubble radius and void fraction distribution for various stand-off distances. Reprint from [67], with permission from Elsevier, 2022, License Number 5830830883340.

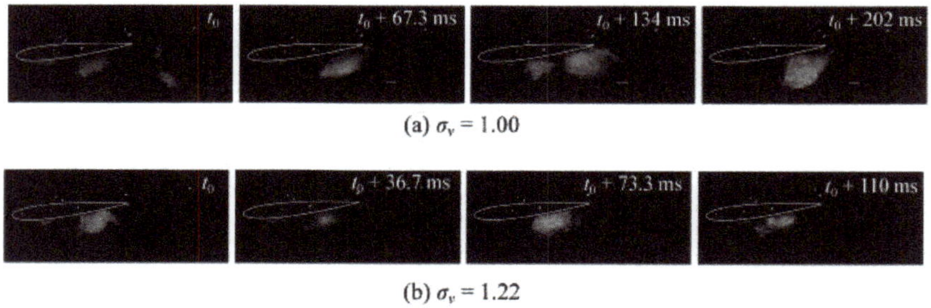

Figure 23. Snapshots of cloud cavitation in four successive shedding cycles. Reprint from [68] with permission from AIP Publishing, 2023, License Number 5830840703844.

Throughout the research above, a recurring theme is the intricate dance of forces and dynamics governing cloud bubble behavior in the vicinity of rigid boundaries. The collective insights from these studies furnish us with a comprehensive understanding of bubble cloud dynamics, from individual bubble behavior to collective cloud phenomena. Through experimental observations and sophisticated numerical simulations, a nuanced picture of bubble–wall interactions has emerged, laying the groundwork for future research and technological innovations in fields as diverse as ultrasonic cleaning, cavitation peening, and cavitation erosion prevention.

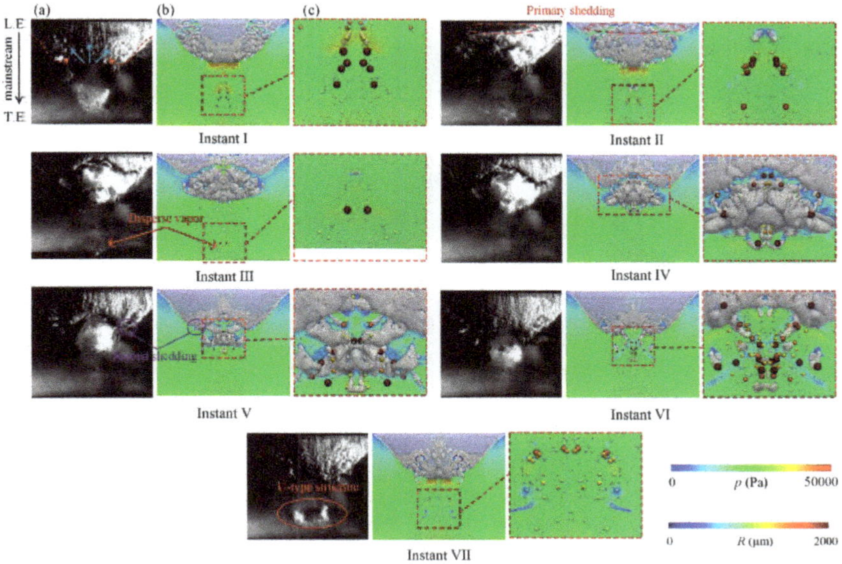

Figure 24. Observed and predicted time evolution of the multi-scale cavitating flow around the twisted hydrofoil: (**a**) experimental pictures, (**b**) global simulation results, and (**c**) partially enlarged simulation results. Reprint from [70], with permission from Elsevier, 2023, License Number 5830841056536.

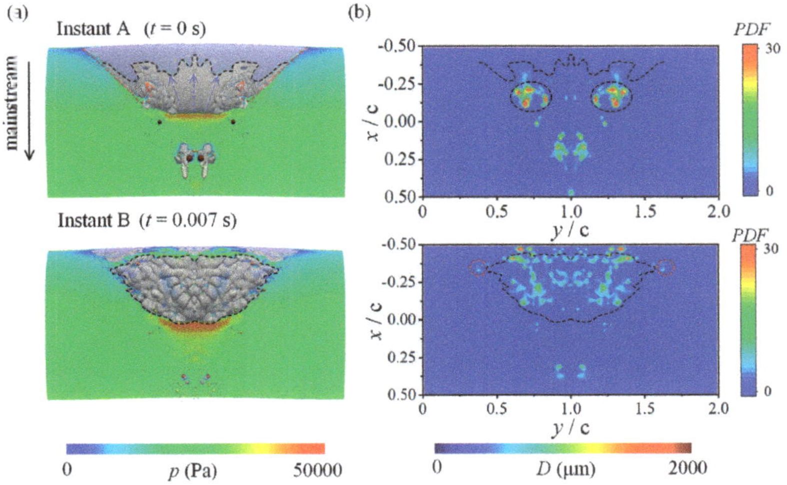

Figure 25. (**a**) Typical cavitating flow fields and (**b**) probability density of discrete bubble numbers at two typical instants during stage I. Reprint from [70], with permission from Elsevier, 2023, License Number 5830841056536.

4. Vortex Cavitation

In this section, we will discuss vortex bubbles' study in research that has applied numerical and experimental analyses combined with hydraulic machinery, mostly on tip vortices. A typical example of vortex cavitation is shown in Figure 26 [71].

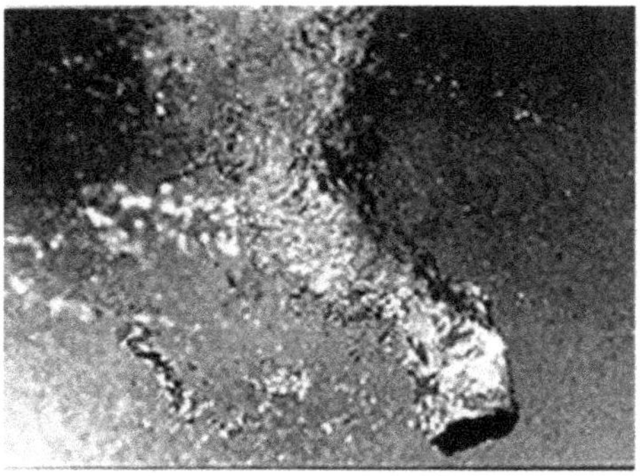

Figure 26. Typical vortex cavitation [71].

In early work, tip vortex cavitation was observed [72]. Figure 27 shows the development of the tip and other forms of cavitation on a hydrofoil. By investigating the characteristics of cavitation near the tips of hydrofoils, valuable information such as Reynold's number and cavitation number was gathered to enhance our understanding of vortex cavitation's inception and development in practical hydrodynamic scenarios.

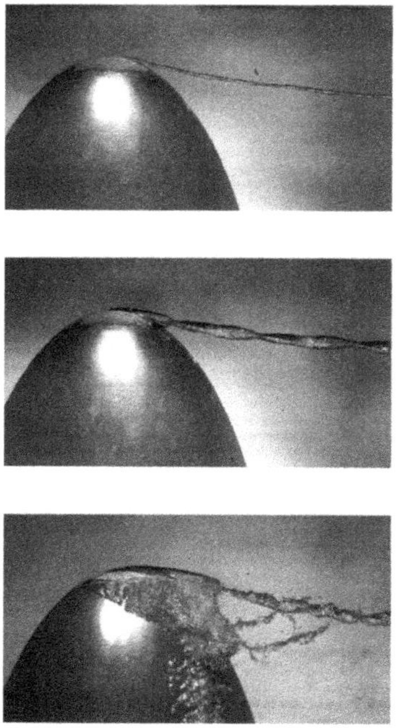

Figure 27. Photographs showing the development of a tip and other forms of cavitation on a hydrofoil, showing the continuous tip vortex cavity, tip vortex cavitation, secondary vortex cavitation, and surface cavitation. Reprint from [72], with permission from Cambridge University Press, 1991, License Number 5860660245448.

A study on the numerical simulation of three-dimensional cavitation shedding dynamics investigated the interaction between cavitation and vortices [73]. This study provides a comprehensive analysis of the formation and evolution of cavitation in vortical fluid flows. Through the mass transfer cavitation model and the modified RNG k-ε model, the shedding behavior of cavitation can be visualized through numerical analysis, as presented in Figure 28. Furthermore, Figure 29 shows the predicted vapor volume fraction and vorticity contours for the middle plane during a typical cavity shedding cycle. Focusing on the dynamic interplay between cavitation bubbles and flow vortices, the study identified three types of flow behaviors along the suction side of the twisted hydrofoil, and demonstrated that cavitation promotes vortex production and increases the boundary layer thickness via local separation and flow unsteadiness. In sum, the numerical simulations presented in this study are noteworthy in capturing the intricate details of cavitation–vortex interactions, deepening our understanding of cavitation dynamics in three-dimensional flows.

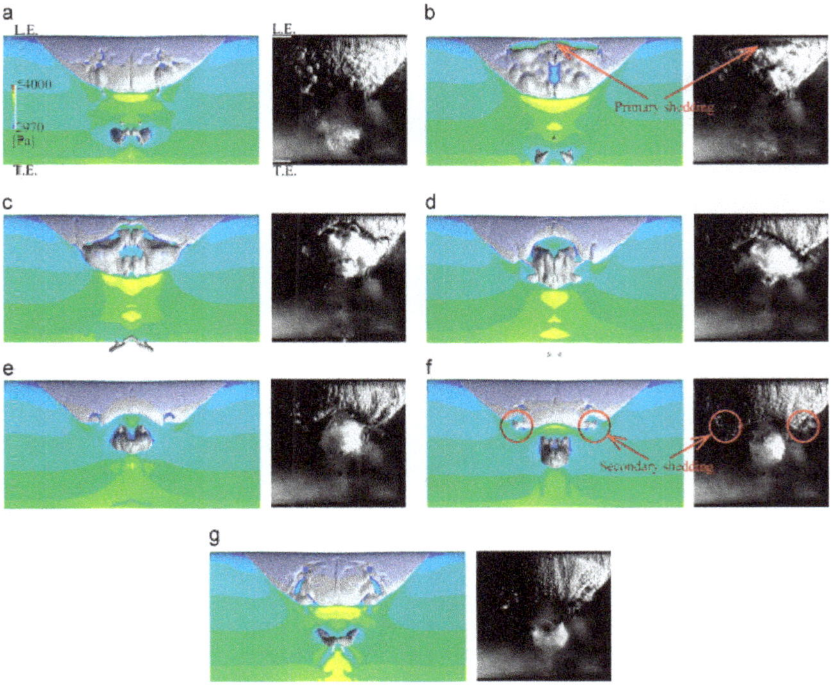

Figure 28. Top view of cavity shedding patterns for $\sigma = 1.07$ during one typical cycle (**a**–**g**). (Left: numerical results. Right: experimental observation by Foeth (2008) [74].) Reprint from [73] with permission from Elsevier, 2014, License Number 5830860095566.

The practical implications of cavitation in jet pump cavitation reactors (JPCRs) have also been explored via an investigation of the flow performance and induced noise [29]. The JPCR is an important object of study in vortex bubble research because it has great potential for future applications. Figure 30 shows its structure.

The important parameters of the JPCR are the flow ratio (q), pressure ratio (h), and efficiency (η), which are defined as follows:

$$q = \frac{Q_s}{Q_d} \tag{4}$$

$$h = \frac{P_0 - P_s}{P_d - P_s} = \frac{\left(\frac{p_0}{\rho_0 g} + \frac{v_0^2}{2g} + z_0\right) - \left(\frac{p_s}{\rho_s g} + \frac{v_s^2}{2g} + z_s\right)}{\left(\frac{p_d}{\rho_d g} + \frac{v_d^2}{2g} + z_0\right) - \left(\frac{p_s}{\rho_s g} + \frac{v_s^2}{2g} + z_s\right)} \tag{5}$$

$$\eta = \frac{\rho_s Q_S (P_0 - P_s)}{\rho_d Q_d (P_d - P_0)} = q\frac{h}{1 - h} \tag{6}$$

where Q represents the flow rate, ρ the density, P the total pressure, p the hydrostatic pressure, z the elevation, and v the averaged flow velocity.

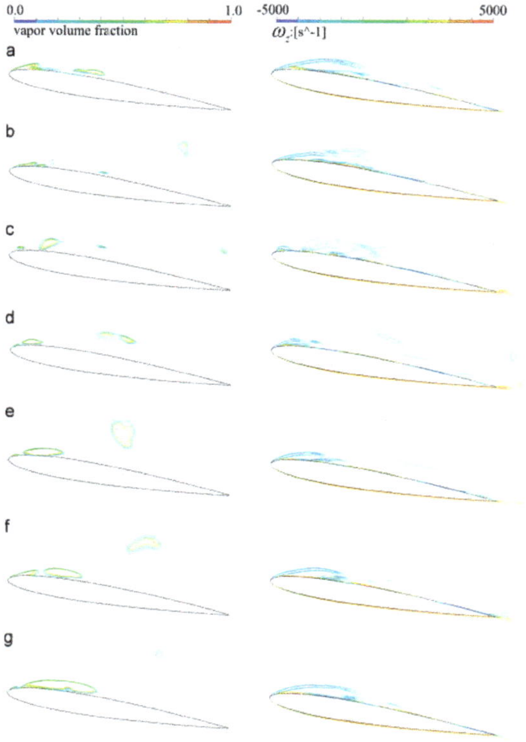

Figure 29. Comparison of vapor volume fraction (**left**) and vorticity (**right**) contours for $\sigma = 1.07$ at the plane with $z = 0$ during one typical cycle (**a–g**). Reprint from [73] with permission from Elsevier, 2014, License Number 5830860095566.

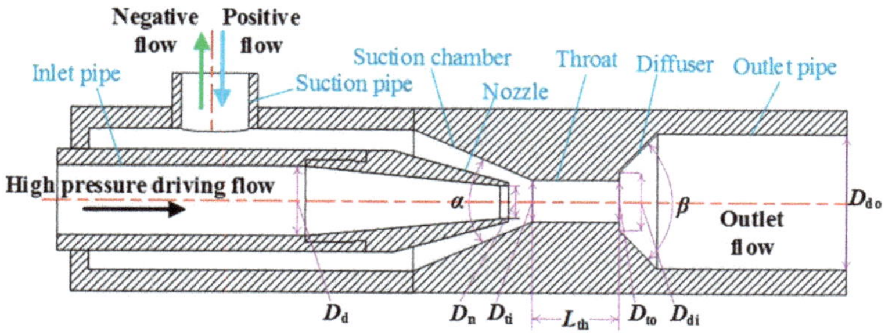

Figure 30. Schematic drawing of JPCR [29].

In Figure 31, we can see the vortex cavitation flow from the JPCR, and in Figure 32, we can see the various cavitation regions of JPCRs with different throat types. This team's research emphasizes the relationship between cavitation stages, noise signals, and flow parameters, which is helpful for JPCR design and application. The findings of their study, obtained by systematically altering the throat length, throat type, and diffuser angle, provide valuable insights into how we may optimize the design of cavitation reactors for various industrial processes, demonstrating that the JPCR has great potential as a piece of equipment for providing vortex cavitation.

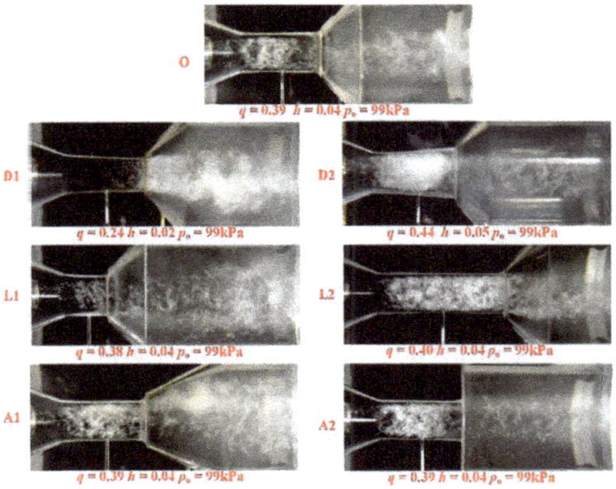

Figure 31. Typical cavitation images at the minimum outlet pressure for different JPCRs [29].

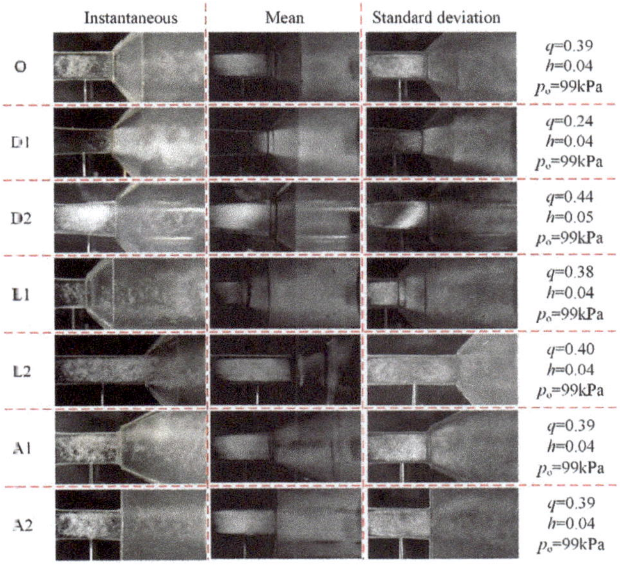

Figure 32. Typical cavitation images at the minimum outlet pressure for JPCRs with different structures [29].

In regard to the application of vortex cavitation, a noteworthy contribution has been made to our understanding of vortex generator effects on propeller cavitation and associated hull pressure fluctuations [75]. The findings of this experimental investigation underscored the role of vortex generators in mitigating the adverse effects caused by propeller cavitation. For instance, installing a vortex generator on a ship can improve sheet cavitation's stability and transparency when installed in the boundary layer and in front of the region of interest. Figure 33 shows cavitation with a vortex generator within a typical cycle. The vortex generator stabilizes cavitation and promotes energy transfer to a larger area on the hull surface. This dual impact of the vortex generator is a significant finding of the study, providing experimental insights into how adjustments in vortex generator configurations can potentially reduce cavitation impacts, offering an opportunity for the application of vortex bubble research.

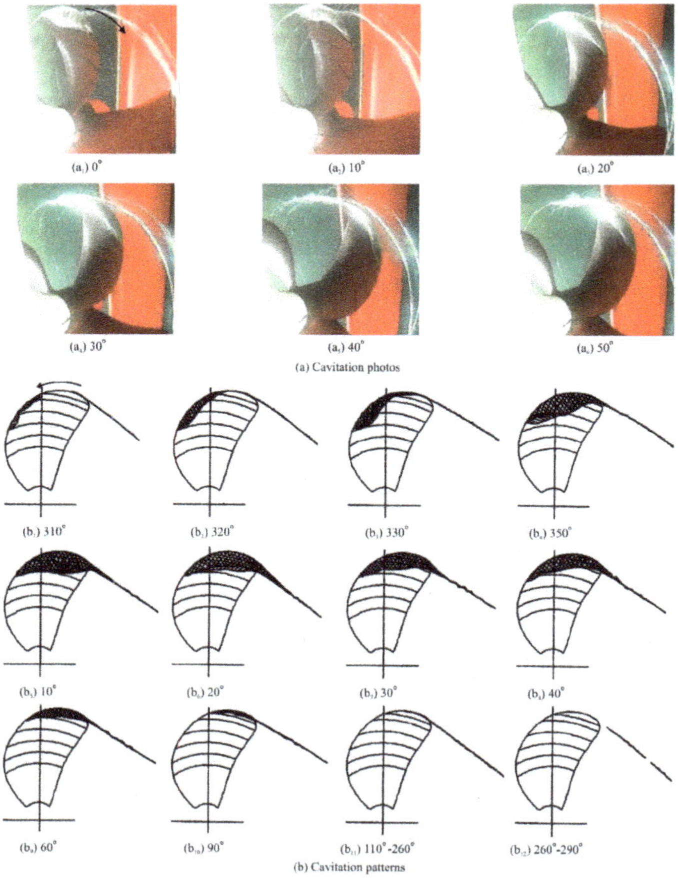

Figure 33. Measured cavitation results with a vortex generator within a typical cycle. Reprint from [75] with permission from Elsevier, 2020, License Number 5831390551079.

Similarly, another cavitation study focused on the acoustic pressures generated by laser-induced cavitation bubble expansion and collapse near rigid walls [76]. Their research revealed that the first and second shock waves released by cavitation bubbles have a depth relationship with γ, the distance ratio between the bubble's nucleation source, and its maximum size. Both pressure waves have the same amplitude for $\gamma = 2.5 \pm 0.1$ and $\gamma > 4$. The pressure and velocity in the numerical analysis also show a peak pressure generated during the bubble's collapse. Simulations predicted the behavior of the bubble, the bubble radius, and the formation and evolution of another vortex cavitation-like vortex. The study underscores the acoustic dimension of cavitation, which is closely related to the vortex bubbles' behavior when collapsing.

Advancing into the characteristics of vortex bubbles, a study applied high-speed photography to investigate the luminescence intensity of vortex cavitation within a Venturi tube and its variation with cavitation number [28]. The cavitation number was defined as follows:

$$\sigma = \frac{p_2 - p_v}{\frac{1}{2}\rho U^2} = \frac{p_2 - p_v}{p_1 - p_2} \tag{7}$$

where p_v and ρ are the vapor pressure and density of the test water, respectively, and U is the velocity at the throat.

As mentioned before, the Venturi tube structure is like that of a JPCR. In the Venturi tube, vortex cavitation can be observed clearly, as shown in Figures 34 and 35, which present the relationship between luminescence intensity and cavitation number. Here, the pressure wave was detected at the collapse of vortex cavitation. Moreover, the sound velocity was inflated, which was attributed to the aggressive intensity of the cavitation. This study explored cavitation's photonic emissions, and the researchers identified a crucial intersection between vortex bubbles and optical phenomena. This offers a new viewpoint on diagnosing and understanding cavitation's energy processes. Such work is integral to applications requiring the precise monitoring and control of cavitation to mitigate destructive effects or harness cavitation energy efficiently. The cited study set the foundation for future vortex bubble research in this regard.

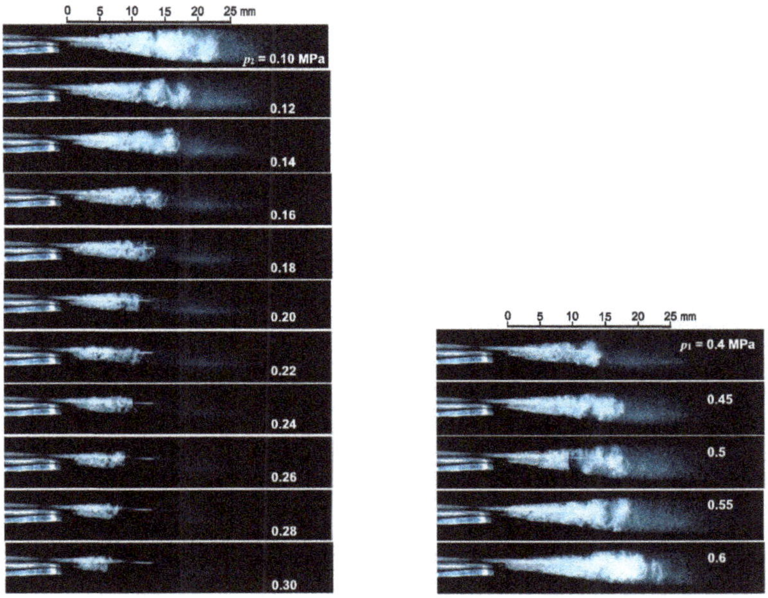

(a) Aspects changing with downstream pressure p_2 (p_1 = 0.6 MPa) (b) Aspect changing with upstream pressure p_1 (p_2 = 0.1 MPa)

Figure 34. Typical aspects of cavitation in a Venturi tube: (**a**) aspects changing with downstream pressure p2 (p1 = 0.6 MPa); (**b**) aspect changing with upstream pressure p1 (p2 = 0.1 MPa) [28].

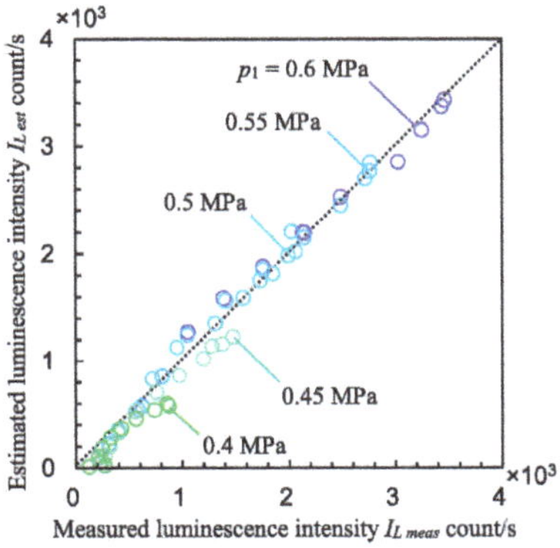

Figure 35. Relationship between measured luminescence intensity and estimated luminescence intensity [28].

A numerical and experimental study on vortex rings extended the discourse to the interaction between shock waves and vortex rings, which has implications for bubble collapse [77]. The vortex ring incubation corresponds to the shock wave when the bubble collapses, as shown in Figure 36. The intricacies of vortex dynamics revealed through this study are instrumental in comprehending how energy is transferred within fluid systems under the influence of shock waves, which directly relates to understanding bubble collapse and related pressure fluctuations.

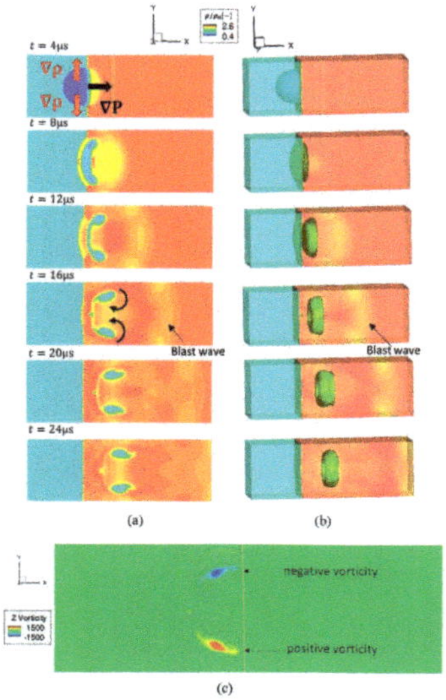

Figure 36. Numerical and experimental study on the behavior of vortex rings generated by shock–bubble interaction. (**a**) two–dimensional views; (**b**) three–dimensional views; (**c**) vorticity contours at t=16 μs. Reprint from [77], with permission from AIP Publishing, 2022, License Number 5831411280013.

A large eddy simulation study focusing on cloud cavitation and wake vortex cavitation around a trailing-truncated hydrofoil provided insights into their influence on the performance and safety of hydraulic machinery [78]. This study sheds light on the intricate flow structures associated with cavitation phenomena, revealing the interactions between cavitation bubbles and the surrounding flow field. The performance of cloud cavitation is shown in Figure 37. The characteristics of cloud bubbles and wake vortex bubbles can be observed. Through simulation, the researchers found that pressure-side bubbles were generally larger than suction-side bubbles in vortex cavitation, and their numerical analysis results matched their experimental results. In sum, these numerical and experimental analyses of cloud and vortex cavitation in hydrofoil machinery equipment help us understand the erosion mechanism at play.

In another large eddy simulation, the mechanisms of tip leakage vortex cavitation development in a cryogenic inducer were researched [79]. By analyzing thermodynamic effects and flow characteristics, this study uncovered how vortex cavitation instabilities, particularly with a tip leakage vortex, can be influenced and optimized in the design of cryogenic pumps. The numerical results agreed with the experimental results, proving that the numerical analysis was reliable. Tip leakage vortex cavitation is especially unstable due to the periodic coupling effect of cavity development, flow rate magnitude, and variation in local incidence angle due to the flow characteristics.

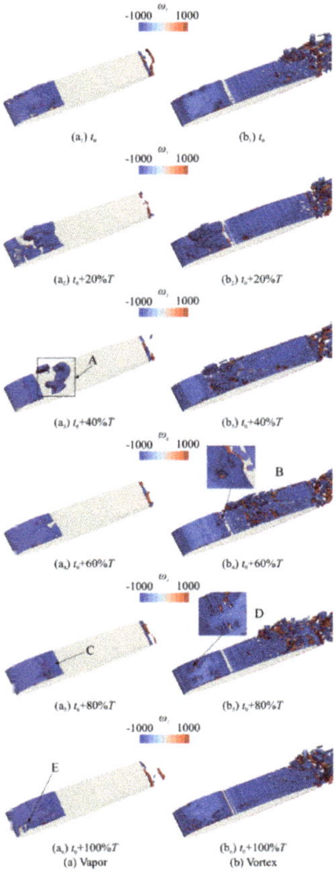

Figure 37. One-cycle shedding process of the cloud cavitation. A: horse-type cavity structure; B: small-scale U-type vortex; C: re-entrant jet; D: vortex tubes; E: non-uniform distribution of cavity and re-entrant jet. The cavity is visualized by the iso-surface with a vapor fraction of 0.1. All the structures are colored in terms of the spanwise vorticity component. Reprint from [78], with permission from Springer Nature, 2022, License Number 5831420313482.

In another area of bubble research, computer fluid dynamics in the Venturi tube have been explored through a numerical experimental analysis, where Eulerian–Lagrangian modeling was applied to the phase transition phenomena in cavitation-driven chemical processes [80]. Figure 38 shows the Venturi tube over time, reporting the axial velocity field with the 10% contour of the volume fraction overlapped. Figure 39 then shows the evolution of the bubbles for all the inlet pressures of 3 to 6 bar. The statistical results showed that the CFD code has an excellent predictive capability and can be used to assess the cavitation behavior in the Venturi tube compared to the experiment. By accurately predicting pressure peak values and bubble evolution using a multi-phase approach, the study demonstrates the efficacy of applying computer analyses in cavitation research. It also offers a reference on carrying out vortex bubble research in the Venturi tube, as well as the potential of cavitation technology in various chemical applications.

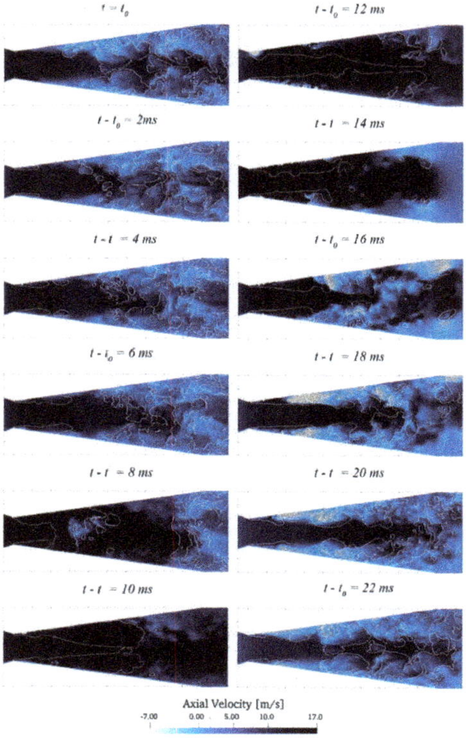

Figure 38. Cross-section of the velocity field with represented volume fraction contour $\gamma = 0.1$ [80].

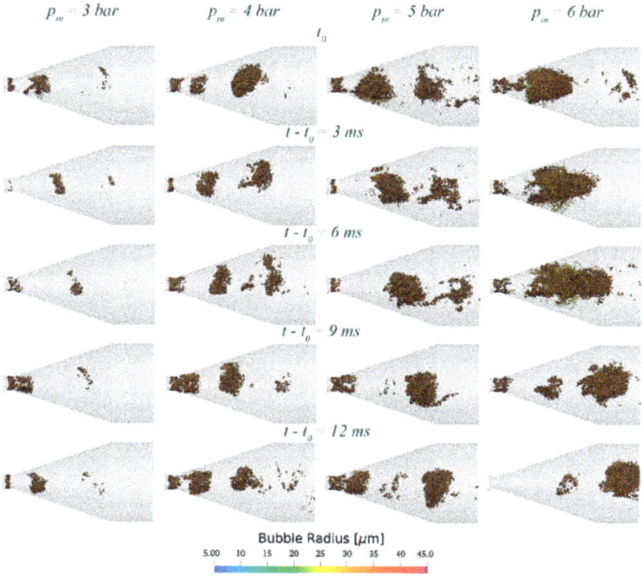

Figure 39. Bubble evolution for p_{in} = 3–6 bar [80].

Moreover, computer simulations were applied for a tip vortex, whereby a numerical investigation was performed of the oscillatory dynamics for different cavitation numbers, focusing on an elliptical hydrofoil section [81]. This study used an effective radius based on the cross-sectional area of the tip vortex cavity, which captured cavity volume variations. Downstream, the study simulated the velocity distribution within the tip vortex region, as shown in Figure 40. Through the detailed simulations and analyses of cavity oscillations at varying cavitation numbers, the study revealed correlations between cavity dynamics, pressure fluctuations, and frequency shifts, contributing to our understanding of tip vortex cavitation behavior.

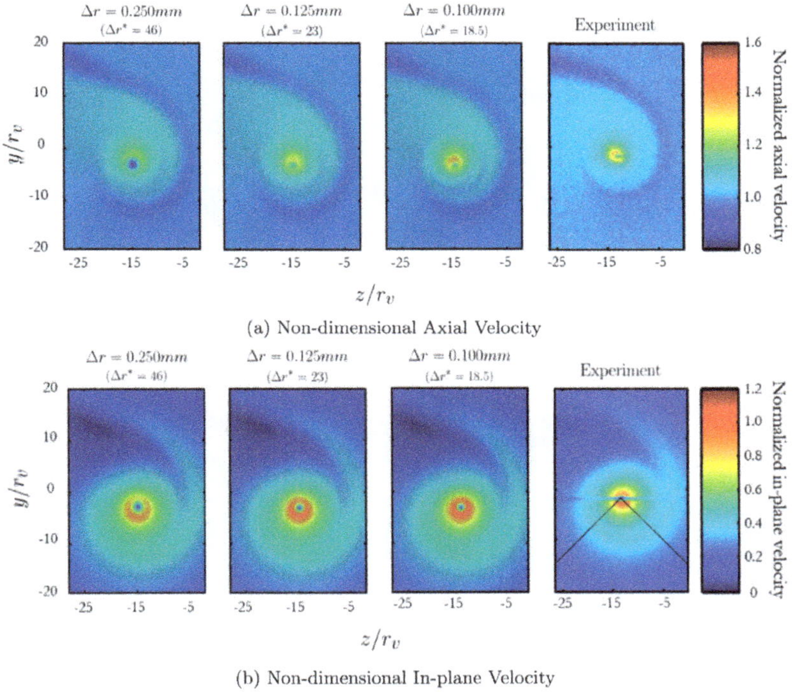

Figure 40. Comparison of nondimensional velocity distribution within the tip vortex region at downstream location $x/c_0 = 1.14$, as obtained from simulations with different mesh sizes, with the experimental measurement [81].

Often, vortex bubbles are investigated through high-speed imaging, but bubbles have been characterized as spherical through numerical analysis. One option to overcome the limitations of these approaches is to instead apply X-ray imaging. Compared to visible light, X-ray imaging offers a different environment in which to observe vortex bubbles [55]. Figures 41 and 42 show the results of such imaging in a Venturi tube. There are many angulated bubbles as well as spherical bubbles, which should be considered when setting the assumptions in future numerical analyses. In addition, the tangential velocity of the surface of vortex cavitation is measured in this approach, along with the relationship of the geometry of the Venturi tube with the shedding frequency in vortex cavitation.

Previously, experimental investigations were performed on tip vortex cavitation (TVC) around a pitching hydrofoil in a cavitation tunnel [82–90]. The experiments used high-speed cameras, torque sensors, and hydrophones to analyze cavitation patterns, the torque exerted on the hydrofoil, and cavitation noise. These research studies focused on understanding the characteristics of TVC induced by the tip vortex and its impact under different flow conditions. After comparing and analyzing the phases of ascent, a static state, and descent, these studies concluded that the tip vortex cavitation intensity is the highest during the descent, moderate during the static state, and the lowest during the ascent.

In summary, the combination of extensive eddy simulations, multi-phase modeling, and advanced numerical techniques showcased in the above studies provides valuable insights into the complex phenomena of vortex cavitation, offering a comprehensive foundation for further research and engineering advancements in fluid dynamics and related fields.

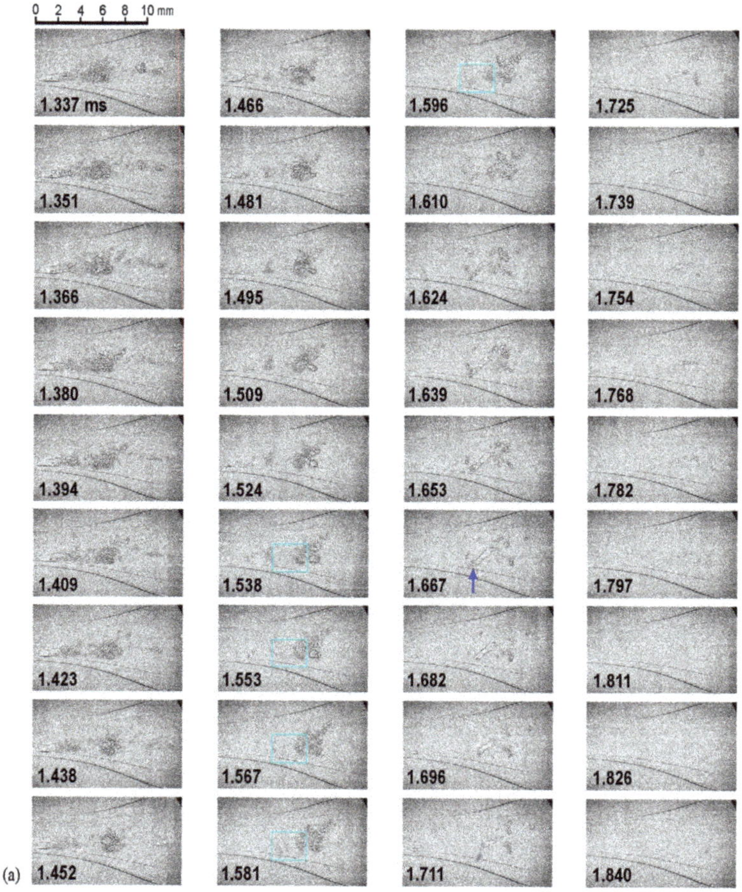

Figure 41. *Cont.*

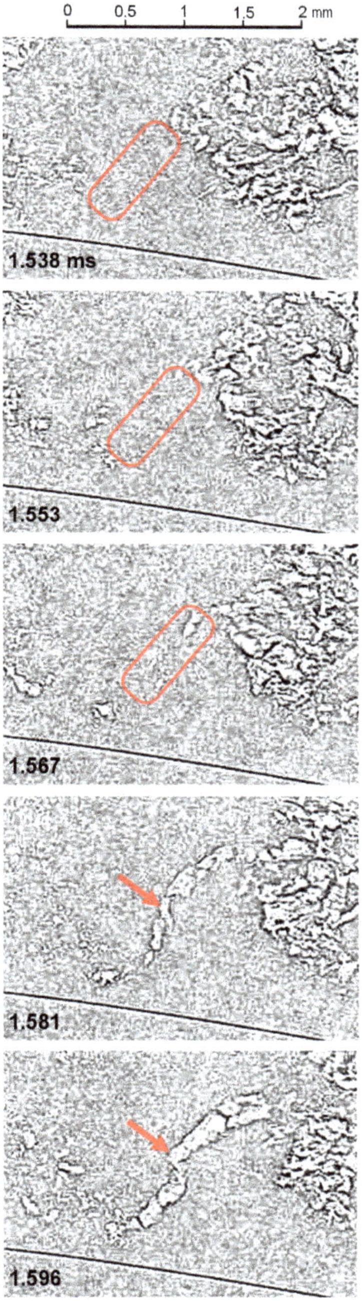

(b)

Figure 41. Vortex cavitation from shedding of trailing edge to collapse, observed by high-speed X-ray imaging. (**a**) Standard view of one cycle. (**b**) Magnified view [55].

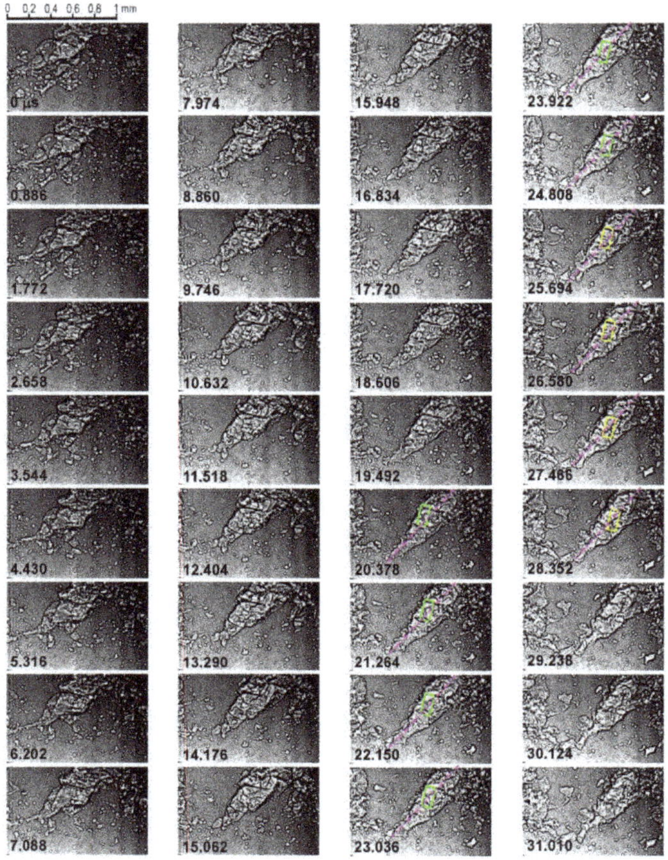

Figure 42. Rotation of vortex cavitation arising in the Venturi tube observed by XFEL [55].

5. Conclusions

This review has covered single cavitation bubbles, cloud cavitation bubbles, and vortex cavitation bubbles. Notably, most research on cavitation bubbles to date has concerned the mechanisms of bubble collapse near a solid boundary. Numerical analyses have been compared with experimental investigations, considering mass transfer, bubble shape, and so on. However, many areas remain to be explored in the research on bubbles, especially concerning their collapse mechanism. The key points highlighted in the present review can be summarized as follows:

(1) Cavitation peening is a new surface modification method used in the production of automobiles and stainless steel. It can improve the life of mechanical parts. However, cavitation bubbles near the solid boundary will cause erosion, such as in pumps and propellers. This highlights the importance of researching cavitation bubbles.

(2) The images obtained to date of cavitation bubbles and the trajectory of the cavitation bubble research field have been summarized, covering single cavitation bubbles, cloud cavitation bubbles, and vortex cavitation bubbles, with a focus on the images of cavitation bubbles near a solid boundary. An important finding is that many studies' numerical analyses of cavitation bubbles have pointed to them having the same characteristics as those depicted in the experimental images. However, recent research has brought to light an angular vortex cavitation bubble, as identified through X-ray imaging, which challenges our prior assumptions on the shapes of bubbles.

(3) The stand-off distance is a primary parameter impacting cavitation bubbles. Numerical and experimental analyses have proven that the stand-off distance is related to the stress and pressure caused by a bubble collapsing. Beyond that, this review has also introduced other parameters that are vital to incorporate when performing simulations, such as the cavitation number and governing equation.

Author Contributions: B.J.: conceptualization, methodology, and writing—original draft preparation. H.S.: visualization, supervision, and writing—review and editing. All authors have read and agreed to the published version of the manuscript.

Funding: This research was partly supported by JSPS KAKENHI (grant numbers 22KK0050 23K25988), JST CREST (JPMJCR2335), and the Suzuki Foundation (5-i19).

Data Availability Statement: The data presented in this study are available upon request from the authors.

Conflicts of Interest: The authors declare no conflicts of interest.

References

1. Soyama, H.; Tanaka, M.; Takiguchi, T.; Yamamoto, M. Development of a Cavitation Generator Mimicking Pistol Shrimp. *Biomimetics* **2024**, *9*, 47. [CrossRef] [PubMed]
2. Lohse, D.; Schmitz, B.; Versluis, M. Snapping shrimp make flashing bubbles. *Nature* **2001**, *413*, 477–478. [CrossRef] [PubMed]
3. Lillis, A.; Perelman, J.N.; Panyi, A.; Aran Mooney, T. Sound production patterns of big-clawed snapping shrimp (*Alpheus* spp.) are influenced by time-of-day and social context. *J. Acoust. Soc. Am.* **2017**, *142*, 3311–3320. [CrossRef] [PubMed]
4. Yang, Y.; Qin, S.; Di, C.; Qin, J.; Wu, D.; Zhao, J. Research on claw motion characteristics and cavitation bubbles of snapping shrimp. *Appl. Bionics Biomech.* **2020**, *2020*, 6585729. [CrossRef]
5. Larsen, O.N.; Reichmuth, C. Walruses produce intense impulse sounds by clap-induced cavitation during breeding displays. *R. Soc. Open Sci.* **2021**, *8*, 210197. [CrossRef]
6. Song, Z.; Salas, A.K.; Montie, E.W.; Laferriere, A.; Zhang, Y.; Aran Mooney, T. Sound pressure and particle motion components of the snaps produced by two snapping shrimp species (Alpheus heterochaelis and Alpheus angulosus). *J. Acoust. Soc. Am.* **2021**, *150*, 3288–3301. [CrossRef]
7. Kingston, A.C.N.; Woodin, S.A.; Wethey, D.S.; Speiser, D.I. Snapping shrimp have helmets that protect their brains by dampening shock waves. *Curr. Biol.* **2022**, *32*, 3576–3583.e3573. [CrossRef]
8. Dinh, J.P.; Patek, S.N. Weapon performance and contest assessment strategies of the cavitating snaps in snapping shrimp. *Funct. Ecol.* **2023**, *37*, 327–342. [CrossRef]
9. Versluis, M.; Schmitz, B.; von der Heydt, A.; Lohse, D. How Snapping Shrimp Snap: Through Cavitating Bubbles. *Science* **2000**, *289*, 2114–2117. [CrossRef]
10. Naude´, C.F.; Ellis, A.T. On the Mechanism of Cavitation Damage by Nonhemispherical Cavities Collapsing in Contact With a Solid Boundary. *J. Basic Eng.* **1961**, *83*, 648–656. [CrossRef]
11. Philipp, A.; Lauterborn, W. Cavitation erosion by single laser-produced bubbles. *J. Fluid Mech.* **1998**, *361*, 75–116. [CrossRef]
12. Brennen, C.E. *Cavitation and Bubble Dynamics*; Cambridge University Press: Cambridge, UK, 2013; pp. 90–95.
13. Gągol, M.; Przyjazny, A.; Boczkaj, G. Wastewater treatment by means of advanced oxidation processes based on cavitation—A review. *Chem. Eng. J.* **2018**, *338*, 599–627. [CrossRef]
14. Zheng, H.; Zheng, Y.; Zhu, J. Recent Developments in Hydrodynamic Cavitation Reactors: Cavitation Mechanism, Reactor Design, and Applications. *Engineering* **2022**, *19*, 180–198. [CrossRef]
15. Bagal, M.V.; Gogate, P.R. Wastewater treatment using hybrid treatment schemes based on cavitation and Fenton chemistry: A review. *Ultrason. Sonochemistry* **2014**, *21*, 1–14. [CrossRef]
16. Wang, B.; Liu, Y.; Zhang, H.; Shi, W.; Xiong, M.; Gao, C.; Cui, M. Hydrodynamic cavitation and its application in water treatment combined with ozonation: A review. *J. Ind. Eng. Chem.* **2022**, *114*, 33–51. [CrossRef]
17. Chen, Y.; Yin, C.; Song, Y. Application of hydrodynamic cavitation in the field of water treatment. *Chem. Pap.* **2023**, *77*, 3521–3546. [CrossRef]
18. Gogate, P.R.; Kabadi, A.M. A review of applications of cavitation in biochemical engineering/biotechnology. *Biochem. Eng. J.* **2009**, *44*, 60–72. [CrossRef]
19. Kelkar, M.A.; Gogate, P.R.; Pandit, A.B. Intensification of esterification of acids for synthesis of biodiesel using acoustic and hydrodynamic cavitation. *Ultrason. Sonochemistry* **2008**, *15*, 188–194. [CrossRef] [PubMed]
20. Innocenzi, V.; Prisciandaro, M. Technical feasibility of biodiesel production from virgin oil and waste cooking oil: Comparison between traditional and innovative process based on hydrodynamic cavitation. *Waste Manag.* **2021**, *122*, 15–25. [CrossRef]
21. Ahmad, A.; Yadav, A.K.; Singh, A.; Pal, A. Optimization of cavitation-assisted biodiesel production and fuel properties from Neochloris oleoabundans microalgae oil using genetic algorithm and response surface methodology. *Proc. Inst. Mech. Eng. Part E J. Process Mech. Eng.* **2024**, *238*, 1821–1835. [CrossRef]

22. Giuliano, A.; Cellamare, C.M.; Chiarini, L.; Tabacchioni, S.; Petta, L. Evaluation of the controlled hydrodynamic cavitation as gas mass transfer system for ex-situ biological hydrogen methanation. *Chem. Eng. J.* **2023**, *471*, 144475. [CrossRef]
23. Deggelmann, M.; Nöpel, J.-A.; Rüdiger, F.; Paustian, D.; Braeutigam, P. Hydrodynamic cavitation for micropollutant degradation in water—Correlation of bisphenol A degradation with fluid mechanical properties. *Ultrason. Sonochemistry* **2022**, *83*, 105950. [CrossRef]
24. Agarkoti, C.; Gujar, S.K.; Gogate, P.R.; Pandit, A.B. Pilot scale degradation of Sulfamerazine using different venturi based hydrodynamic cavitation and ultrasound reactors in combination with oxidation processes. *J. Environ. Chem. Eng.* **2023**, *11*, 109857. [CrossRef]
25. Feng, Z.; Wang, F.; Zhu, K.; Wang, Z.; Ning, J. Degradation of ammonia nitrogen by an economic combined hydrodynamic cavitation method. *Environ. Sci. Pollut. Res.* **2023**, *30*, 72782–72792. [CrossRef]
26. Min Oo, Y.; Somnuk, K. Investigation of free fatty acid reduction from mixed crude palm oil using 3D-printed rotor–stator hydrodynamic cavitation: An experimental study of geometric characteristics of the inner hole. *Ultrason. Sonochemistry* **2023**, *98*, 106472. [CrossRef] [PubMed]
27. Xu, S.; Wang, J.; Chen, W.; Ji, B.; Yan, H.; Zhang, Z.; Long, X. Removal of field-collected Microcystis aeruginosa in pilot-scale by a jet pump cavitation reactor. *Ultrason. Sonochemistry* **2022**, *83*, 105924. [CrossRef] [PubMed]
28. Soyama, H. Luminescence intensity of vortex cavitation in a venturi tube changing with cavitation number. *Ultrason. Sonochemistry* **2021**, *71*, 105389. [CrossRef] [PubMed]
29. Wang, J.; Cheng, H.; Xu, S.; Ji, B.; Long, X. Performance of cavitation flow and its induced noise of different jet pump cavitation reactors. *Ultrason. Sonochemistry* **2019**, *55*, 322–331. [CrossRef]
30. Liu, L.; Yang, X.; Guo, Y.; Li, B.; Wang, L.-P. Reactive mixing performance for a nanoparticle precipitation in a swirling vortex flow reactor. *Ultrason. Sonochemistry* **2023**, *94*, 106332. [CrossRef]
31. Song, Y.; Hou, R.; Liu, Z.; Liu, J.; Zhang, W.; Zhang, L. Cavitation characteristics analysis of a novel rotor-radial groove hydrodynamic cavitation reactor. *Ultrason. Sonochemistry* **2022**, *86*, 106028. [CrossRef]
32. Soyama, H.; Korsunsky, A.M. A critical comparative review of cavitation peening and other surface peening methods. *J. Mater. Process. Technol.* **2022**, *305*, 117586. [CrossRef]
33. Soyama, H. Improvement of Fatigue Strength of 3D-Metal by Combined Process of Blasting and Cavitation Peening. In Proceedings of the 3rd International Conference on Advanced Surface Enhancement (INCASE) 2023, Singapore, 25–27 September 2023.
34. Soyama, H.; Shimizu, M.; Hattori, Y.; Nagasawa, Y. Improving the fatigue strength of the elements of a steel belt for CVT by cavitation shotless peening. *J. Mater. Sci.* **2008**, *43*, 5028–5030. [CrossRef]
35. Soyama, H.; Sekine, Y. Sustainable surface modification using cavitation impact for enhancing fatigue strength demonstrated by a power circulating-type gear tester. *Int. J. Sustain. Eng.* **2010**, *3*, 25–32. [CrossRef]
36. Zhang, Y.; Zhang, Y.; Qian, Z.; Ji, B.; Wu, Y. A review of microscopic interactions between cavitation bubbles and particles in silt-laden flow. *Renew. Sustain. Energy Rev.* **2016**, *56*, 303–318. [CrossRef]
37. Wang, T.; Chen, L. Thermodynamic Behavior and Energy Transformation Mechanism of the Multi-Period Evolution of Cavitation Bubbles Collapsing near a Rigid Wall: A Numerical Study. *Energies* **2023**, *16*, 1048. [CrossRef]
38. Yu, J.; Yan, J.; Jiang, L.; Li, J.; Guo, H.; Qiao, M.; Qu, L. Fluorescence enhancement of organic dyes by femtosecond laser-induced cavitation bubbles for crystal imaging. *Nanoscale* **2023**, *15*, 8730–8739. [CrossRef] [PubMed]
39. Subramanian, R.K.; Coutier-Delgosha, O. Intrusive temperature measurement of the cavitation bubbles using cold wire. *J. Acoust. Soc. Am.* **2022**, *152*, A249. [CrossRef]
40. Reuter, F.; Cairós, C.; Mettin, R. Vortex dynamics of collapsing bubbles: Impact on the boundary layer measured by chronoamperometry. *Ultrason. Sonochemistry* **2016**, *33*, 170–181. [CrossRef]
41. Yasui, K. Production of O Radicals from Cavitation Bubbles under Ultrasound. *Molecules* **2022**, *27*, 4788. [CrossRef]
42. Izak Ghasemian, S.; Reuter, F.; Ohl, C.D. High-speed ultrasound imaging of laser-induced cavitation bubbles. *Appl. Phys. Lett.* **2021**, *119*, 114101. [CrossRef]
43. Krella, A.K. Degradation and Protection of Materials from Cavitation Erosion: A Review. *Materials* **2023**, *16*, 2058. [CrossRef] [PubMed]
44. Reuter, F.; Gonzalez-Avila, S.R.; Mettin, R.; Ohl, C.-D. Flow fields and vortex dynamics of bubbles collapsing near a solid boundary. *Phys. Rev. Fluids* **2017**, *2*, 064202. [CrossRef]
45. Iga, Y.; Sasaki, H. Relationship between a non-spherical collapse of a bubble and a stress state inside a wall. *Phys. Fluids* **2023**, *35*, 023312. [CrossRef]
46. Vogel, A.; Lauterborn, W. Acoustic transient generation by laser-produced cavitation bubbles near solid boundaries. *J. Acoust. Soc. Am.* **1988**, *84*, 719–731. [CrossRef]
47. Hickling, R.; Plesset, M.S. Collapse and Rebound of a Spherical Bubble in Water. *Phys. Fluids* **1964**, *7*, 7–14. [CrossRef]
48. Benjamin, T.B.; Ellis, A.T. The Collapse of Cavitation Bubbles and the Pressures thereby Produced against Solid Boundaries. *Philos. Trans. R. Soc. London. Ser. A Math. Phys. Sci.* **1966**, *260*, 221–240.
49. Chapman, R.B.; Plesset, M.S. Collapse of an initially spherical vapour cavity in the neighbourhood of a solid boundary. *J. Fluid Mech.* **1971**, *47*, 283–290.

50. Brujan, E.; Keen, G.; Vogel, A.; Blake, J. The final stage of the collapse of a cavitation bubble close to a rigid boundary. *Phys. Fluids* **2002**, *14*, 85–92. [CrossRef]
51. Brujan, E.A.; Ikeda, T.; Yoshinaka, K.; Matsumoto, Y. The final stage of the collapse of a cloud of bubbles close to a rigid boundary. *Ultrason. Sonochemistry* **2011**, *18*, 59–64. [CrossRef]
52. Colonius, T.I.M.; Johnsen, E. Numerical simulations of non-spherical bubble collapse. *J. Fluid Mech.* **2009**, *629*, 231–262.
53. Li, S.; Zhang, A.M.; Han, R.; Liu, Y. Experimental and numerical study on bubble-sphere interaction near a rigid wall. *Phys. Fluids* **2017**, *29*, 092102. [CrossRef]
54. Fujikawa, S.; Akamatsu, T. Experimental Investigations of Cavitation Bubble Collapse by a Water Shock Tube. *Bull. JSME* **1978**, *21*, 223–230. [CrossRef]
55. Soyama, H.; Liang, X.; Yashiro, W.; Kajiwara, K.; Asimakopoulou, E.M.; Bellucci, V.; Birnsteinova, S.; Giovanetti, G.; Kim, C.; Kirkwood, H.J.; et al. Revealing the origins of vortex cavitation in a Venturi tube by high speed X-ray imaging. *Ultrason. Sonochemistry* **2023**, *101*, 106715. [CrossRef]
56. Soyama, H.; Iga, Y. Laser Cavitation Peening: A Review. *Appl. Sci.* **2023**, *13*, 6702. [CrossRef]
57. Soyama, H. Development of Laser Cavitation Peening Using a Normal-Oscillation Nd:YAG Laser. *Coatings* **2023**, *13*, 1395. [CrossRef]
58. Kuji, C.; Soyama, H. Mechanical Surface Treatment of Titanium Alloy Ti6Al4V Manufactured by Direct Metal Laser Sintering Using Laser Cavitation. *Metals* **2023**, *13*, 181. [CrossRef]
59. Soyama, H.; Kuji, C.; Liao, Y. Comparison of the effects of submerged laser peening, cavitation peening and shot peening on the improvement of the fatigue strength of magnesium alloy AZ31. *J. Magnes. Alloys* **2023**, *11*, 1592–1607. [CrossRef]
60. Sasaki, H.; Ochiai, N.; Iga, Y. Numerical Analysis of Damping Effect of Liquid Film on Material in High Speed Liquid Droplet Impingement. *Int. J. Fluid Mach. Syst.* **2016**, *9*, 57–65. [CrossRef]
61. Aganin, A.A.; Davletshin, A.I. A particle model of interaction between weakly non-spherical bubbles. *Appl. Math. Model.* **2024**, *126*, 185–205. [CrossRef]
62. Chahine, G.L.; Duraiswami, R. Dynamical interactions in a multi-bubble cloud. *J. Fluids Eng.* **1992**, *114*, 680. [CrossRef]
63. Reisman, G.E.; Wang, Y.C.; Brennen, C.E. Observations of shock waves in cloud cavitation. *J. Fluid Mech.* **1998**, *355*, 255–283. [CrossRef]
64. Ma, J.; Hsiao, C.-T.; Chahine, G.L. Euler–Lagrange Simulations of Bubble Cloud Dynamics Near a Wall. *J. Fluids Eng.* **2015**, *137*, 041301. [CrossRef]
65. Ma, J.; Hsiao, C.T.; Chahine, G.L. Dynamics of a Bubble Cloud near a Rigid Wall. *J. Phys. Conf. Ser.* **2015**, *656*, 012153. [CrossRef]
66. Ma, J.; Hsiao, C.-T.; Chahine, G.L. Numerical study of acoustically driven bubble cloud dynamics near a rigid wall. *Ultrason. Sonochemistry* **2018**, *40*, 944–954. [CrossRef] [PubMed]
67. Okita, K.; Miyamoto, Y.; Furukawa, T.; Takagi, S. Numerical study on stress in a solid wall caused by the collapse of a cavitation bubble cloud in hydraulic fluid. *Int. J. Multiph. Flow* **2022**, *150*, 103965. [CrossRef]
68. Liu, Y.; Zhang, H.; Zhang, W.; Wang, B. Bubble Size Distribution at Early Stage of Hydrodynamic Cloud Cavitation. *SSRN Electron. J.* **2023**, *35*, 063305. [CrossRef]
69. Zhang, H.; Liu, Y.; Wang, B. Spatial-temporal features of the coherent structure of sheet/cloud cavitation flows using a frequency-weighted dynamic mode decomposition approach. *Phys. Fluids* **2021**, *33*, 053317. [CrossRef]
70. Wang, Z.; Cheng, H.; Ji, B.; Peng, X. Numerical investigation of inner structure and its formation mechanism of cloud cavitating flow. *Int. J. Multiph. Flow* **2023**, *165*, 104484. [CrossRef]
71. Soyama, H.; Ohba, K.; Takeda, S.; Oba, R. High-speed observations of highly erosive vortex cavitation around butterfly valve. *JSME* **1994**, *60*, 1133–1138. [CrossRef]
72. Arndt, R.; Arakeri, V.; Higuchi, H. Some observations of tip-vortex cavitation. *J. Fluid Mech.* **1991**, *229*, 269–289. [CrossRef]
73. Ji, B.; Luo, X.; Arndt, R.E.; Wu, Y. Numerical simulation of three dimensional cavitation shedding dynamics with special emphasis on cavitation–vortex interaction. *Ocean Eng.* **2014**, *87*, 64–77. [CrossRef]
74. Foeth, E.-J.; van Terwisga, T.; van Doorne, C. On the Collapse Structure of an Attached Cavity on a Three-Dimensional Hydrofoil. *J. Fluids Eng.* **2008**, *130*, 071303. [CrossRef]
75. Huang, H.-b.; Long, Y.; Ji, B. Experimental investigation of vortex generator influences on propeller cavitation and hull pressure fluctuations. *J. Hydrodyn.* **2020**, *32*, 82–92. [CrossRef]
76. Gonzalez-Avila, S.R.; Denner, F.; Ohl, C.-D. The acoustic pressure generated by the cavitation bubble expansion and collapse near a rigid wall. *Phys. Fluids* **2021**, *33*, 032118. [CrossRef]
77. Kitamura, K.; Yue, Z.; Fujimoto, T.; Asai, H.; Kubota, A.; Myokan, M.; Ichihara, D.; Sasoh, A. Numerical and experimental study on the behavior of vortex rings generated by shock–bubble interaction. *Phys. Fluids* **2022**, *34*, 046105. [CrossRef]
78. Yin, T.; Pavesi, G.; Pei, J.; Yuan, S.Q.; Gan, X. Large eddy simulation of cloud cavitation and wake vortex cavitation around a trailing-truncated hydrofoil. *J. Hydrodyn.* **2022**, *34*, 893–903. [CrossRef]
79. Chen, T.; Mu, Z.; Chen, J.; Tan, S.; Fan, Y. Numerical investigations on the mechanisms of the tip leakage vortex cavitation development in a cryogenic inducer with large eddy simulation. *Phys. Fluids* **2023**, *35*, 073328. [CrossRef]
80. Duronio, F.; Di Mascio, A.; De Vita, A.; Innocenzi, V.; Prisciandaro, M. Eulerian–Lagrangian modeling of phase transition for application to cavitation-driven chemical processes. *Phys. Fluids* **2023**, *35*, 053305. [CrossRef]
81. Lak, S.; Jaiman, R. A numerical study on the oscillatory dynamics of tip vortex cavitation. *arXiv* **2023**, arXiv:2312.15579.

82. Wang, L.; Wang, P.; Wu, K.; Wang, H.; Huang, B.; Wu, D. Experiment investigation of the tip vortex cavitation around a pitching hydrofoil. *Phys. Fluids* **2023**, *35*, 103313. [CrossRef]
83. Amini, A.; Reclari, M.; Sano, T.; Iino, M.; Farhat, M. Suppressing tip vortex cavitation by winglets. *Exp. Fluids* **2019**, *60*, 159. [CrossRef]
84. Chahine, G.L.; Hsiao, C.-T.; Raju, R. Scaling of cavitation bubble cloud dynamics on propellers. In *Advanced Experimental and Numerical Techniques for Cavitation Erosion Prediction. Fluid Mechanics and Its Applications*; Springer: Dordrecht, The Netherlands, 2014; Volume 106; pp. 345–372.
85. Khoo, M.T.; Venning, J.A.; Pearce, B.W.; Brandner, P.A. Nucleation and cavitation number effects on tip vortex cavitation dynamics and noise. *Exp. Fluids* **2021**, *62*, 216. [CrossRef]
86. Lee, J.-H.; Han, J.-M.; Park, H.-G.; Seo, J.-S. Application of signal processing techniques to the detection of tip vortex cavitation noise in marine propeller. *J. Hydrodyn.* **2013**, *25*, 440–449. [CrossRef]
87. Park, I.; Kim, J.; Paik, B.; Seol, H. Numerical Study on Tip Vortex Cavitation Inception on a Foil. *Appl. Sci.* **2021**, *11*, 7332. [CrossRef]
88. Park, K.; Seol, H.; Choi, W.; Lee, S. Numerical prediction of tip vortex cavitation behavior and noise considering nuclei size and distribution. *Appl. Acoust.* **2009**, *70*, 674–680. [CrossRef]
89. Peng, X.-X.; Zhang, L.-X.; Wang, B.-L.; Xu, L.-H.; Song, M.-T.; Cao, Y.-T.; Liu, Y.-W.; Hong, F.-W.; Yan, K. Study of tip vortex cavitation inception and vortex singing. *J. Hydrodyn.* **2019**, *31*, 1170–1177. [CrossRef]
90. Zhang, L.-X.; Zhang, N.; Peng, X.-X.; Wang, B.-L.; Shao, X.-M. A review of studies of mechanism and prediction of tip vortex cavitation inception. *J. Hydrodyn.* **2015**, *27*, 488–495. [CrossRef]

fluids

Article

Effect of Dissolved Carbon Dioxide on Cavitation in a Circular Orifice

Sina Safaei * and Carsten Mehring

Institute of Mechanical Process Engineering, Böblinger Str. 72, 70199 Stuttgart, Germany; carsten.mehring@imvt.uni-stuttgart.de
* Correspondence: sina.safaei@imvt.uni-stuttgart.de

Abstract: In this work, we investigate the effect of dissolved gas concentration on cavitation inception and cavitation development in a transparent sharp-edged orifice, similar to that previously analyzed by Nurick in the context of liquid injectors. The working liquid is water, and carbon dioxide is employed as a non-condensable dissolved gas. Cavitation inception points are determined for different dissolved gas concentration levels by measuring wall-static pressures just downstream of the orifice contraction and visually observing the onset of a localized (vapor) bubble cloud formation and collapse. Cavitation onset correlates with a plateau in wall-static pressure measurements as a function of a cavitation number. An increase in the amount of dissolved carbon dioxide is found to increase the cavitation number at which the onset of cavitation occurs. The transition from cloud cavitation to extended-sheet or full cavitation along the entire orifice length occurs suddenly and is shifted to higher cavitation numbers with increasing dissolved gas content. Volume flow rate measurements are performed to determine the change in the discharge coefficient with the cavitation number and dissolved gas content for the investigated cases. CFD analyses are carried out based on the cavitation model by Zwart et al. and the model by Yang et al. to account for non-condensable gases. Discharge coefficients obtained from the numerical simulations are in good agreement with experimental values, although they are slightly higher in the cavitating case. The earlier onset of fluid cavitation (i.e., cavitation inception at higher cavitation numbers) with increasing dissolved carbon dioxide content is not predicted using the employed numerical model.

Keywords: cavitation number; cavitation inception; outgassing; dissolved carbon dioxide; discharge coefficient

1. Introduction

When the pressure p in a liquid falls below a certain threshold or critical pressure p_{crit}, vapor bubbles are generated [1–3]. This abrupt phase-change process, which is associated with the generation of noise and vibration in the system, is called cavitation or, more precisely, vapor cavitation. When the pressure drop is a result of pressure reduction due to an acceleration of the liquid, the cavitation phenomenon is referred to as hydrodynamic cavitation. On the other hand, pressure reduction in a liquid can also result in a degassing process, i.e., the evolution of gases previously dissolved in the liquid phase. This process is frequently referred to as gas cavitation. Vapor cavitation can have severe adverse effects on the processes within which it occurs. For instance, on the efficiency and service life of ship propellers and hydrofoils, since it causes disturbances of the flowfield around the propeller blades and foils, energy dissipation, and the collapse of the generated vapor bubbles (once they reach areas of higher pressure in the flow) can significantly damage the blade/foil surfaces and lead to material erosion or failure. The cavitation phenomenon has also found broad technical applications. Acoustic cavitation (where pressure variations are induced via ultrasonication) is used to erode and break kidney stones [4] or to clean the surfaces of root canals, filter membranes, surgical equipment, or electronic substrates [5].

Hydrodynamic cavitation has found applications in water purification, the recycling of waste, and process water and material recycling [6–10]. As gases dissolved in the liquid phase evolve, they can affect cavitation inception [11,12] and the subsequent dynamics of the cavitation bubbles, including, for example, peak pressures during a bubble collapse. Control of this latter and other parameters using the amount of dissolved gas present in the liquid phase can be used in cavitation reactors [13] to control the cleaning action exerted on dispersed matter present within the liquid phase or treat its surface in a controlled way.

The present investigation aims to provide additional information on the influence of dissolved gases on cavitation onset by systematically varying the dissolved gas (i.e., CO_2) fluid content for a well-known orifice flow case [14]. The newly obtained experimental results augment the fairly small database on orifice cavitation in the presence of dissolved gases, improving the ability to test and validate new cavitation models, which include the effect of non-condensable gases on hydrodynamic cavitation.

1.1. Experimental Studies on Orifice Cavitation

Prediction and/or sensing of cavitation inception is an important topic for operational safety (w.r.t process performance and system hardware failure) [15] and its promotion or control in related applications. Observing the onset of cavitation and its transition to fully cavitating flow in simple hydrodynamic systems has also been a useful benchmark case for evaluating the accuracy of CFD analysis tools used for the design of hydrodynamic cavitation reactors or other processes where cavitation phenomena play an important role. Nurick [14] was one of the first to study cavitation in straight cylindrical sharp-edged orifices, specifically in context with their relevance to so-called unlike (fuel-oxidizer) impinging doublet injector elements for rocket propulsion. Nurick measured static pressures at two locations in close vicinity and downstream of the contraction as a function of orifice inlet pressure (at fixed downstream pressure). He also provided flow visualizations for a nozzle made of Lucite (PMMA), visualizing the cavitation region from the onset of cavitation to full cavitation. However, Nurick did not investigate the effect of dissolved gases in his work. Nurick's early contribution to the topic remains a frequently used benchmark for CFD-based cavitation models, even though he later also investigated flow transition (from non-cavitating to cavitating flow) for other geometries, e.g., in a 90-degree sharp-edge orifice [16].

The effect of dissolved gas on the onset of cavitation in a liquid flowing through an orifice was already discussed by Yan and Thorpe [17] and later investigated by Mishra and Peles [18]. They found that cavitation inception occurred at higher values of cavitation numbers with increasing amounts of oxygen dissolved in the liquid. They also observed a faster transition between cavitation onset and full cavitation with increasing dissolved oxygen content (see also Nurick [16] in this context). On the other hand, the dissolved gas content did not have an effect on the discharge coefficient in their experiments. The influence of dissolved gas content on hydrodynamic cavitation was also considered by Zhang, Zuo, and Liu [19], finding only little influence of the dissolved gas (oxygen) content on the (mean) length of the cavitation zone downstream of the ramp.

Li, Gu, and Chen [11] studied the influence of dissolved gases on the tensile strength of liquids and cavitation inception under acoustic cavitation conditions. They found that with increased gas concentration (nitrogen or oxygen), the tensile strength of water decreases, whereby the effect is stronger for oxygen in comparison to nitrogen as dissolved gas. The effect of dissolved gases on liquid cavitation has also been analyzed by Rooze [20], focusing on radical production in ultrasonic cavitation. The author also analyzed hydrodynamic cavitation for a similar configuration as Misha and Peles [18] but limited to the range of full cavitation without including dissolved gases. Freudigmann et al. [21] and Kowalski et al. [22] studied the evolution of dissolved gases out of the liquid phase as a result of vapor cavitation. The latter authors identify three regimes for cavitation-induced gas release with an exponential increase in evolved gas volume fraction with a decreasing cavitation number at high cavitation intensities.

Duke et al. [23–25] presented a series of papers investigating cavitating nozzle flow via X-ray radiography with and without dissolved gases. In order to determine the vapor and evolved gas phase fractions, they doped the liquid phase in [24] with an X-ray fluorescent bromine tracer and substituted the dissolved air with krypton. They observed that vapor cavitation is found only in an annular wall region beginning at the edge of the contraction. For the case of dissolved gas, they found a core region with increased, evolved gas content and a reduction in the true vapor cavitation zone. The latter is explained by the appearance of gas bubbles (appearing prior to vapor cavitation), thereby displacing the cavitation zones that would otherwise appear at the respective locations. In [25], the authors use a nozzle with a similar geometry as in [24] but slightly different transition sections. In addition, the nozzle is made of beryllium alloy, whereas in prior investigations, polycarbonate plastic nozzles were used. Quantitative measurements were provided for (total) vapor distribution, area concentration, and discharge coefficients for a broad range of Reynolds numbers and cavitation numbers and for the case with and without dissolved nitrogen in the liquid phase. Selected results were compared to those obtained previously under the same operating conditions for nozzles made of plastic. In the latter case, the authors found considerable amounts of vapor/gas in the center region of the constriction for the case with dissolved gas but not for the case without dissolved gas (see above). However, for the same case with a beryllium alloy nozzle, no differences were found between the cases with and without dissolved gas (nitrogen) present. The authors attributed this difference to gas evolving out of the liquid phase at imperfections on the nozzle walls in the case of the plastic material, which is not found in the nozzles made of beryllium. The authors found that the physical area contraction coefficient (i.e., the ratio of liquid flow area and geometric area) is independent of the amount of dissolved gas in the beryllium nozzle but decreases slightly when moving from the degassed to dissolved gas (i.e., nitrogen) condition in case of the plastic nozzle. The discharge coefficients of either the nozzle, plastic, or beryllium alloy were not affected by the amount of dissolved gases in the liquid and were found to follow the Nurick solution for sharp-edged orifices. For the plastic nozzle, this suggests that the evolving gas displaces the cavitation voids rather than the liquid phase such that the void distribution and velocity field yield the same effective discharge. Measurements on the effect of dissolved gas content on the onset of cavitation were not provided by Duke et al. [25].

1.2. Numerical Studies on Orifice Cavitation and the Effect of Dissolved Gases on Cavitation

Cavitating flows pertain to the class of multiphase flows with a mass transfer between the phases, namely evaporation (vapor cavitation) and condensation. If the liquid phase contains non-dissolved gases, this gas can also evolve out of or re-dissolve into the liquid phase, thereby impacting the cavitation phenomena. Numerous physical and mathematical models were developed to describe the multiphase flow and the phase-transition/change aspects of cavitating flows. The approaches for modeling multiphase flows can be categorized based on how the various phases are being described and the type of equations that are being solved for each phase, namely the Euler–Lagrange (see, for example [26]) and the Euler–Euler approach.

In the latter case, both liquid and vapor phases are considered as interpenetrating continua for which proper field or transport equations are being solved. Here, one distinguishes between mixture models, interface-tracking models, and multifluid Eulerian models.

Within multifluid Eulerian models, momentum and continuity equations are being solved for each phase while they share the same pressure field. The interaction between each phase depends on the type of phase involved and phase concentrations.

In mixture models, the various phases are considered to be a single continuous mixture. Proper transport equations for the mixture and continuity equations for the dispersed phases are solved together with algebraic equations for the relative velocities between the dispersed phase(s) and the continuous phase. In the homogeneous mixture model, no slip exists between the phases. In interface-tracking methods such as volume of fluid (VOF) or level-set (LS) methods, the phase transport (of all the phases) is solved using a single momentum equation, while the interface between the phases is tracked explicitly.

For the modeling of the mass transfer between the phases, the most frequently employed (macroscopic) models pertain to two major groups, i.e., equation of state (EOS) models and bubble dynamics models.

The latter group calculates the rate of mass transfer based on bubble dynamics governed using the Rayleigh–Plesset (R–P) equation or a modified form of the RP equation (without the influence of thermal effects). The pressure difference between the pressure inside the bubble (containing saturated vapor at the given temperature) and the surrounding liquid represents a mechanical driving force for mass transfer. Examples of this type of cavitation model are the Schnerr–Sauer model [27], the Kunz et al. model [28], the Zwart–Gerber–Belamri model [29], and the model by Singhal et al. [30]. Within the EOS-type models, mass transfer is based on thermodynamic relations for the phase mixture, with pressure representing one of the thermodynamic properties governing the phase-change process [31]. The so-called homogeneous equilibrium model (HEM) assumes a thermal equilibrium between the liquid and vapor phase, whereas in barotropic models, density only depends on pressure, and an energy equation does not have to be considered. In contrast, the homogeneous relaxation model (HRM) determines the mass transfer rate based on the thermal nonequilibrium between the phases [31]. For a more detailed and complete list of cavitation models and their classification, the reader is referred to [32].

The presence of non-condensable gas(es) and its release/evolution from or re-adsorption/dissolution into the liquid phase affects the overall flowfield and can thereby influence flow cavitation; see, for example, [12,33] discussed below. In addition, similar to impurities on the wetted surfaces of wall-bounded flows and suspended solid particles in the flow, microscopic amounts of non-dissolved gas serve as cavitation nuclei. Increasing the amount of dissolved gases leads to an increase in the nucleation rate and results in an elevation of the liquid's vapor pressure and a reduction in its tensile strength for cavitation inception, see [34–36].

Since the gas evolution or dissolution process is diffusion-based and, therefore, slow in comparison to the phase-change process in vapor cavitation, some cavitation models limit the dynamics of the non-condensable gas (originally present in the flow in the form of nuclei or microbubbles) to its expansion or contraction based on the flowfield conditions, see [30,31,33]. Other models capture the co-existence of both vapor and gaseous phases in a single bubble, thereby being able to predict phenomena such as cavitation-induced gas release; see, for example, [21,34,37–39]. Note that an arbitrary combination of the prescribed multiphase and mass transfer models is not always possible.

For completeness, some of the above-mentioned references are briefly addressed in terms of their stated objectives.

Battistoni et al. [31] compared the results of two different simulation models for cavitating flow through a test nozzle (under full cavitation conditions) with experimental data by Duke et al. [23]. The first simulation model uses a mixture model in conjunction with a VOF formulation (as implemented in the CFD solver CONVERGE (2.1.0) [40]) together with an HRM cavitation model. The second model approach uses a multifluid Eulerian approach 1 in conjunction with a simplified R-P equation for calculating mass transfer due to vapor cavitation, as implemented in the solver AVL FIRE ^{TM}. The model also accounts for vapor bubble coalescence and disintegration by tracking the number density and interfacial area density of vapor bubbles. The non-condensable gas phase (air), on the other hand, was assumed to be a dispersed phase of bubbles with a constant diameter of 1 μm. In both models, the gaseous phase was assumed to be free (non-dissolved) at all times and was taken into account in the mixture density; volume or mass fraction equations were solved for all three phases (liquid, vapor, and gas).

Comparison with the experimental data showed that the multifluid Eulerian model more accurately predicts mass flow rate, total void fraction, and liquid stream reattachment to the wall than the mixture model. In a later study, Battistoni et al. [33] provided an in-depth numerical analysis based on the prescribed mixture model for the same geometry under standard and de-gassed conditions. They demonstrated good agreement with experimental data [23] for these two test cases. However, experimental data collected later for the same test cases [25] but using a nozzle made of beryllium alloy, rendered the physical arguments arising from the numerical results questionable.

Motivated by the data presented in [25], Yang and Habchi [34] proposed a fully compressible two-phase flow model based on phase-equilibrium theory and non-linear real fluid equation of state for liquid and gas phases, whereby both phases are considered as multicomponent systems containing either vapor and free gas or liquid and dissolved gas. The authors implemented their real fluid-phase model into the flow-solver IFP-C3D [41] but with turbulence being modeled via LES and the Smagorinsky subgrid-scale model. While their simulations for the non-degassed case agreed well with experimental data, results for the degassed case were predicted to be much more fragmented in the occurrence of cavitation than found experimentally.

Assuming thermodynamic equilibrium and neglecting thermal effects, Iben et al. [37,38] conducted numerical simulations on a planar micro-nozzle using a mixture model in conjunction with a HEM cavitation model. The liquid phase was considered weakly compressible, and the vapor and released gas were compressible using suitable equations of state. Vapor mass fraction was obtained directly as a function of the prevailing fluid pressure; the evolution of dissolved gas was limited to its release into existing vapor bubbles or cavities and governed using a combination of Henry's law valid at the bubble or cavity interface and Fick's law governing the transport of dissolved gas within the liquid phase. The observed cavitation-induced gas release was similar in nature to that observed by Yang and Habchi [34].

Numerical simulations of the cavitation test geometry previously investigated by Nurick [14] and also presented as part of the present investigation have been analyzed by several authors. Singhal et al. [30] were the first to consider this case as a test case for their full cavitation model and its implementation in the commercial flow-solver CFD-ACE+. Their results for the discharge coefficient compared well with Nurick's correlation (derived from test data), even though values are consistently slightly higher than Nurick's correlation curve for the cavitating flow regime. The onset of cavitation was reported to occur at a cavitation number of 1.7, analogous to correlation data. Numerical simulations of this case based on a homogenous mixture model and considering a three-phase (liquid, vapor, and non-condensable gas) system were also carried out by Darbandi et al. [42]. They employed the full cavitation model by Singhal et al. [30], even though there are some inconsistencies in their reported formulation of the vapor condensation source term. Their results for the discharge coefficient as a function of a cavitation number agree well with the experimental measurements by Nurick and the simulation results reported by Singhal et al. [30], even though their values were consistently below those predicted by the latter authors. Both Singhal et al. [30] and Darbandi et al. [42] assumed the incoming flow to contain non-condensable gas with a mass fraction of 15 ppm dissolved in the liquid phase (i.e., water). Note that, in his experiments, Nurick [14] did not report the amount of dissolved gas contained in their water supply system. He did, however, report the aspiration of air between the test orifice and the measurement system under some conditions. Later, Darbandi et al. [43] compared their prior simulation results [42] with predictions from a mixture model for a two-fluid (liquid and vapor) system without dissolved gases and based on a cavitation model similar to that by Schnerr and Sauer [27]. They found that results for the discharge coefficient from the three-phase model agreed better with experimental results than the two-phase model results. Cavitation inception, i.e., the onset of cavitation, is found to occur at a higher cavitation number for the case with the dissolved gases considered in the model (i.e., 1.69 versus 1.51).

Nurick's nozzle geometry (made of Lucite) was also analyzed by Lifante and Frank [44] in the context of a validation effort regarding their implementation of the Zwart–Gerber–Belamri model [29] for vapor cavitation and the dissolved non-condensable gas model by

Yang et al. [45] in the commercial flow-solver ANSYS CFX. Their results for the discharge coefficient agree better with Nurick's correlation than those by Singhal et al. [30] in the cavitating regime, but underpredict its value in the non-cavitating region. As part of an investigation on cavitating aircraft ejector pumps, Mehring [12] also analyzed Nurick's Lucite nozzle geometry for both water and jet-A, both with dissolved air in the liquid phase (based on equilibrium conditions) and free air in the form of bubble nuclei in the upstream flow. Later, Coussirat et al. [46] also studied this orifice geometry using ANSYS Fluent with a focus on the effect of different turbulence models, their calibration and by employing the Zwart–Gerber–Belamri model [29] and Singhal et al. [30] model.

The model equations employed within the numerical analysis conducted as part of the present work closely relate to those employed by Lifante and Frank [44] and are presented in Appendix A. Specifically included herein are also the model equations for the dissolution and evolution of non-condensable gases, which have previously been detailed and explained only in [45]; a reference which is not readily available.

Effects of varying the amount of dissolved gas in the flow were not investigated in any of the prescribed investigations based on Nurick's nozzle geometry, even though Coussirat et al. [46] considered small amounts of dissolved gases for calibrating their model(s). To the authors' knowledge, Coussirat et al. were also the only ones to compare, aside from the predictions for the discharge coefficient, predicted local wall-pressure data with the respective data measured by Nurick. The present work rectifies these shortcomings by providing experimental data and numerical analysis for Nurick's orifice geometry under the systematic variation of the amount of dissolved gas (CO_2) in the upstream flow.

2. Basic Definitions for Orifice Flows

The likelihood for the flow through a contracting orifice, as depicted in Figure 1, to develop cavitation can be characterized by the cavitation number, CN, which is commonly defined as:

$$CN = \left(\frac{p_1 - p_v}{p_1 - p_2}\right), \tag{1}$$

where p_1 and p_2 are the pressures upstream and downstream of the orifice (see Figure 1) and p_v denotes the (equilibrium) vapor pressure of the liquid.

The dimensionless cavity pressures at the position of the pressure taps shown in Figure 1 are defined as:

$$P_{cav1,2} = \frac{p_{cav1,2}}{p_1 - p_2}, \tag{2}$$

where p_{cav1} and p_{cav2} denote the dimensional wall-static pressures measured at a distance $d/4$ and $d/2$ downstream of the contraction (see Figure 1). The actual mass flow rate through an orifice can be characterized using the discharge coefficient, which is defined as:

$$C_d = \frac{\dot{m}_a}{\dot{m}_t}, \tag{3}$$

where $\dot{m}_a$ and $\dot{m}_t$ represent the actual and the theoretical mass flow rates, respectively. The theoretical mass flow rate describes the mass flow in the absence of any losses. In our case, it can be written as the mass flow rate through a straight tube of diameter d with the same pressure conditions as for the orifice case and without any frictional losses, i.e.,

$$\dot{m}_t = \frac{\pi d^2}{4}\sqrt{2\rho_l(p_1 - p_2)}, \tag{4}$$

where ρ_l denotes the fluid density (liquid water). Various expressions can be found in the literature for determining the discharge coefficient. A vital consideration in this context is the presence or absence of cavitation in the nozzle. Since the cylindrical orifice geometry investigated here (and as shown in Figure 1) conforms to the Lucite orifice considered by Nurick [14], except for a scale-down factor of $(d/d_{Nurick}) = 2.7/7.6$, we use Nurick's analytical expressions for the cavitating flow in the nozzles as a reference.

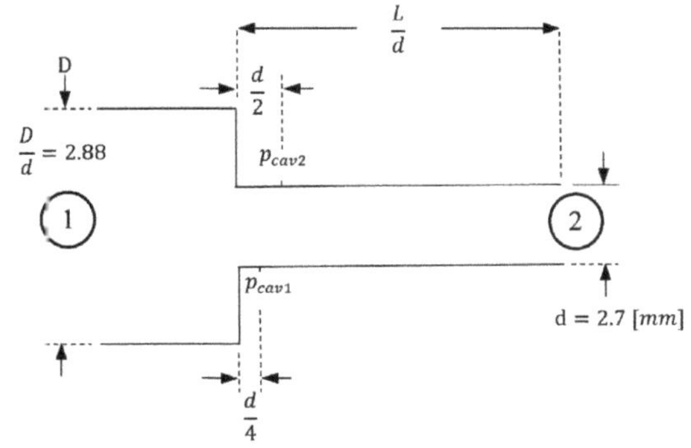

Figure 1. Geometry and dimensions of the investigated sharp-edge cylindrical test orifice and positions of cavity pressure taps 1 and 2 downstream of the sudden contraction. A scaled-down version of Nurick's Lucite orifice [14].

3. Experimental Test Setup

3.1. Water-Preparation Cycle

In order to investigate the effect of dissolved gases on cavitation inception and its evolution, the amount of non-condensable dissolved gases has to be adjustable and controllable in the experimental setup. In an upfront water-preparation cycle, water is pumped from the bottom of a tank to a membrane contactor, where the absorption and desorption of gases in or out of the liquid takes place, see Figure 2 in this context.

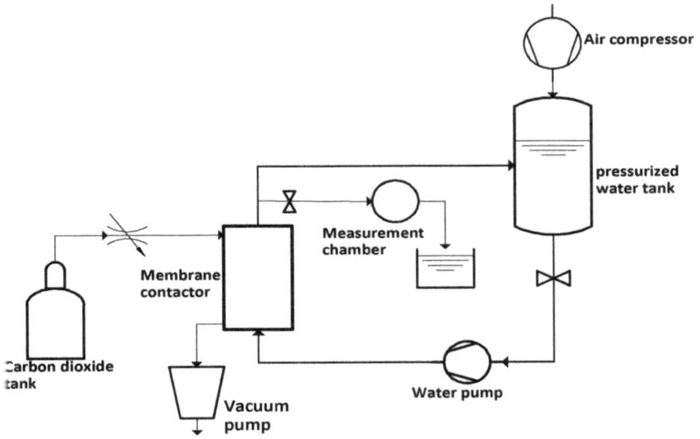

Figure 2. Schematic diagram of the water-preparation cycle in the experimental setup.

The contactor contains inlet and outlet ports for water as well as for the gas. For the degasification of water, a vacuum pump is connected to the corresponding gas port of the contactor while the "sweep" gas port remains closed. In order to dissolve non-condensable gases in water, the contactor is used in sweep mode. For this purpose, the "sweep" gas is introduced to the contactor at the corresponding gas port while the vacuum gas port remains closed. At suitable time intervals water samples are taken and guided to the measurement chamber at suitable time intervals, where the gas sensors measure the dissolved gas concentration. Dissolved oxygen is measured using an optical oxygen sensor (Orbisphere M1100 Luminescent oxygen sensor, Hach Lange GmbH, Loveland, CO,

USA), and nitrogen and carbon dioxide concentrations are measured using two thermal conductivity gas sensors (the Orbisphere 31590HP nitrogen sensor and Orbisphere 31490TC carbon dioxide sensor Hach Lange GmbH). To reach higher gas concentrations, an air compressor is utilized to increase the pressure level within the system. In this work, we use carbon dioxide as a test gas because of its higher solubility in comparison to nitrogen and oxygen. Water with five different dissolved carbon dioxide concentration levels is prepared for testing purposes. Table 1 summarizes the dissolved gas concentrations for the five investigated cases.

Table 1. In water, dissolved gas concentrations for the five investigated cases GC1–GC5.

Gas Type	Mass Concentration [ppm]				
	GC1	GC2	GC3	GC4	GC5
Carbon dioxide	0	318	707	1060	1493
Oxygen	0.85	1.09	0.42	1.02	0.52
Nitrogen	2.03	2.41	1.30	2.28	1.67

3.2. *Cavitating Orifice Flow–Flow Measurements*

Cavitation inception is investigated for water flowing through the sharp-edged orifice illustrated in Figure 1. The test section (i.e., drilled and polished) was made out of PMMA for flow visualization purposes. The orifice is a scaled-down version (scale factor $2.7/7.6 \approx 35.5\%$) of the Lucite orifice employed by Nurick [14]. The scale-down was necessary due to flow rate limitations based on the available water tank volume and the required testing time to assure steady-state conditions for each measurement. For cavity pressure measurements, two wall pressure taps have been precision drilled and are located at a $d/2$ and $d/4$ distance from the sharp-edged contraction across which the cylindrical cross-section reduces from diameter D to d. Figure 3 depicts our experimental setup for the cavitation inception measurement in an orifice flow. In the employed flow system setup, water flow is maintained by pressurizing the water tank using an air compressor, providing tank pressures of up to 10 bar. The orifice inlet pressure must be precisely regulated to resolve the inception point accurately. This has been accomplished using a back-pressure regulator. The water mass flow rate is measured upstream of the orifice using a magnetic mass flow meter (Promag P 10, 5PBB15, Endress + Hauser Group Services AG, Reinach, Basel, Switzerland.). Experiments and respective measurements were conducted for orifice inlet absolute pressures between 0.96 bar and 3.45 bar. A high-pressure build-up in the tank (10 bar) just behind the pressure valve guarantees that the orifice inlet pressure remains constant during the experiments. No free or evolved gas was observed at the orifice inlet within our experiments. Experiments were performed in batch mode; that is, data for each operating point were collected separately. Measurements of cavity pressures $p_{cav1,2}$, pressures $p_{1,2}$ upstream and downstream of the orifice, and the mass flow rate $\dot{m}_a$ were taken simultaneously for a duration of 5 s after quasi-steady flow conditions through the orifice were reached. The temporal resolution of the employed mass flow meter is 1 Hz, that of the wall pressure tap sensors 20 Hz, and for upstream and downstream pressure sensors 1 Hz. The accuracy of the latter sensors is $\pm 0.5\%$ of the full-scale measurement range, and that of the sensors employed for the pressure taps is $\pm 0.35\%$. The accuracy of the mass flow meter is $\pm 0.5\%$ of the actual measurement value.

Note that in the experiments conducted by Nurick [14], the flow (after leaving the orifice) was directly discharged into a pressure-regulated tank filled with nitrogen, allowing the study of the hydraulic flip. In contrast, in the present work, the orifice is connected to a larger diameter pipe section, allowing the flow to expand before discharging into a filled water tank. For some experiments, this pipe was made of PMMA as part of the test orifice. This allowed the observation of gas bubbles downstream as a consequence of outgassing or gas cavitation throughout the orifice flow section.

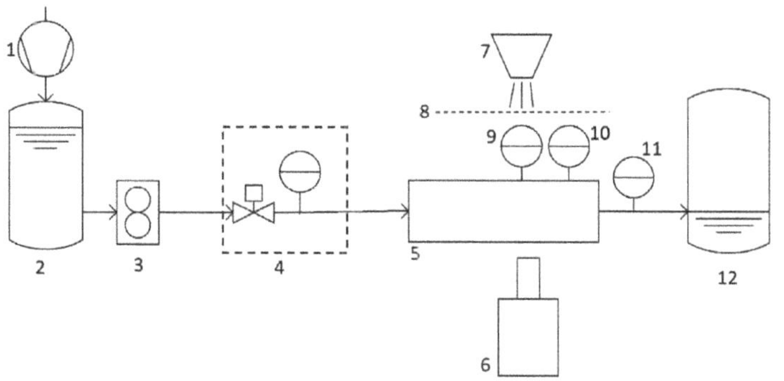

Figure 3. Flow-system and measurement setup for cavitation inception experiments. 1: air compressor, 2: pressurized water tank, 3: flow meter, 4: pressure regulator unit, 5: PMMA orifice, 6: high-speed camera, 7: LED light, 8: light diffuser, 9: cavity pressure sensor, 10: cavity pressure sensor, 11: downstream pressure sensor, and 12: water tank.

3.3 *Cavitating Orifice Flow–Flow Visualization*

Diffuse backlight illumination was used for flow visualization. A 50W LED chip was used as a light source and placed behind the test article. To ensure uniform illumination of the observation window and prevent any bright spots, a diffuser sheet was placed between the test article and the light source. This was necessary for imaging the orifice with a relatively large field of view (see, for example, Figures 10 and 11 below). Flow visualization for a small field of view is exemplary, shown in Figure 4, for an area just after the contraction. Here, imaging was performed without a diffuser because the light distribution on that field of view was reasonably homogenous. Images were captured using a high-speed camera (Photron FASTCAM SA-Z) equipped with a zoom system lens to achieve the desired magnification for three different sizes of the field of view (see, for example, Figure 4, Figure 10, and Figure 11). The Images in Figure 4 were captured at a frame rate of 4000 frames per second (fps) to capture more details of the bubble dynamics at the contraction. Removing the diffuser cover, in this case, was also necessary in order to reach the higher light intensity necessary for imaging at high magnification and high frame rates. Images such as those shown in Figures 10 and 11 were captured at a frame rate of 1000 fps to show the overall flow characteristics in the orifice and to observe the fluid flow over a longer time period, especially in the context of the unsteady flow at transition points (See for example Figure 10, GC5, CN = 1.80). It is important to note that cavitation bubbles or cavitation clouds for the attached cavitation illustrated in Figure 10 appear blurry, i.e., not sharp, compared to the images in Figure 4 because the focal planes in both sets of images are different. The focal plane of the images in Figure 10 was set to coincide with the position of the pressure taps, i.e., the plane through the centerline of the orifice, allowing the observation of the flow near the walls (top and bottom region in the images) and also in the center region of the orifice. Cavitation bubbles or cavitation clouds do not appear sharp in those images because they are generated at the walls around the entire circumference, which includes out-of-focus areas between the focal plane and the camera. In Figure 4, the focal plane was adjusted so that bubbles generated in the wall near the region closest to the camera appear exactly on the focal plane, unobstructed from any other flow features. In this case, we can even see the scratches located on the inner wall surfaces of the test article. The prescribed optical interrogation technique follows the work of others; see, for example, Sou et al. [25] and Stanley et al. [47], and differs from standard shadowgraphy or Schlieren photography frequently used for investigating cavitating nozzle flow [48]. The present method solely aims to identify the overall flow structure without aiming to extract details of that structure or quantitative information such as local density. Other optical methods in the visible spectrum, which provide, for example, information about local vapor fraction or information on fluid velocity, do exist but either limit themselves to 2D orifices, a scaled-up flow geometry, or require additional particle seeding, thereby modifying the flow behavior, especially for small-scale orifices. See also [49].

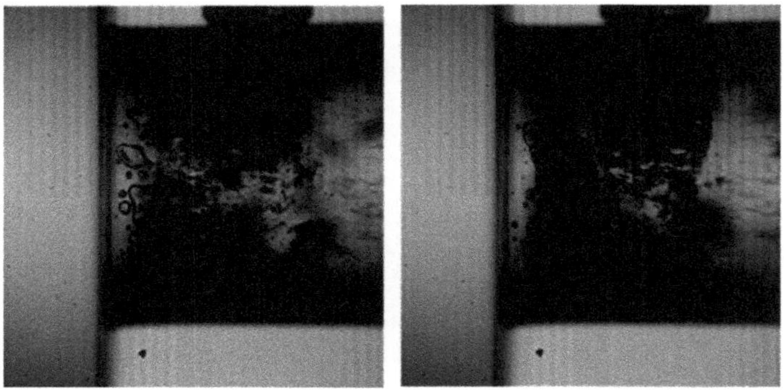

Figure 4. High-speed images of the flowfield just downstream of the orifice contraction show cavitation clouds at two different instances in time. The focal plane is on the orifice wall closest to the camera. Flow from right to left; backlight images without diffuser cover. Images taken with Photron FASTCAM SA-Z at a frame rate of 4000 fps.

4. Numerical Analysis Model

Numerous modeling approaches have been developed to predict cavitation and outgassing phenomena in the framework of CFD simulations; see also Section 1.2 in this context. To compare our experimental results with numerical model predictions, we employ the Zwart et al. model [29] for the mass transfer between the liquid and vapor phase in combination with the outgassing model by Yang et al. [45], the latter of which is also employed in the so-called "full cavitation model" by Singhal [30]. The model employed here and the slightly different "full cavitation model" are still in widespread use in the industry today (see, for example, Mehring [12]), which motivated the use of one of these models in the present investigation. The model considers a three-phase mixture consisting of a gaseous phase of non-condensable gas, water vapor, and liquid water, which itself is considered a two-component liquid composed of pure water and dissolved carbon dioxide. The homogenous mixture assumption is employed, and all phases share the same velocity and pressure field. Here, we make use of the commercial flow-solver CFX v19.1 by ANSYS Inc., in which the prescribed models are available: see also Lifante and Frank [44] in this context. For a detailed explanation of the phase change models, the reader is referred to Zwart et al. [29], Yang et al. [45], and Lifante and Frank [44]. Values for model parameters were chosen identical to those in [44] for the same flow configuration (see Section 1.2). The k-ω SST turbulence model [50] with standard model coefficients was used for turbulence closure. The overall governing equations solved as part of the present investigation are provided in Appendix A. Only carbon dioxide is considered a dissolved non-condensable gas for simulating the orifice flow. The mass fraction of dissolved CO_2 at the inlet boundary was specified according to the experimentally investigated test cases summarized in Table 1. Analyses were only carried out for cases GC1 through GC3 due to (known) convergence issues of the outgassing model for mass fractions beyond 1000 ppm. The Henry constant for carbon dioxide was chosen from the literature to be $H_v^{px} = 1.6 * 10^8 \left[Pa \frac{mol_{water}}{mol_{co2}}\right]$. The vapor volume fraction and free non-condensable gas volume fraction were both set to zero at the inlet boundary, while the nuclei volume fraction was set to 0.01247 with a nuclei radius of 2 μm. Liquid or water vapor pressure was set to 2680 [Pa]. Figure 5 provides a summary of all employed model parameters and boundary conditions. For the simulations, a sharp-edge orifice was assumed. To model the 2D axisymmetric problem within the 3D CFX solver environment, a narrow one-degree wedge of the configuration with symmetry conditions in a circumferential direction was considered. Only a single computational cell was used in that direction. In CFX, hydrodynamic equations (for velocity and pressure) and other transport equations are solved via a pressure-based coupled solver and based on an element-based finite-volume approach with a co-located (non-staggered) grid and an interpolation for pressure and velocity via FE shape functions. Steady-state solutions for the prescribed problem have been produced via pseudo-transient time-stepping with a special high-resolution upwind-type scheme for the advection terms. The solution of the linearized system of equations is obtained via an algebraic multigrid (AMG) solver. A block-structured rectilinear mesh consisting of five blocks was used in the (r-z) plane. The mesh count for the inflow and contracted flow section (three blocks) was 20,000, and the downstream expanded flow section was 21,000 (two blocks). The respective local mesh resolution is greater than that employed by Lifante and Frank [44] and Singhal et al. [30] for a similar flow configuration. Further mesh refinement did not provide any meaningful changes (<1%) in the flow properties of interest. Numerical solutions for the various cases will be presented and compared to experimental results in Section 5.2.

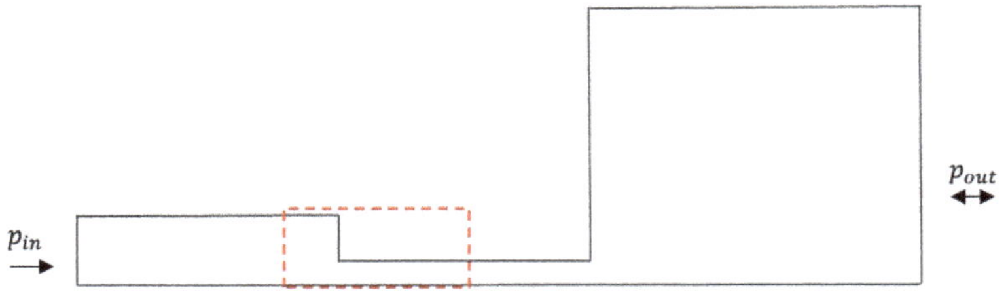

Fluid Properties			
Fluid	Molar Mass $Kg/_{Kmol}$	Density $Kg/_{m^3}$	Dynamic viscosity $Kg/_{ms}$
Water	18.02	997	8.899×10^{-4}
Water vapor	18.02	0.02308	9.862×10^{-6}
CO2	44.01	1.87	1.49×10^{-5}

Dissolved gas Modeling Parameters		
Absorption Coefficient	0.1	$m.s/_{kg}$
Desorption Coefficient	2	$m.s/_{kg}$
Henry's Law Constant	1.6×10^{8}	Pa
Maximum Solubility	0.001	-

Inlet BC:
Type: Inlet with
Total Pressure p_t =1.75 – 3.45 bar
Flow Dir: Normal to the boundary
Turbulence intensity = medium
CO2 Volume Fraction = 0
Liquid Volume Fraction = 1
Dissolved CO2 mass fraction according to table 1

Outlet BC:
Type: Opening Pressure with
Static pressure p =1.05 – 1.10 bar

All Fluid walls are no-slip walls
Heat transfer model: none
Reference pressure: 0
Turbulence Model: k-ω SST

Cavitation Model:
Vapor Pressure (p_v): 2680 pa
Nuclei mean diameter: 2×10^{-6} m

Figure 5. Summary of all employed model parameters and boundary conditions for the simulations. The dashed line illustrates the flow area displayed in Figure 14 below.

5. Results

5.1. Experimental Results

5.1.1. Comparison of Measured Cavity Pressures with Results from Nurick Experiments

Figure 6 illustrates the dimensionless cavity pressures measured at tap positions 1 and 2 (as specified in Figure 1) dependent on the cavitation number *CN* for water with an aeration level GC1 (i.e., no dissolved CO_2). Also shown are the results reported by Nurick for the similar (scaled-up) orifice made of Lucite [14]. Note that the latter study did not measure or report the amount of dissolved gas contained in the water supply system. Cavitation onset in the present case is found to occur at $CN = 2.02$ and coincides with a plateau in the pressure measured at tap position 2. Full cavitation is found to occur at cavitation numbers below $CN = 1.73$ when both pressure tap measurements are identical. The prescribed observation in regards to cavitation onset has also been made by Nurick in his original paper [14] and later again in the "Authors' Closure" comment contained in the commentary by Peterson [51]. However, in Nurick's experiment, the cavity pressure reaches the vapor pressure at a higher cavitation number, i.e., CN = 2.5, and the plateau formation reported by Nurick is less pronounced; see Figure 6.

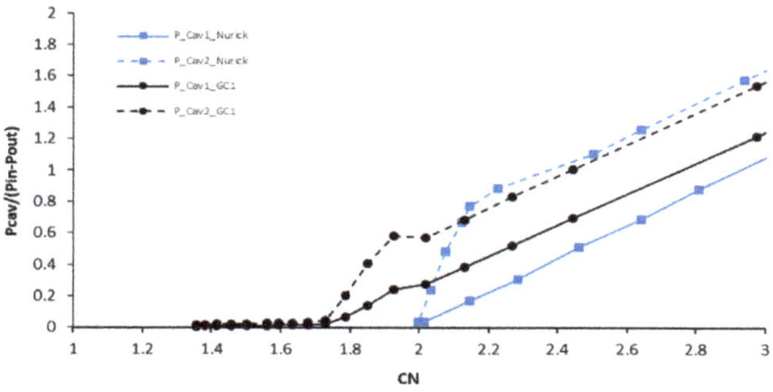

Figure 6. Comparison of dimensionless cavity pressures $P_{cav1,2}$ at tap positions 1 and 2 (according to Figure 1), with results presented by Nurick [14] (Figure 5).

Deviations from Nurick results might be due to scale differences and differences in entrance roundness r in relation to the orifice diameter(s). As already reported by Nurick, the roundness of the sharp-edged entrance region has a significant influence on cavity pressure and, consequently, on cavitation inception [14]. In addition to entrance roundness, imperfections along the edge of the contraction caused by drilling and polishing can affect the flow through the contraction and, consequently, cavity pressures as well. These effects are expected to have a stronger impact on the cavity pressure at tap 1, which is located closer to the contraction. As noted earlier, our orifice geometry is approximately a factor 3 smaller than Nurick's Lucite orifice. However, the corner radius of our test article was indeed quite small as well. Differences in the amount of dissolved air in both cases will also impact cavitation onset, as will be discussed in detail later in the context of dissolved carbon dioxide. Assuming equilibrium conditions in Nurick's experiment, Henry's law yields a dissolved air mass fraction of 15 ppm, whereas in Case GC1, it is approximately 3 ppm (see Table 1). An explanation for the variation of pressure at the two pressure taps, including the prescribed plateau formation, at tap position 2 was already given by Nurick [14]. As the cavitation number is decreased, i.e., by increasing the upstream pressure, the pressure at both taps initially drops linearly, with the pressure at tap 2 being higher, as the latter is located in the recompression zone. It is interesting to note that Nurick explains the plateau formation in the pressure for Tap 2 as being the result of a variation of the reattachment phenomenon with cavitation onset. However, in the present investigation, the pressure curve for tap 1 also exhibits the plateau phenomena to an extent similar to that observed by Nurick [14] for the pressure at tap 2 only. This suggests that cavitation onset also changes the flow dynamics around the vena contracta, thereby initially delaying a further drop in pressure, also at tap 1. A possible explanation might be an initial decrease in shear stress in the shear layer between the flow recirculation zone next to the wall and the main flow since cavitation inception and initial vapor formation is found to occur in that shear layer; see, for example, [52]. See also Figure 14 below in this context.

Figure 7 shows the post-test CT scans of our test orifice. The test article was cut in order to allow a better resolution for measuring the corner radius of the orifice. The measured mean corner radius was $r = 9$ µm, i.e., $r/d = 0.003$. The CT scans also reveal some local topological non-uniformities along and near the edge. In the context of quantitative differences to Nurick's data observed in Figure 6, we note that aside from the prescribed geometric scale differences, there are other scaling effects that need to be considered to ensure similarity between hydrodynamically cavitating flows [53], which might have contributed to the observed differences between the present results and those reported by Nurick [14].

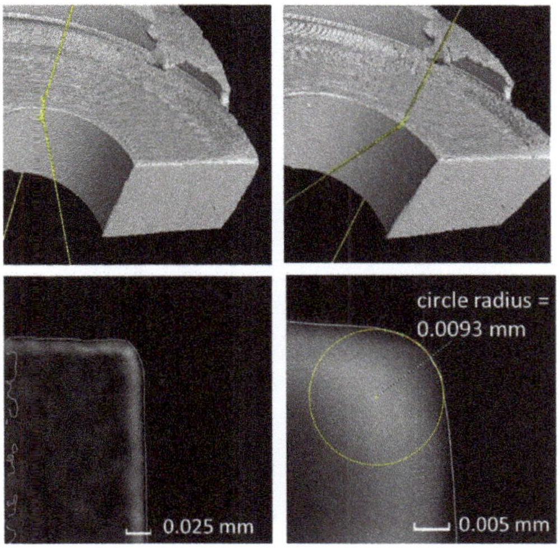

Figure 7. Top: Post-test CT scan of the orifice section with some topological anomalies along or near the edge. **Bottom**: Zoom-in (with different magnifications) on two edge locations identified via the yellow crosshairs in the top pictures.

5.1.2. Influence of Dissolved Gas on Cavitation Inception

In the work presented by Nurick [14], cavitation inception was characterized by the formation of a fuzzy region just downstream of the contraction. Within the present study, flow visualization at inception using a high-speed camera (Photron FASTCAM SA-Z) shows that the observed fuzzy region consists of tiny vapor bubbles that disappear further downstream due to pressure recovery. Figure 8 illustrates the first appearance of vapor bubbles at the inception point for case GC1. The images were captured at a frame rate of 100 k [fps].

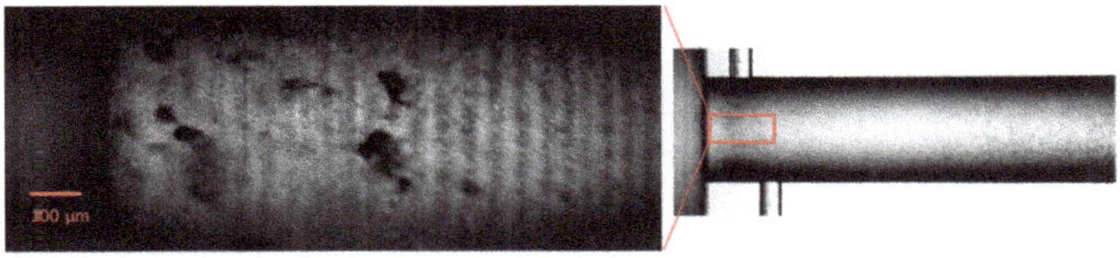

Figure 8. Right: Fuzzy region just downstream of the sharp-edge orifice at cavitation inception (flow from left to right). **Left**: Magnified view of cavitation onset indicating the presence of single cavitation bubbles (and striped structures from machining grooves on orifice walls).

Figure 9 depicts the measured change in pressures $p_{cav1,2}$ at pressure taps 1 and 2 as a function of the cavitation number for all five CO_2 aeration levels (i.e., GC1 through GC5) listed in Table 1. Each data point in Figure 9 was determined based on the time averages of cavity pressures $p_{cav1,2}$ and upstream/downstream pressures $p_{1,2}$ recorded during the test time of 5 s. As already noted in the previous section, the instance of cavitation inception is readily recognized by a plateau in the plot for cavity pressure p_{cav1} at tap 1. This plateau is less pronounced with increasing dissolved CO_2 content; in fact, for the highest aeration level (case GC5), a plateau in the pressure curve is no longer observed. We also note that for cases GC1 through GC4, a less pronounced pressure plateau at the onset of cavitation is observed for the cavity pressure at tap 2, i.e., p_{cav2}, as already discussed earlier in the context of Figure 6. Again, such behavior was not reported by Nurick [14]. Cavitation numbers at cavitation inception are 1.93, 2.02, and 2.14 for aeration levels GC1, GC2, and

GC3, respectively. Visual observations of inception (via the appearance of the prescribed fuzzy region) confirm these values; see the photographic inserts in Figure 9 as well as Figure 10. It can be deduced that an increase in dissolved gas concentration leads to higher cavitation numbers at inception. For fully cavitating flow, characterized here by a cavitation or multiphase cloud that extends over the entire length of the orifice, pressure values at both taps are nearly the same and always higher for higher aeration levels. See Figure 10 for $CN < 1.73$.

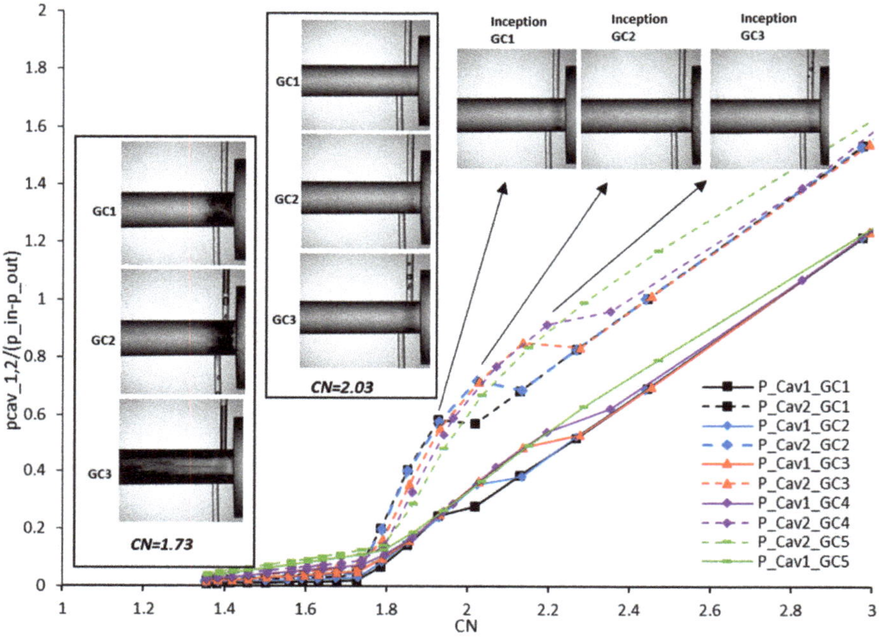

Figure 9. Cavitation inception and evolution as a function of dissolved CO_2 content. Shown are normalized pressure values at tap positions 1 and 2 (see Figure 1) as a function of cavitation number. An increase in dissolved gas concentration leads to cavitation onset at higher cavitation numbers. Accuracy < ±0.35% FSO according to IEC 60770 [54]. Flow direction in the images from right to left.

We now consider the flow images at cavitation number $CN = 2.03$ inserted in Figure 9; for aeration level GC1, i.e., no CO_2 dissolved, and no cavitation occurs. This is in agreement with the information from the cavity pressure curves, according to which inception, in this case, occurs at $CN = 1.94$. Note that the size of the cavitation cloud for aeration level GC3 is bigger than that for aeration level GC2 at that same cavitation number, even though the cavity pressures do not vary. This indicates that the potential of cavitation generation improves with an increase in dissolved gas concentration. The effect of dissolved gas concentration on vapor cavitation is twofold: increasing the amount of dissolved gases (1) aids in the formation of cavitation nuclei and (2) increases the vapor/saturation pressure and reduces the tensile strength of the liquid [34–37]. From the images, we also note that, for case GC3, there is a notably higher amount of CO_2 gas bubbles being released into the access port of pressure tap 1 than in the GC2 case. In the present investigation, outgassing occurs parallel to vapor cavitation. Depending on the dissolved gas content, more or less gaseous microbubbles are being released in the flow, indicated by a slight darkening of the flow region downstream of the contraction. This is not readily seen in the photographs of Figure 10 but can be observed in the figure of Appendix B, e.g., for GC2 and $CN = 1.86$ in comparison to GC5 and $CN = 1.98$. Clear evidence of these bubbles is found in the expansion section (with the exception of case GC1, due to its very little dissolved gas content). See Figures 11 and 12 in this context. The onset of outgassing or gas cavitation is governed by the threshold or equilibrium pressure given by the sum of the liquid's vapor pressure and (CO_2) gas-phase partial pressure, the latter being determined (under equilibrium conditions) by Henry's law [55] and the amount of dissolved gas (CO_2) molar fraction in the liquid. In contrast, vapor cavitation occurs if the local fluid pressure drops below the critical pressure given by the sum of liquid vapor pressure augmented by the maximum principal component of the strain-rate tensor [53], including the effects of turbulent fluctuations. As water containing noticeable amounts of dissolved (non-condensable) gases passes through the flow contraction, the static pressure drops below the prescribed equilibrium pressure, leading to gas desorption. At the same time, vapor bubbles appearing in the low-pressure zone due to cavitation will collapse as soon as they migrate to areas where the fluid pressure has recovered. In contrast to vapor bubbles, gas bubble formation and dissolution occur via diffusion. Once formed, gas bubbles will persist due to the large characteristic time scale associated with this process compared to the other relevant processes. Accordingly, gas bubbles will predominantly be found well downstream of the contraction zone or in the recirculation region next to the vena contracta. Evidence of the latter is given via the gas bubbles detected in the pressure tap access ports, especially at high-dissolved CO_2 concentrations. Evidence of the former is given by observations of the downstream flow for a PMMA test article, which features a transparent, extended outlet section; see Figures 11 and 12. A similar observation can also be made for aeration with air, as illustrated in Figures 17–19 of Section 5.2 in the context of a comparison to numerical analysis results.

Figure 10 summarizes the visual observations from our test campaign for all investigated CO_2 aeration levels. Here, we observe again that cavitation inception shifts to higher cavitation numbers as the amount of dissolved carbon dioxide increases. Comparing fluid images for cavitation number $CN = 1.93$, we observe that the size of the cavitation cloud generated after the contraction is bigger or more prominent for GC2 versus GC1 (without increasing its axial length), while no free gas can be observed in the pressure ports in either case. Accordingly, while the gas content furthers cavitation, it has not yet produced considerable amounts of free carbon dioxide. Also, the cloud vanishes even before reaching the position of the second pressure tap, which demonstrates that it mainly consists of water vapor, which disappears as the static pressure in the flow recovers. By further increasing the gas content, significant amounts of gas evolve from the liquid phase, the extent of the cavitation cloud increases, and free gas is observed in the pressure ports; see Figure 10 for GC3 through GC5 at $CN = 1.93$ or 1.94. Note that the axial extent of the cavitation cloud or multiphase region/zone has been noted in Figure 10 below each photograph. The length was determined as the time-mean value of the maximum zone length determined using high-speed videography and after suitable image post-processing (i.e., increasing image brightness and saturation, rendering the entire image white except for the wall-near the cavitation zone). Time-mean values were determined based on an imaging frame rate of 1000 fps for the entire test duration of 5 s with an image resolution of 1024×1024 at a pixel size of 10 μm. Also, to determine the maximum zone length in a single image, a reference flow image without cavitation was used and subtracted from the image to be evaluated. A quantification of how far vapor cavitation influences the gas evolution rate (as described by Kowalski et al. [22]) was not possible within the present study.

Figure 10 also shows that, depending on the dissolved gas concentration, the length of the cavitation zone increases with a decreasing cavitation number up to a length of about 1.1–1.6 times the orifice diameter. With a further decrease in the cavitation number, the cavitation zone suddenly expands over the entire length of the orifice and up to the downstream expansion zone; a transition occurs from cloud or bubble cavitation to extended-sheet cavitation or full cavitation, as discussed in the context of Figure 12 below.

Cavitation Number, CN	1.68	1.73	1.79	1.85	1.93	2.02	2.13	2.27
GC1	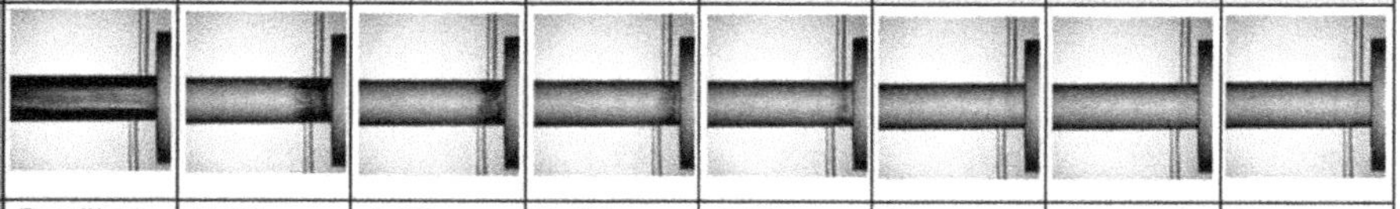							
length of cavitation zone [mm]	Transition to full cavitation	3.00	1.95	1.55	1.31	0	0	0
Cavitation Number, CN	1.68	1.73	1.79	1.85	1.93	2.02	2.13	2.27
GC2	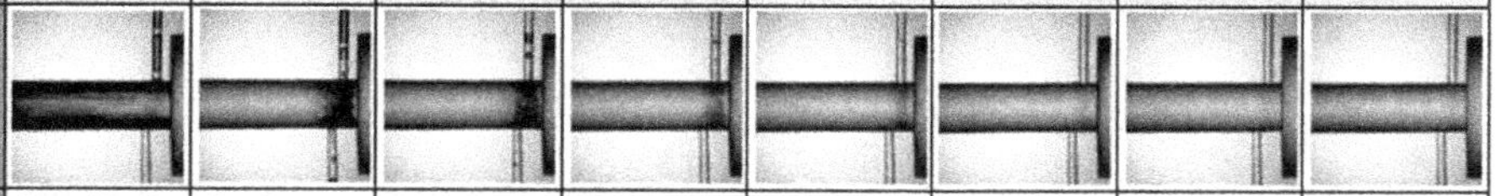							
length of cavitation zone [mm]	Transition to full cavitation	3.42	2.48	2.13	1.72	1.23	0	0
Cavitation Number, CN	1.68	1.73	1.79	1.86	1.93	2.03	2.14	2.28
GC3	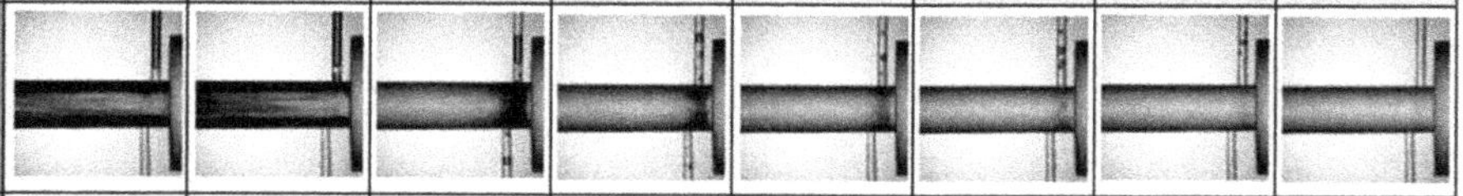							
length of cavitation zone [mm]	Full cavitation	Transition to full cavitation	3.41	2.53	1.97	1.57	1.26	0
Cavitation Number, CN	1.69	1.74	1.80	1.86	1.94	1.96	2.07	2.20
GC4	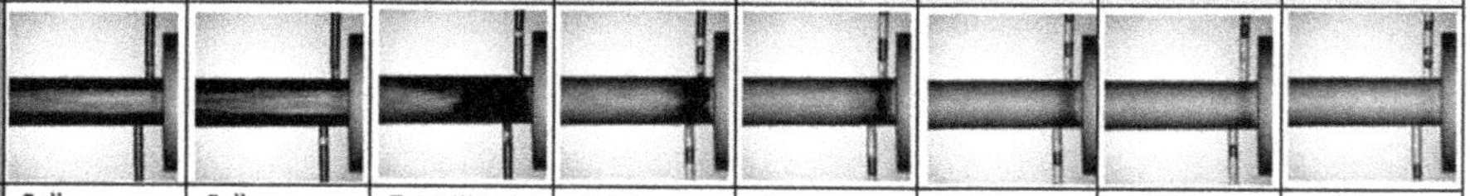							
length of cavitation zone [mm]	Full cavitation	Full cavitation	Transition to full cavitation	4.20	3.43	2.41	1.72	1.35
Cavitation Number, CN	1.69	1.74	1.80	1.87	1.94	2.04	2.15	2.29
GC5	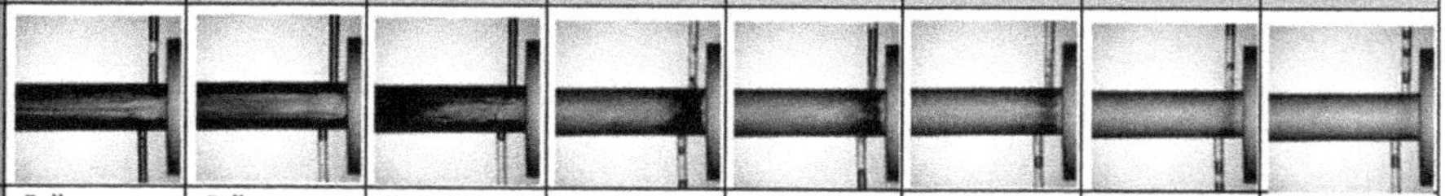							
length of cavitation zone [mm]	Full cavitation	Full cavitation	Transition to full cavitation	4.70	4.15	3.61	2.08	1.40

Figure 10. Summary of flow observations for broad range of cavitation numbers CN and 5 different mass fractions of dissolved CO_2 , i.e,. GC1–GC5 according to Table 1. Presence of dissolved CO_2 accelerates cavitation development. Accumulation of evolved gas in pressure ports with decreasing cavitation number indicates that outgassing, i.e., gas cavitation, occurs parallel to vapor cavitation. Transition from confined cloud/bubble cavitation to extended-sheet cavitation, i.e., fully caviating flow, is also influenced by dissolved gas content. Flow direction from right to left.

In the cloud or bubble cavitation stage, the cavitation zone is filled with highly dispersed vapor or vapor/gas bubbles influenced in part by the dynamics of the downstream flow reattachment and re-entrant jet dynamics. The zone appears dark in the photographs due to excessive light scattering at the many phase interfaces in the annular flow section occupied by vapor/gas bubbles. At the fully cavitating stage, core flow re-attachment does not take place. A continuous cavitation sheet filled with vapor and gas extends from the contraction throughout the entire length of the orifice. In this case, light is merely scattered at the orifice walls and the surface of the continuous annular gas/vapor cavity surrounding the liquid core flow region.

For cases with high dissolved gas concentrations and prior to cavitation onset, enough CO_2 evolves in the recirculation zone around the vena contracta, such that it can be observed in the pressure tap connection channels; see Figure 10 for cases GC4 and GC5 at $CN = 2.2$ and 2.29, respectively. In Appendix B, we include a second set of experimental data from our test campaign for similar conditions as presented in Figure 10 but with a frame of view more focused on the flow area after the contraction. This set of figures also supports the prescribed observations and illustrates the repeatability of our results.

To readily observe the outgassing process ahead of the onset of vapor cavitation and to illustrate the entire flowfield under sheet- or full-cavitation conditions, a test article was produced without pressure tabs and with both the orifice and the downstream expanded outlet section made of PMMA, i.e., acrylic glass. Figure 11 illustrates results obtained for both test articles under non-cavitating conditions (cavitation number $CN = 2.5$) for two specific cases: (1) with little or no dissolved gas (corresponding to case GC1) and (2) with a large amount of dissolved CO_2 (corresponding to Case GC5) in the upstream flow. For the "degassed" case, no bubbles or microbubbles are visible within the entire flow domain. In contrast, for the highly aerated case, CO_2 microbubbles grow, potentially coalesce, and become observable in areas where their flow residence time has become large enough, i.e., in the pressure port(s) connected to the flow recirculation zone and the downstream expanded outlet section.

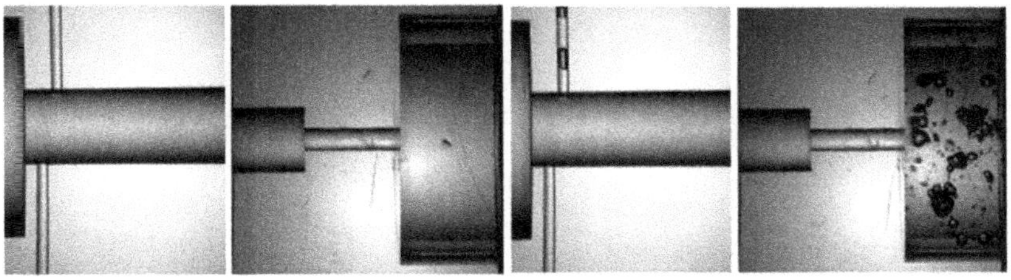

Figure 11. Visualization of CO_2 outgassing for non-cavitating flow at $CN = 2.5$ for a test article with pressure ports and a test article w/o pressure ports but with a transparent, extended outlet section. Left: degassed water, right: highly aerated water. Accumulation of evolved CO_2 gas in the access port of pressure tap 2 shows that outgassing occurs near the contraction or vena contracta in the form of microbubbles expanding in zones with slow flow velocity, including the extended outlet zone. Flow direction from left to right.

Figure 12 visualizes the evolution of the multiphase flow in the orifice and the extended outlet zone for highly aerated water (dissolved CO_2 concentration similar to case GC5) with changes in the cavitation number, i.e., from $CN = 2.13$ (early cavitation), $CN = 1.85$ (transition to full cavitation), $CN = 1.55$ (full/sheet cavitation). At $CN = 2.13$, a small cavitation region establishes after the contraction indicated by the darkened area. After the pressure recovers, the downstream orifice retains a darkened color, indicative of the formation of gaseous microbubbles. The patchy appearance might be indicative of non-uniform bubble growth and/or nuclei distribution associated with the flow dynamics in the cavitation region; it might also be related to the complexity of the turbulent flow with microbubble accumulation correlating to turbulent flow structures. At $CN = 1.85$, a continuous gas/vapor annular core has formed almost along the entire length of the orifice, characterized by a smoother streaky appearance; the extended darkened region prior to the flow expansion into the downstream section is indicative of flow reattachment and re-entrant jet dynamics. At $CN = 1.55$, the flow reattachment zone has reached the downstream edge of the orifice, although the gas/vapor annulus does not extend meaningfully into the expansion section. With continuous operation, more and larger bubbles start to form in the recirculation region surrounding the free jet in the downstream expanded flow section.

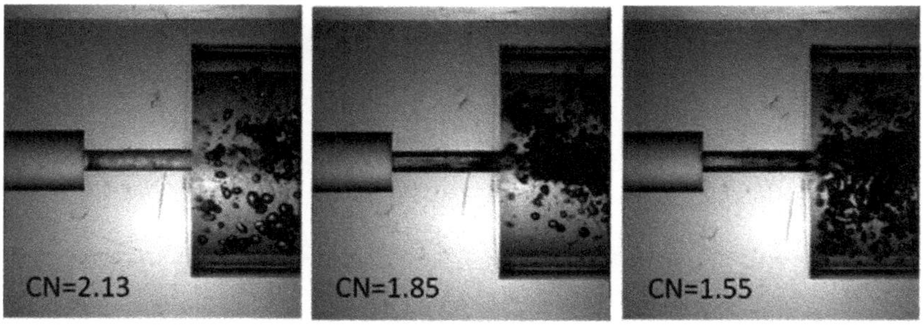

Figure 12. Cavitation development for highly aerated water with a reduction in cavitation number. Due to high amounts of dissolved carbon dioxide, outgassing occurs at higher rates. At cavitation number $CN = 1.85$, we observe full cavitation. Flow direction from left to right.

5.1.3. Measurements of the Discharge Coefficient

Figure 13 shows a comparison of the orifice discharge coefficients C_d calculated according to the equations presented in Section 2 and based on the present experimental results for the five investigated CO_2 aeration levels (see Table 1). Each data point in Figure 13 was determined based on the time averages of upstream/downstream pressures $p_{1,2}$ and mass flow rates recorded during the test time of 5 s. Also shown is the correlation presented by Nurick for this orifice geometry [14] without considering dissolved gas content (thick solid black line) together with a linear approximation of the experimental data reported by the same author at higher cavitation numbers under different back-pressure conditions (thick dashed black line). See Figure 8 in [14] for $L/d = 6$.

We note that for low-dissolved CO_2 content, discharge coefficients from the current measurements agree reasonably well with those presented by Nurick, although measurement uncertainties for the lowest dissolved gas content (GC1) were considerable. Also, even though the amount of dissolved CO_2 changes the cavitation number CN at which cavitation first occurs (see the discussion in the prior section), it does not notably change the variation of discharge coefficient C_d with CN for lower aeration levels, i.e., cases GC1 and GC2. At higher dissolved gas content (i.e., for cases GC3–GC5), gas evolution, i.e., gas cavitation, is increased at the same CN number, which might also significantly enhance vapor cavitation (possibly also vice versa) (see Figure 10) and the evolved gas volume fraction leads to further blockage of the flow cross-section, thereby leading to a notable drop in the discharge coefficient; see Figure 13. Despite variations in some of the instantaneously measured values, illustrated by the error bars in Figure 13, the measured average mass flow rates at a given cavitation number clearly decrease as the amount of dissolved carbon dioxide increases. Note that variations in instantaneous mass flow measurements were smallest for the cases with larger dissolved gas contents, i.e., GC3–GC5. The prescribed differences in mass flow rates for different dissolved gas content continuously decrease once the cavitation number drops below the value where full cavitation (as discussed earlier) occurs in all cases, i.e., below $CN = 1.73$. As the cavitation number continues to decrease below this value, the vapor-to-gas volume fraction continuously increases, diminishing the effect of the evolved gas on mass flow rate, as well as the pressure conditions at the two pressure taps. Below $CN = 1.35$, the discharge coefficient and pressure values at the two pressure taps become independent of the amount of dissolved gas originally contained in the liquid.

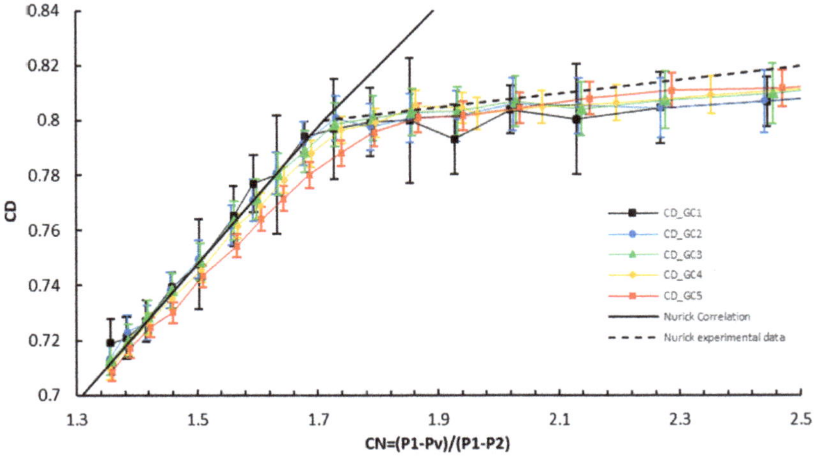

Figure 13. Discharge coefficient C_d as a function of cavitation number *CN* and dissolved CO_2 gas content (GC1–GC5 according to Table 1) together with data from Nurick [14]. Increasing gas content leads to a decrease in the discharge coefficient, especially near the onset of full cavitation. Here, evolved gas contributes meaningfully to r liquid phase displacement resulting in lower mass flow rates.

5.2. Numerical Analysis Results

Figure 14 illustrates simulation results based on the numerical analysis model described in Section 4 for the three experimentally investigated dissolved gas configurations GC1, GC2, and GC3 at cavitation number $CN = 1.85$. (As mentioned already in Section 4, convergence issues with the outgassing model for high mass fractions prevented the analysis of cases GC4 and GC5.)

One observes that the velocity fields for the three cases do not differ significantly, even though the predicted minimum absolute pressure found near the wall just after the contraction is larger if the dissolved gas concentration of the upstream flow is larger. Correspondingly, the size of the low-pressure zone is predicted to decrease with increasing dissolved gas concentration. As a result, we observe less cavitation (a lower vapor production rate) with increased dissolved gas concentration. In fact, for the highest dissolved gas content (case G3), the flow is not cavitating at all. The numerical results suggest that cavitation inception occurs at lower cavitation numbers for higher dissolved gas content. This is in contrast to our experimental observations (see the discussion in Section 5.1.2) and will be further addressed later on.

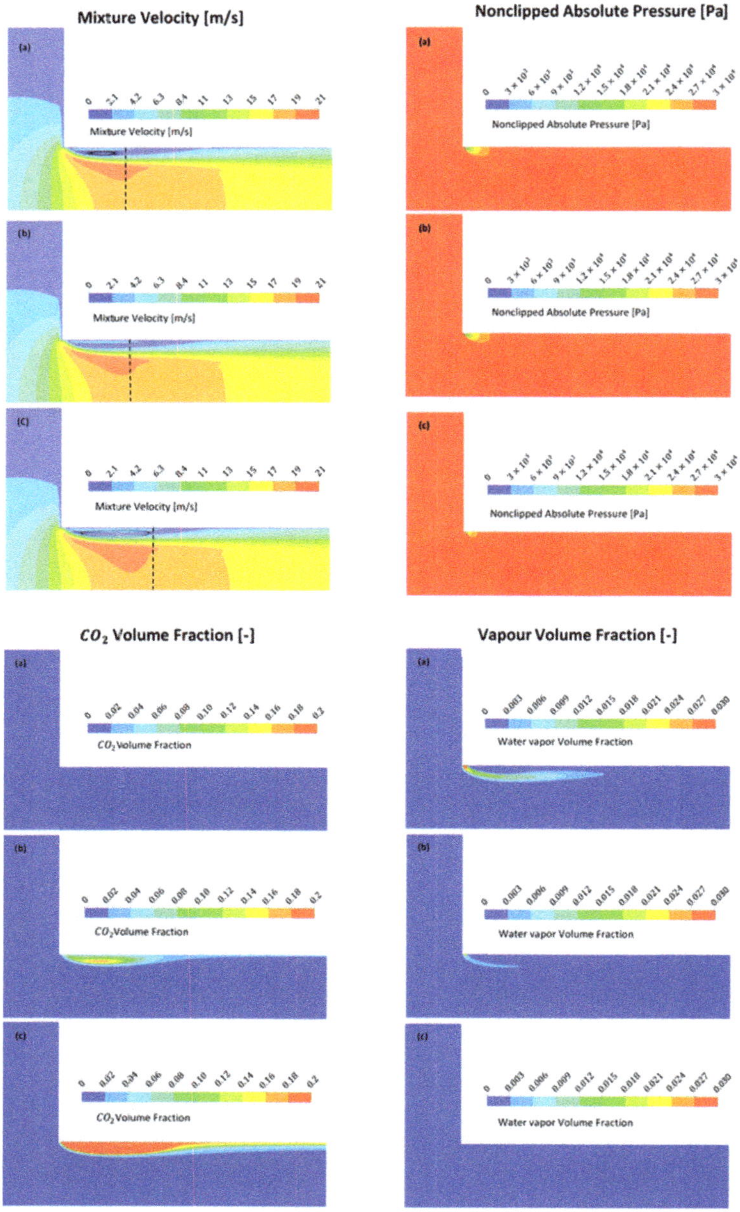

Figure 14. Contours of predicted absolute velocity (top, left), absolute pressure (top, right), gaseous CO_2 volume fraction (bottom, left) and water vapor volume fraction (bottom, right) near the sharp-edged contraction for CN = 1.88 and different values for dissolved-gas content in the upstream flow, i.e., (**a**–**c**) corresponding to G1, G2, and G3 according to Table 1. The axial length of the recirculation zone measured from the contraction is 1.35 mm (**a**), 1.47 mm (**b**), and 1.94 mm (**c**), as indicated by dash marks in the velocity contour plots. The area experiencing cavitation shrinks with an increase in gas content. Cavitation occurs in the shear layer between the flow recirculation zone and the core flow; evolved CO_2 remains highest in the flow recirculation zone and, for high dissolved CO_2 concentrations in the near wall region further downstream. Flow direction from left to right.

The results presented in Figure 14 also show that with increasing dissolved CO_2 concentration in the upstream flow, the amount of gaseous CO_2 released out of a solution near or around the vena contracta increases and accumulates in the recirculation zone. Experimental evidence for this gas release is given by gas bubbles in the pressure ports at these operating conditions, as shown in Figure 10. Taking the dividing streamline between core flow and recirculation zone as a reference, the simulations show that with increasing the dissolved gas content from GC1 to GC3, the depth of the recirculation zone at the positions of pressure taps 1 and 2 increases by about 10% and 15%, respectively. For case GC3, its depth at Tap 2 is about 10% of the orifice diameter. To distinguish between the onset of outgassing versus the onset of cavitation, we observe Henry's constant discussed in Section 4 and the resulting equilibrium pressure based on Henry's law. Because the local equilibrium pressure of the dissolved CO_2 is higher than the water vapor pressure, the onset of CO_2 outgassing occurs prior to fluid cavitation. The reader is referred to Figures 9 and 11 in this context.

Figure 11 visualizes flow characteristics in the orifice and the extended outlet section of the test setup for degassed and highly aerated water at cavitation number $CN = 2.5$. As illustrated in Figure 9, no fluid cavitation occurs at this cavitation number, and no fuzzy region can be found near the contraction. However, gas bubbles can be observed in the extended outlet section in Figure 11 and also in the pressure port at tap 2 in Figure 9 for this case, which is indicative of gas release.

Despite the prescribed differences, computed mass flow rates through the orifice for the three cases GC1–GC3 at the investigated cavitation number ($CN = 1.85$) differ by less than 0.5%. The mass flow rate predicted for case GC3 is $\dot{m} = 0.0735$ kg/s and only 2% above the experimentally measured value. The discharge coefficient calculated from the measured data for this case is $C_d = 0.792$. This value is slightly below the experimental data reported by Nurick at the same cavitation number (see Figure 13).

Figure 15 illustrates computed results for non-dimensional wall-static pressures at the location of pressure tap 1 and 2 (denoted by subscripts "$Cav1$" and "$Cav2$") as a function of the cavitation number CN for cases GC1 and GC3, whereby the former case corresponds to no dissolved CO_2 (see Table 1). Computed results for case GC2 have been excluded here since they compare closely to those for case GC1. Cavitation onset is predicted to occur at $CN = 1.85$ for case GC1 and at $CN = 1.79$ for GC3, respectively. Also included in the diagrams are experimentally measured values, as already presented in Figure 9, with the plateau formation indicative of cavitation onset.

An initial increase in the amount of dissolved CO_2 (i.e., from GC1 to GC2) does not have a noticeable effect on computed cavity pressures at tap 1 and tap 2. However, a higher amount of dissolved CO_2 (case GC3) leads to slightly higher pressures at the position of pressure tap 1 and a reduction in pressure at tap 2. The predicted rate in pressure change at taps 1 and 2 with changes in the cavitation number is in good agreement with experimental data in the non-cavitating range; however, absolute pressure values at the upstream tap (tap 1) are overpredicted. The experimentally observed formation of a pressure plateau at the cavitation onset is not captured by the numerical results. The numerically predicted pressure variations illustrated in Figure 15 provide no indication of the onset of cavitation. More importantly, cavitation onset is found to occur at lower cavitation numbers than observed experimentally, i.e., for case CG1 at $CN = 1.85$, whereas $CN = 2$ in the experimental results and for case CG3 at $CN = 1.79$, whereas $CN = 2.28$ experimentally. This demonstrates that the employed physical-numerical model does not properly predict the effect of the dissolved CO_2 gas content; i.e., while in the experiment, the onset of cavitation is moved to higher cavitation numbers with an increase in dissolved gas content, the numerical results predict the cavitation onset to occur at lower cavitation numbers with an increase in dissolved gas content. From the discussion on numerical studies of cavitating orifice flows and dissolved gas effects (see Section 1.2) and the various existing modeling approaches, it is clear that the currently employed analysis model(s) have considerable limitations, leading to the prescribed differences between numerical and experimental results. These limitations include: (1) no slip between the phases based on the homogenous mixture assumption, (2) thermodynamic relations for the phase mixture are not considered within the employed (R-P equation-based) liquid-vapor mass transfer model, (3) cavitation-induced gas release is not captured, since only pure vapor and gas bubbles are being modelled, (4) gas desorption and absorption is modeled ad hoc based on the difference between local partial gas pressure and its equilibrium pressure based on the amount of dissolved gas, and (5) the critical pressure of the liquid is set to equal its vapor pressure without considering the influence of viscous normal stresses. To demonstrate the consistency of our numerical results with prior numerical work on this specific orifice geometry, Figure 15 also includes results presented by Coussirat et al. [46] based on a SST k-ω turbulence model and the cavitation model by Singhal et al. [30] calibrated for this flow configuration. Note that Coussirat et al. did not find meaningful differences in their results when using the Singhal et al. [30] and the Zwart–Gerber–Belamri model [29]. They did not provide the results of their mesh sensitivity study for this model combination since the effects were reported to be small. We note that their data agree closely with our numerical results, even though their prediction for the static pressure variation at tap 1 does show a slight "kink" at the cavitation number where cavitation onset is observed in our experiments.

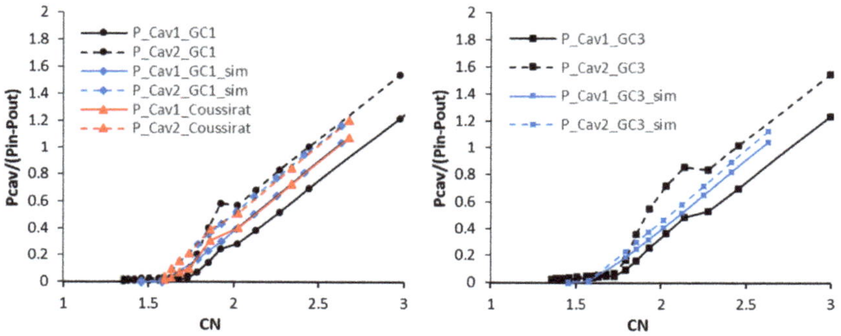

Figure 15. Dimensionless cavity pressures $P_{Cav1,2}$ at pressure taps 1 and 2 as a function of the cavitation number, for two dissolved gas concentrations, i.e., GC1 (**left**) and GC3 (**right**). Comparison between experimental (black) and numerical (blue) results. Also shown are simulation results by Coussirat et al. [46].

Figure 16 shows a comparison of the orifice discharge coefficients C_d predicted using our CFD simulations for the three considered dissolved gas contents (i.e., GC1, GC2, and GC3) with those obtained from our experiments (see also Figure 13). Also presented are Nurick's data [14], which are already included in Figure 13. Simulation results are found to be in good agreement with our experimental results and Nurick's data in the non-cavitating case (high CN values) and also agree well with Nurick's correlation in the full cavitation zone ($CN < 1.3$). However, the simulations predict cavitation onset at smaller cavitation numbers (see also Figure 15 in this context), resulting in an overprediction of the discharge coefficient in the respective range of cavitation numbers. Similar observations were also made by Singhal et al. [30] and Lifante and Frank [44]. Darbandi et al. [42] and Coussirat et al. [46], using the cavitation model by Singhal et al. [30] and a k-ω and SST k-ω turbulence model, respectively, do not find this overprediction; however, their C_d values at higher cavitation numbers are underpredicted.

Similar to our experimental observations, the simulations do not show a meaningful reduction in the discharge coefficient with an increase in dissolved CO_2 content from GC1 to GC2. For case GC3, a reduction in the discharge coefficient is only predicted for the data point at $CN = 1.57$, whereas in the experimental data, a reduction is observed over a broader CN value range; see Figure 13.

Since most prior published work on this flow geometry was carried out with water containing dissolved air (or untreated water) in lieu of dissolved CO_2, numerical simulations and experimental flow visualizations were also carried out for three select cases, representative of three cavitation numbers (at or near the onset of cavitation and full cavitation) and with a dissolved air mass fraction of 15 ppm. This latter value corresponds to the value employed in the analyses by Singhal et al. [30], Lifante and Frank [44], and Darbandi et al. [42]. Findings for these cases with dissolved air were found to correspond to those discussed in the context of our study of dissolved CO_2 and specifically with low amounts of dissolved CO_2 (i.e., cases GC1 and GC2 according to Table 1).

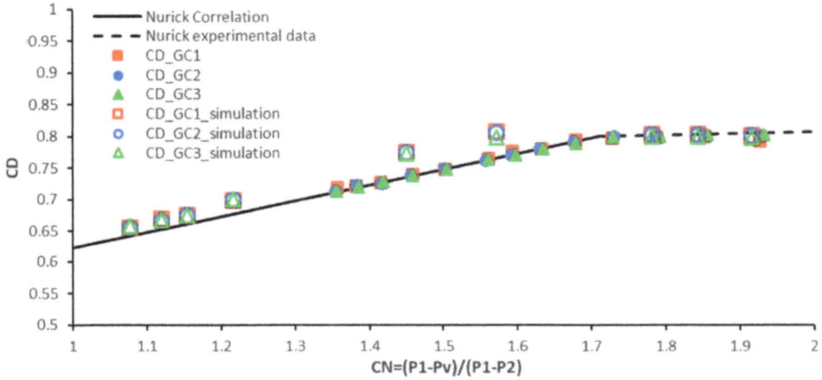

Figure 16. Discharge coefficient as a function of the cavitation number based on experimental results, correlation by Nurick [14], and CFD simulation results based on the full cavitation model.

Figures 17–19 illustrate computed results for the prescribed cases with dissolved air together with the respective experimental observations over the entire flow domain, i.e., including both the orifice region and the downstream expanded flow section. The overall development of the cavitation and outgassing phenomena is captured reasonably well using the employed model (see Section 4) at low dissolved gas content, even though quantitative differences exist. Figure 17 illustrates the point of cavitation inception within the experiment, taking place at cavitation number $CN = 2.13$. In the respective simulation, cavitation has not yet occurred. However, air is evolving out of the liquid phase, albeit only at a small rate. This conforms to the experimental result, which does not yet show the appearance of air bubbles in the expanded downstream section. It is important to note that the zero values of dissolved gas content in the expanded downstream section result from the established large recirculation zone surrounding the discharging orifice jet (see the streamline plot in Figure 17) and the imposed outlet pressure condition, allowing only pure liquid water, i.e., with zero dissolved air content, to re-enter the domain. With an increase in inlet pressure (i.e., a decrease in cavitation number), cavitation becomes stronger.

Figure 18 illustrates the case for $CN = 1.85$; this represents the value at which cavitation inception is predicted using the simulation model (see also Figure 15 for case CG1). Cavitation takes place in the shear layer between the core flow and the flow recirculation zone after the contraction. Surprisingly, the axial extent of the predicted cavitation zone is close to that found experimentally. Parallel to vapor cavitation, outgassing is predicted to occur now at a higher rate, whereby the free air remains confined to the recirculation zone near the vena contracta, the downstream orifice wall, and the shear layer of the discharging orifice jet. This conforms to the appearance of microbubbles in the expanded downstream section within the experiment. With a further reduction in the cavitation number, the rate of vapor cavitation increases, and the cavitating flow zone starts to expand downstream while the air outgassing rate also increases. At $CN = 1.55$, as illustrated in Figure 19, full cavitation is observed in the experiment, i.e., the cavitating flow region extends over the entire length of the orifice and up to the downstream expansion zone. The streaky bright and dark areas are indicative of the existence of single-phase areas containing pure gas or vapor and areas with finely dispersed gas/vapor bubbles. In addition, significantly more bubbles are observed in the expanded flow region downstream. In the respective numerical solution, the cavitating flow region has not yet reached the entire length of the orifice. However, for a length of about two orifice diameters downstream of the contraction, the wall region is filled with only vapor (vapor volume fraction equal to 1). Further downstream, the wall region is occupied by air evolved out of the liquid phase, which subsequently discharges with the core jet into the expanded downstream flow region. The air outgassing rate has now grown to approximately three times that predicted for $CN = 2.13$. A quantitative comparison with experimentally determined released air volume rates has not been carried out.

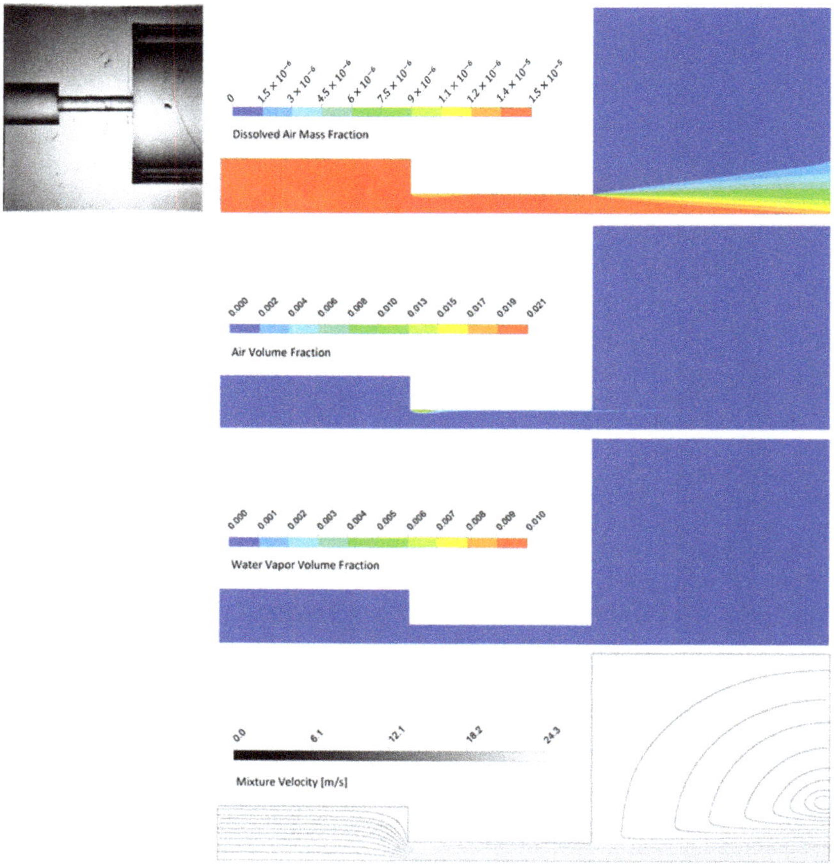

Figure 17. Comparison of simulation results with experimental observations at $CN = 2.13$, corresponding to the experimentally observed cavitation onset. Operating fluid: tap water with 15 ppm dissolved air mass fraction. Also shown are backward-traced streamlines beginning at the flow outlet. Gas content in the recirculation region of the extended outlet is zero since inflow from that boundary is set to be pure water. Flow direction from left to right.

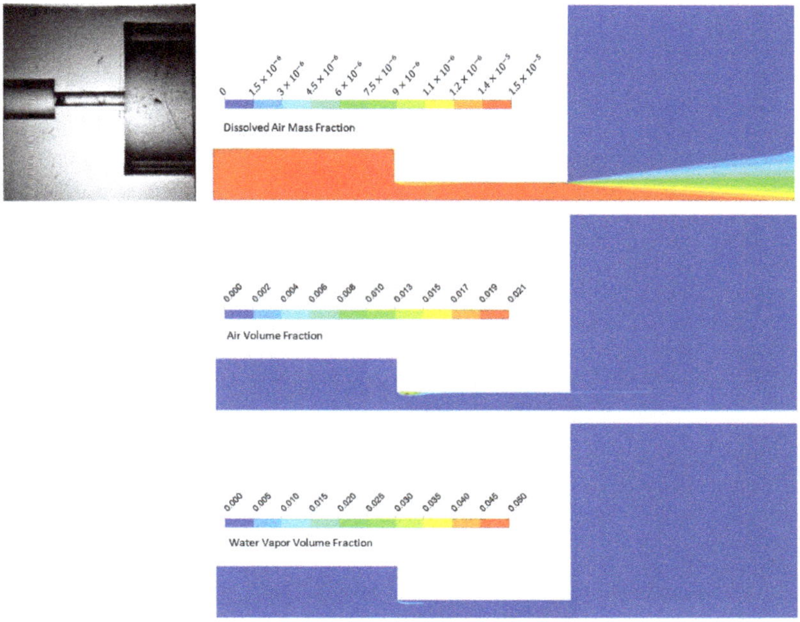

Figure 18. Comparison of simulation results with experimental observations for a similar case as in Figure 17 but at $CN = 1.85$, corresponding to cavitation onset in the numerical prediction. Flow direction from left to right.

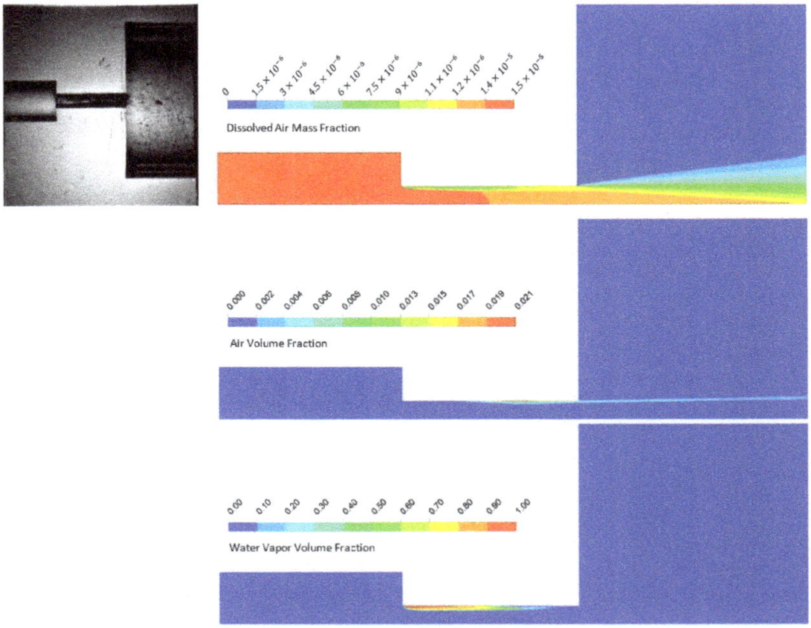

Figure 19. Comparison of simulation results with experimental observations for a similar case as in Figure 17 but at $CN = 1.55$, corresponding to experimentally observed full cavitation. Flow direction from left to right.

6. Summary

The effect of dissolved gases on the inception and evolution of vapor cavitation was investigated for the well-known sharp-edged orifice geometry already analyzed by Nurick [14], who did not consider the influence of dissolved gases. Water with five different concentrations of dissolved CO_2 was prepared. In the present setup and in contrast to Nurick's configuration, the flow does not discharge into a gas-filled vessel but discharges into an expanding flow section from which it continues to flow downstream. Mass flow through the orifice was controlled by adjusting the pressure in the upstream vessel containing the prepared water volume while the downstream pressure was held constant. The test article was made of PMMA, and backlighting was used to observe the orifice flow using a high-speed camera. Cavitation inception for five different carbon dioxide contents was detected by visual observation and by measuring changes in wall-static pressures (at distances $d/4$ and $d/2$ downstream of the orifice edge) with changes in the cavitation number. Comparing visual observations with recorded pressure data shows that, except for the case with the highest dissolved CO_2 content onset of cavitation coincides with a plateau formation in the variation of wall pressures with the cavitation number. A similar observation was already made by Nurick [14]; however, in the present case, the plateau formation is more pronounced and is also found at the pressure tab closest to the orifice edge.

The experimental results show that an increase in dissolved CO_2 concentration leads to higher cavitation numbers at which cavitation inception occurs. Outgassing, i.e., the desorption of dissolved gas from the liquid phase, was found to occur at higher cavitation numbers and before the onset of vapor cavitation. Evidence of outgassing was provided by the appearance of gas bubbles in the downstream expanded flow section and in the channels of the wall-static pressure tabs, the latter being situated in the recirculation region near the vena contracta.

For low dissolved CO_2 content, discharge coefficients from the current measurements agree well with those presented by Nurick [14], even though the amount of dissolved CO_2 changes the cavitation number CN at which cavitation inception occurs. At higher dissolved gas content, gas evolution is increased at the same CN number (which might also significantly enhance vapor cavitation), and the evolved gas (and additional vapor) fraction leads to a further blockage of the flow cross-section for the liquid phase, thereby leading to a notable drop in discharge coefficient. At low CN numbers, i.e., at full cavitation conditions, the amount of dissolved CO_2 is found to have no impact on the discharge coefficient.

CFD analyses for the orifice flow have been carried out using the cavitation model by Zwart et al. [29] and the model by Yang et al. [45] to account for non-condensable gas. Numerical results for three dissolved CO_2 mass fractions, as well as a selected number of cases with dissolved air in lieu of CO_2 (at a mass fraction of 15 ppm) were obtained. Predictions for cases with a low CO_2 mass fraction and those with dissolved air follow the results predicted by other authors using similar models and 15 ppm of dissolved air. An agreement was found in terms of discharge coefficient as a function of cavitation number CN and in the case of Coussirat et al. [46] also in terms of dimensionless wall pressure predictions with variation in CN values.

In the numerical simulations, the onset of vapor cavitation is predicted to occur at lower cavitation numbers than found experimentally, irrespective of the amount of dissolved CO_2 gas. The experimental observation that the onset of cavitation occurs at larger cavitation numbers with increased dissolved CO_2 content is not reflected in the numerical results. On the contrary, the simulations predict that cavitation occurs at smaller cavitation numbers with an increase in dissolved CO_2 content. Also, the plateau observed in the variation of static wall pressure tab values with changes in the cavitation number found at the onset of cavitation is not captured using the present numerical simulations.

The present investigation has augmented the experimental data set for the orifice geometry previously investigated by Nurick by including information on the effect of non-condensable dissolved gas (specifically carbon dioxide) on flow cavitation. The new data can serve for the validation of advanced cavitation models, which are able to capture the complex interaction of the vapor cavitation and outgassing (i.e., gas cavitation) phenomena.

Author Contributions: Conceptualization, C.M.; Methodology, S.S.; Software, S.S.; Validation, S.S.; Formal analysis, C.M.; Investigation, S.S.; Writing—original draft, S.S.; Writing—review & editing, C.M.; Visualization, S.S.; Supervision, C.M.; Project administration, C.M.; Funding acquisition, C.M. All authors have read and agreed to the published version of the manuscript.

Funding: This research received no external funding.

Data Availability Statement: Data are contained within the article.

Conflicts of Interest: The authors declare no conflict of interest.

Appendix A

In this appendix, we briefly review the model equations employed to produce the numerical results presented in this work. Numerical simulations were carried out using the commercial flow-solver ANSYS CFX together with the mass transfer models for vapor (i.e., vapor cavitation) and non-condensable gases (i.e., pseudo cavitation) as previously presented by Zwart et al. [29] and Yang et al. [45] and later referenced by Lifante and Frank [44]. The working fluid is a three-phase mixture consisting of gaseous carbon dioxide, water vapor, and liquid water, which itself is considered a two-component liquid composed of pure water and dissolved carbon dioxide. The homogenous mixture assumption is employed, and all phases share the same velocity and pressure field. The mass transport equation for each phase k, i.e., liquid water ($k = l$), water vapor ($k = v$), and gaseous CO_2 ($k = g$) reads:

$$\frac{\partial(\alpha_k \rho_k)}{\partial t} + \frac{\partial(\alpha_k \rho_k u_i)}{\partial x_i} = \dot{S}_k \tag{A1}$$

in the absence of a relative velocity between phase k and the mixture [56]; where α_k denotes the phase volume fraction of phase k, ρ_k its density, u_i is the velocity component of the mixture and $\dot{S}_k$ the mass transfer rate of phase k due to phase transition. For the dissolution of CO_2 in water, it is assumed that the liquid phase volume does not change despite increasing its mass [50].

Based on conservation principles, we have:

$$\sum_k \alpha_k = 1. \tag{A2}$$

In addition, $Y_w + Y_{D,g} = 1$ for the liquid phase where Y_w and $Y_{D,g}$ denote the mass fractions of pure water and dissolved CO_2, respectively. For determining the local mass fraction of dissolved CO_2 in the liquid phase, an additional mass transport equation is solved, i.e.,

$$\frac{\partial}{\partial t}(\alpha_l \rho_l Y_{D,g}) + \nabla \bullet ((\alpha_l u\, \rho_l Y_{D,g}) - (\alpha_l\, \rho_l D_g (\nabla Y_{D,g}))) = -\dot{S}_g \tag{A3}$$

where D_g denotes the diffusivity of dissolved CO_2 in pure water. Conservation of momentum equations for the mixture is given by

$$\frac{\partial(\rho_m u_i)}{\partial t} + \frac{\partial(\rho_m u_i u_i)}{\partial x_j} = -\frac{\partial p}{\partial x_i} + \frac{\partial(\tau_{ij})}{\partial x_j} + \rho_m g_i\,, \tag{A4}$$

where p is the pressure, τ_{ij} the components of the stress tensor, and g_i the directional component of gravitational acceleration. Mixture density ρ_m is given by

$$\rho_m = \sum_k \alpha_k \rho_k, \tag{A5}$$

where the density of each phase was assumed to be constant in the present investigation (and the liquid density was assumed to be that of pure water irrespective of the amount of dissolved gas); in this case, the velocity divergence balances the volume generation due to phase change (see [29]). Components of the stress tensor are evaluated based on Stokes' hypothesis, i.e.,

$$\tau_{ij} = \mu_m \left[\left(\frac{\partial u_i}{\partial x_j} + \frac{\partial u_j}{\partial x_i} \right) - \frac{2}{3} \delta_{ij} \frac{\partial u_k}{\partial x_k} \right] \tag{A6}$$

and with the molecular viscosity of the mixture μ_m calculated as a volume-weighted average analogous to the mixture density, i.e.,

$$\mu_m = \sum_k \alpha_k \mu_k. \tag{A7}$$

For turbulent flows, as considered here, the velocity components in the above equations represent time-averaged quantities, i.e., after averaging the local velocity on the time scale of the turbulent fluctuations (Reynolds averaging). Based on Boussinesq's hypothesis, the additional stress terms (Reynolds stresses) appearing in Equation (A4), in this case, can be modeled by adding an eddy viscosity μ_t to the value of the molecular viscosity μ_m in Equation (A7). Here, we employ the two-equation k-ω SST turbulence model to express μ_t based on the density of the mixture; see also [50].

For calculating the mass transfer source terms due to cavitation and condensation in Equation (A1), the model by Zwart et al. [29] is employed. The model is derived from a simplified version of the Rayleigh–Plesset equation (neglecting second-order terms and surface tension in the equation). It neglects bubble interaction but considers the effect of vapor volume fraction on nucleation site density. The respective source terms for the liquid and vapor phase in Equation (A1) are

$$\dot{S}_v = -\dot{S}_l = \begin{cases} F_{vap} \frac{3\alpha_{nuc}(1-\alpha_v)\,\rho_v}{R_{nuc}} \sqrt{\frac{2}{3}\left(\frac{p_v - p}{\rho_l}\right)} & for \quad p < p_v \\ F_{cond} \frac{3\alpha_v \rho_v}{R_{nuc}} \sqrt{\frac{2}{3}\left(\frac{p_v - p}{\rho_l}\right)} & for \quad p > p_v \end{cases} \tag{A8}$$

with the model parameters $F_{vap} = 50$ and $F_{cond} = 0.01$ and where R_{nuc} is the nuclei radius set to be 2×10^{-6} m in the present investigation.

The model assumes that the breaking strength of the liquid is reached when the pressure (also referred to as the critical pressure p_{cr}) reaches the vapor pressure p_v. However, viscous normal stresses might have to be considered [57] in determining p_{cr} [58] and improved results might be obtained by replacing p_v in Equation (A8) above by the respective p_{cr} value, i.e.,

$$p_{cr} = p_v + 2\,(\mu_m + \mu_t)\, T_{ij}^{max}, \tag{A9}$$

where T_{ij}^{max} denotes the maximum principal component of the strain-rate tensor. However, the present investigation assumes $p_{cr} = p_v$.

Following Yang et al. [45], the mass transfer source term $\dot{S}_g$ due to gas desorption and absorption in Equations (A1) and (A3), can be written as the sum of a desorption term R_{des} and an absorption term R_{abs}, i.e.,

$$\dot{S}_g = R_{des} - R_{abs} \tag{A10}$$

with absorption and desorption being driven by the difference between the local partial pressure of the non-condensable gas in its gaseous sate and its equilibrium pressure based on the amount of gas dissolved in the liquid. This is different from Rayleigh–Plesset-equation-based models, where the phase change rates are governed by a square-root dependence of pressure difference. Considering that no desorption is possible if there is no liquid phase locally (i.e., $\alpha_g = 1$) or no dissolved gas content (i.e., $Y_{D,g} = 0$) and similarly considering that no gas absorption is possible if there is no free gas locally (i.e., $\alpha_g = 0$) or the liquid phase is saturated with dissolved gas already; the formulas for evaporation and condensation rates according to Yang et al. [45] can be written as follows:

$$R_{abs} = C_a \rho_g \left(p - p_{equil}\right) \left(f_{g,\ l,\ lim} \ - Y_{D,g} \frac{\rho_l \alpha_l}{\rho_m}\right) \frac{\rho_g \alpha_g}{\rho_m}, \tag{A11}$$

$$R_{des} = C_d \rho_g \left(p_{equil} - p\right) \left(1.0 - \frac{\rho_g \alpha_g}{\rho_m}\right) Y_{D,g} \frac{\rho_l \alpha_l}{\rho_m}. \tag{A12}$$

Here, p_{equil} denotes the sum of liquid/water vapor pressure and the equilibrium gas-phase pressure p_g based on the dissolved gas fraction in the liquid, i.e.,

$$p_{equil} = p_v + p_g \tag{A13}$$

with p_g evaluated using Henry's law

$$p_g = H_v^{px} X_{D,g}. \tag{A14}$$

In this equation, H_v^{px} denotes the Henry volatility constant and

$$X_{D,g} = \left(\frac{M_l}{M_g}\right) Y_{D,g} \tag{A15}$$

is the mole fraction of dissolved gas. C_a and C_D in the prior equations are empirical constants and $f_{g,\ l,\ lim}$ denotes the maximum solubility value of the non-condensable gas (i.e., CO_2 in the present investigation) written in terms of its local overall mass fraction. Here, we have chosen C_a = 0.1 and C_D = 2, corresponding to the values proposed by Yang et al. [45] for orifice flows. In order to consider the effect of turbulent pressure fluctuations on the threshold pressure below which phase change phenomena occur, the value of vapor pressure p_v in the above model equations was increased by the local turbulent pressure fluctuation [45], i.e.,

$$p_v = p_{sat} + p'_{turb} \tag{A16}$$

where p_{sat} refers to the vapor (or saturation) pressure without turbulence effects and

$$p'_{turb} = 0.39(1 - \alpha_v)\rho_l k \tag{A17}$$

with k denoting the local turbulent kinetic energy. The expression for p'_{turb} corresponds to that proposed by Yang et al. but with the corrective factor $(1 - \alpha_v)$ as proposed by Lifante and Frank [44].

Appendix B

The second set of test data and respective visual observations from the experimental test campaign for similar conditions is presented in Figure 10 but with a frame of view more focused on the flow area after the contraction. Flow direction from right to left.

Cavitation Number, CN	1.68	1.73	1.79	1.85	1.93	2.02	2.13	2.27
GC1	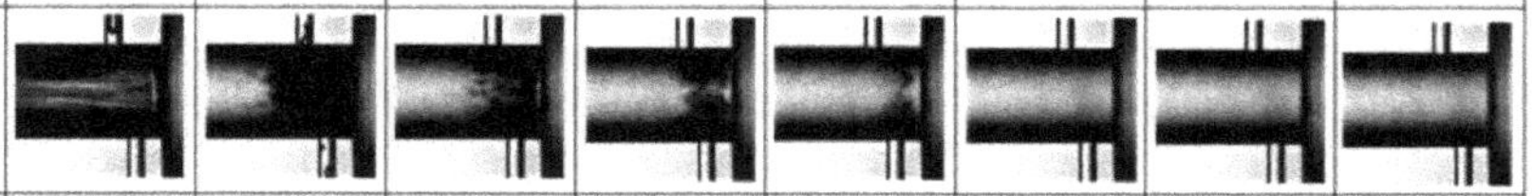							
Length of cavitation zone [mm]	Transition to full cavitation	3.22	2.23	1.64	1.31	0	0	0

Cavitation Number, CN	1.68	1.73	1.79	1.85	1.93	2.02	2.13	2.27
GC2	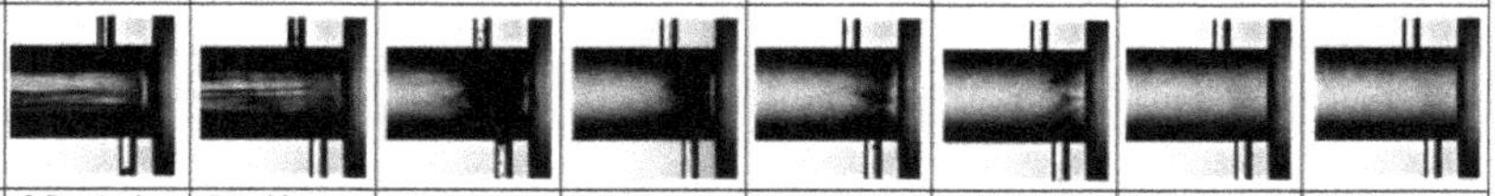							
Length of cavitation zone [mm]	full cavitation	Transition to full cavitation	2.92	2.06	1.63	1.35	0	0

Cavitation Number, CN	1.68	1.73	1.79	1.86	1.93	2.03	2.14	2.28
GC3	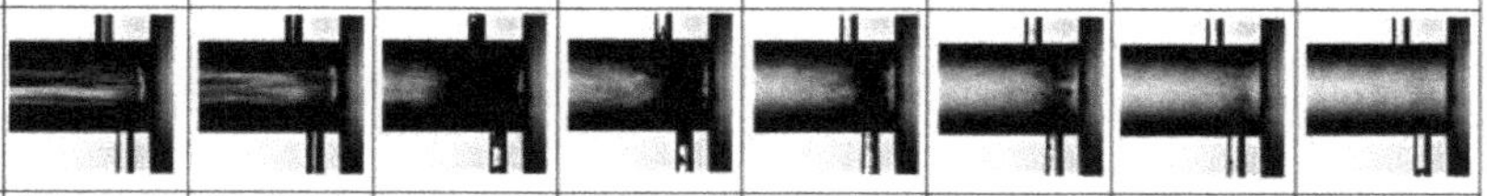							
Length of cavitation zone [mm]	Full cavitation	Transition to full cavitation	4.22	2.92	1.96	1.50	1.33	0

Cavitation Number, CN	1.69	1.74	1.80	1.86	1.94	1.96	2.07	2.20
GC4	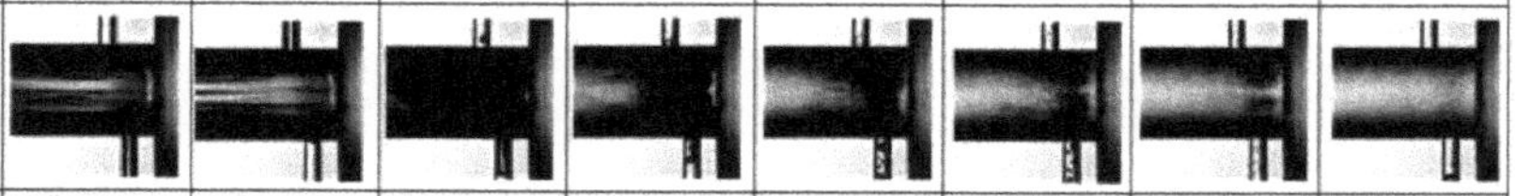							
Length of cavitation zone [mm]	Full cavitation	Full cavitation	Transition to full cavitation	4.10	3.27	2.22	1.56	1.31

Cavitation Number, CN	1.69	1.74	1.80	1.87	1.94	2.04	2.15	2.29
GC5	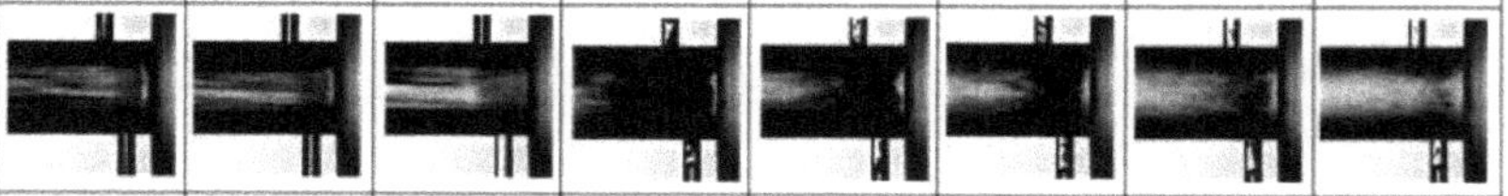							
Length of cavitation zone [mm]	Full cavitation	Full cavitation	Full cavitation	4.90	4.44	3.69	2.82	2.05

Figure A1. Flow observations from second set of test data for broad range of cavitation numbers *CN* and 5 different mass fractions of dissolved CO_2, i.e., GC1–GC5 according to Table 1. Flow direction from right to left. Results are comparable to those presented in Figure 10.

References

1. Brennen, C. *Cavitation and Bubble Dynamics*; Cambridge University Press: Cambridge, UK, 2013.
2. Brennen, C. *Fundamentals of Multiphase Flow*; Cambridge University Press: Cambridge, UK, 2005.
3. Franc, J.-P.; Michel, J.-M. Fundamentals of Cavitation. In *Fluid Mechanics and Its Applications*; Moreau, R., Ed.; Springer Science & Business Media: Dordrecht, The Netherlands, 2005.
4. Perk, O.Y.; Şeşen, M.; Gozuacik, D.; Koşar, A. Kidney stone erosion by micro scale hydrodynamic cavitation and consequent kidney stone treatment. *Ann. Biomed. Eng.* **2012**, *40*, 1895–1902. [CrossRef]
5. Zijlstra, A.G. Acoustic Surface Cavitation. Ph.D. Dissertation, University of Twente, Enschede, The Netherlands, 2011.
6. Vasile, E.; Ciocanea, A.; Ionescu, V.; Lepadatu, I.; Diac, C.; Stamatin, S.N. Making precious metals cheap: A sonoelectrochemical–Hydrodynamic cavitation method to recycle platinum group metals from spent automotive catalysts. *Ultrason. Sonochem.* **2021**, *72*, 105404. [CrossRef]

7. Kim, M.; Kim, D.; Cho, J.M.; Nam, K.; Lee, H.; Nayak, M.; Han, J.; Oh, H.-M.; Chang, Y.K. Hydrodynamic cavitation for bacterial disinfection and medium recycling for sustainable Ettlia sp. Cultivation. *J. Environ. Chem. Eng.* **2021**, *9*, 105411. [CrossRef]
8. Kosel, J.; Šuštaršič, M.; Petkovšek, M.; Zupanc, M.; Sežun, M.; Dular, M. Application of (super) cavitation for the recycling of process waters in paper producing industry. *Ultrason. Sonochemistry* **2020**, *64*, 105002. [CrossRef] [PubMed]
9. Song, Y.; Hou, R.; Zhang, W.; Liu, J. Hydrodynamic cavitation as an efficient water treatment method for various sewage: A review. *Water Sci. Technol.* **2022**, *86*, 302–320. [CrossRef]
10. Gogate, P.R.; Pandit, A.B. Cavitation generation and usage without ultrasound: Hydrodynamic cavitation. *Theor. Exp. Sonochem. Involv. Inorg. Syst.* **2011**, 69–106. [CrossRef]
11. Li, B.; Gu, Y.; Chen, M. An experimental study on the cavitation of water with dissolved gases. *Exp. Fluids* **2017**, *58*, 164. [CrossRef]
12. Mehring, C. Liquid-Fuel Ejector Pump under Multiphase-Flow Conditions. In Proceedings of the 25th European Conference on Liquid Atomization and Spray Systems, Chania, Greece, 1–4 September 2013.
13. Zheng, H.; Zheng, Y.; Zhu, J. Recent developments in hydrodynamic cavitation reactors: Cavitation mechanism, reactor design, and applications. *Engineering* **2022**, *19*, 180–198. [CrossRef]
14. Nurick, W.H. Orifice cavitation and its effect on spray mixing. *J. Flusids Eng.* **1976**, *98*, 681–687. [CrossRef]
15. Arakeri, V.H. Cavitation inception. *Proc. Indian Acad. Sci. Sect. C Eng. Sci.* **1979**, *2*, 149–177. [CrossRef]
16. Nurick, W.H.; Ohanian, T.; Talley, D.G.; Strakey, P.A. Impact of L/D on 90 Degree Sharp-Edge Orifice Flow with Manifold Passage Cross Flow (Preprint). *J. Artic.* **2007**, *30*, 2.
17. Yan, Y.; Thorpe, R.B. Flow regime transitions due to cavitation in the flow through an orifice. *Int. J. Multiph. Flow* **1990**, *16*, 1023–1045. [CrossRef]
18. Mishra, C.; Peles, Y. Cavitation in flow through a micro-orifice inside a silicon microchannel. *Phys. Fluids* **2005**, *17*, 013601. [CrossRef]
19. Zhang, H.; Zuo, Z.; Liu, S. Influence of Dissolved Gas Content on Venturi Cavitation at Thermally Sensitive Conditions. In Proceedings of the 10th International Symposium on Cavitation (CAV2018), Baltimore, MD, USA, 14–16 May 2018; pp. 546–550.
20. Rooze, J. Cavitation in Gas-Saturated Liquids. Ph.D. Thesis, Eindhoven University of Technology, Eindhoven, The Netherlands, 2012.
21. Freudigmann, H.A.; Dörr, A.; Iben, U.; Pelz, P.F. Modeling of cavitation-induced air release phenomena in micro-orifice flows. *J. Fluids Eng.* **2017**, *139*, 111301. [CrossRef]
22. Kowalski, K.; Pollak, S.; Skoda, R.; Hussong, J. Experimental study on cavitation-induced air release in orifice flows. *ASME J. Fluids Eng.* **2018**, *140*, 061201. [CrossRef]
23. Duke, D.; Kastengren, A.L.; Tilocco, F.Z.; Swantek, A.B.; Powell, C.F. X-ray radiography measurements of cavitating nozzle flow. *At. Sprays* **2013**, *23*. [CrossRef]
24. Duke, D.J.; Swantek, A.B.; Kastengren, A.L.; Powell, C.F. X-ray diagnostics for cavitating nozzle flow. *J. Phys. Conf. Ser.* **2015**, *656*, 012110. [CrossRef]
25. Duke, D.J.; Matusik, K.E.; Kastengren, A.L.; Swantek, A.B.; Sovis, N.; Payri, R.; Viera, J.P.; Powell, C.F. X-ray radiography of cavitation in a beryllium alloy nozzle. *Int. J. Engine Res.* **2017**, *18*, 39–50. [CrossRef]
26. Sou, A.; Biçer, B.; Tomiyama, A. Numerical simulation of incipient cavitation flow in a nozzle of fuel injector. *Comput. Fluids* **2014**, *103*, 42–48. [CrossRef]
27. Schnerr, G.H.; Sauer, J. Physical and Numerical Modeling of Unsteady Cavitation Dynamics. In Proceedings of the Fourth International Conference on Multiphase Flow (Vol. 1), New Orleans, LO, USA, 27 May–1 June 2001.
28. Kunz, R.F.; Boger, D.A.; Stinebring, D.R.; Chyczewski, T.S.; Lindau, J.W.; Gibeling, H.J.; Venkateswaran, S.; Govindan, T.R. A preconditioned Navier–Stokes method for two-phase flows with application to cavitation prediction. *Comput. Fluids* **2000**, *29*, 849–875. [CrossRef]
29. Zwart, P.J.; Gerber, A.G.; Belamri, T. A Two-Phase Flow Model for Predicting Cavitation Dynamics. In Proceedings of the Fifth International Conference on Multiphase Flow, Yokohama, Japan, 30 May–4 June 2004; p. 152.
30. Singhal, A.K.; Athavale, M.M.; Li, H.; Jiang, Y. Mathematical basis and validation of the full cavitation model. *ASME J. Fluids Eng.* **2002**, *124*, 617–624. [CrossRef]
31. Battistoni, M.; Som, S.; Longman, D.E. Comparison of mixture and multifluid models for in-nozzle cavitation prediction. *J. Eng. Gas Turbines Power* **2014**, *136*, 061506. [CrossRef]
32. Li, W.; Yu, Z. Cavitation models with thermodynamic effect for organic fluid cavitating flows in organic Rankine cycle systems: A review. *Therm. Sci. Eng. Prog.* **2021**, *26*, 101079. [CrossRef]
33. Battistoni, M.; Duke, D.; Swantek, A.B.; Tilocco, F.Z.; Powell, C.F.; Som, S. Effects of noncondensable gas on cavitating nozzles. *At. Sprays* **2015**, *25*. [CrossRef]
34. Yang, S.; Habchi, C. Real-fluid phase transition in cavitation modeling considering dissolved non-condensable gas. *Phys. Fluids* **2020**, *32*, 032102. [CrossRef]
35. Amini, A.; Reclari, M.; Sano, T.; Farhat, M. Effect of Gas Content on Tip Vortex Cavitation. In Proceedings of the 10th International Symposium on Cavitation (CAV2018), Baltimore, MD, USA, 14–16 May 2018.
36. Mukherjee, S.; Gomez, H. Effect of dissolved gas on the tensile strength of water. *Phys. Fluids* **2022**, *34*, 126112. [CrossRef]
37. Iben, U.; Makhnov, A.; Schmidt, A. Numerical study of the effects of dissolved gas release in cavitating flow. *AIP Conf. Proc.* **2018**, *2027*, 030128.

38. Iben, U.; Makhnov, A.; Schmidt, A. Numerical study of dissolved gas release induced by cavitation in a high speed channel flow. *J. Phys. Conf. Ser.* **2019**, *1400*, 077037. [CrossRef]
39. Malekshah, E.H.; Wróblewski, W. Merging theory-based cavitation model adaptable with non-condensable gas effects in prediction of compressible three-phase cavitating flow. *Int. J. Heat Mass Transf.* **2022**, *196*, 123279. [CrossRef]
40. Richards, K.; Senecal, P.; Pomraning, E. *Converge 2.1.0 Theory Manual*; Convergent Science Inc.: Middleton, WI, USA, 2013.
41. Bohbot, J.; Gillet, N.; Benkenida, A. IFP-C3D: An unstructured parallel solver for reactive compressible gas flow with spray. *Oil Gas Sci. Technol.-Rev. De L'ifp* **2009**, *64*, 309–335. [CrossRef]
42. Darbandi, M.; Sadegh, H.; Schneider, G.E. Simulating Orifice Cavitation Using the Full Cavitation Model. In Proceedings of the 17th Annual Conference of the CFD Society of Canada, Ottawa, ON, Canada, 3–5 May 2009.
43. Darbandi, M.; Sadeghi, H. Numerical simulation of orifice cavitating flows using two-fluid and three-fluid cavitation models. *Numer. Heat Transf. Part A Appl.* **2010**, *58*, 505–526. [CrossRef]
44. Lifante, C.; Frank, T. *Investigation of Higher Order Pressure Fluctuations and Its Influence on Ship Stern, Taking into Account Cavitation at Propeller Blades*; TR-08-04; ANSYS Germany GmbH: Berlin, Germany, 2008.
45. Yang, H.Q.; Singhal, A.K.; Megahed, M. Industrial two-phase flow CFD—The Full Cavitation Model. *Von Karman Inst. Fluid Dyn. Lect. Ser.* **2005**, *4*.
46. Coussirat, M.; Moll, F.; Cappa, F.; Fontanals, A. Study of Available Turbulence and Cavitation Models to Reproduce Flow Patterns in Confined Flows. *ASME. J. Fluids Eng.* **2016**, *138*, 091304. [CrossRef]
47. Stanley, C.; Barber, T.; Milton, B.; Rosengarten, G. Periodic cavitation shedding in a cylindrical orifice. *Exp. Fluids* **2011**, *51*, 1189–1200. [CrossRef]
48. Mauger, C.; Méès, L.; Michard, M.; Azouzi, A.; Valette, S. Shadowgraph, Schlieren and interferometry in a 2D cavitating channel flow. *Exp. Fluids* **2012**, *53*, 1895–1913. [CrossRef]
49. Saurabh, A. Cavitation in Injectors: A Brief Review of the Diagnostics of Liquid-Vapour Flow Within Injector Nozzles. Simulations and Optical Diagnostics for Internal Combustion Engines. *Curr. Status Way Forw.* **2020**, 85–93. [CrossRef]
50. Menter, F. Zonal Two Equation kw Turbulence Models for Aerodynamic Flows. In Proceedings of the 23rd Fluid Dynamics, Plasmadynamics, and Lasers Conference, Orlando, FL, USA, 6–9 July 1993; p. 2906.
51. Peterson, F.B. Discussion: "Orifice Cavitation and Its Effect on Spray Mixing" (Nurick, W.H., 1976, ASME J. Fluids Eng., 98, pp. 681–687). *ASME. J. Fluids Eng.* **1977**, *99*, 426. [CrossRef]
52. Winklhofer, E.; Kull, E.; Kelz, E.; Morozov, A. Comprehensive Hydraulic and Flow Field Documentation in Model Throttle Experiments under Cavitation Conditions. In Proceedings of the ILASS-Europe Conference, Zurich, Switzerland, 2–6 September 2001; pp. 574–579.
53. Martynov, S.; Mason, D.; Heikal, M.R.; Sazhin, S.S.; Gorokhovski, M. Modelling of Cavitation Flow in a Nozzle and Its Effect on Spray Development. In Proceedings of the International Heat Transfer Conference 13, Sydney, Australia, 13–18 August 2006; Begel House Inc.: New York, NY, USA, 2006.
54. *IEC 60770*; Transmitters for Use in Industrial-Process Control Systems—Part 1: Methods for Performance Evaluation. International Electrotechnical Commission: Geneva, Switzerland, 2010.
55. Battino, R.; Clever, H.L. The solubility of gases in liquids. *Chem. Rev.* **1966**, *66*, 395–463. [CrossRef]
56. Manninen, M.; Taivassalo, V.; Kallio, S. *On the Mixture Model for Multiphase Flow*; Technical Research Center of Finland, VIT Publications 288: Espoo, Finland, 1996.
57. Joseph, D.D. Cavitation and the state of stress in a flowing liquid. *J. Fluid Mech.* **1998**, *366*, 367–378. [CrossRef]
58. Martynov, S.B.; Mason, D.J.; Heikal, M.R. Effect of Viscous Stress on Cavitation Flow in Nozzles. *J. Fluids Eng.* **2006**, *9*, 2006.

fluids

Article

Application of Central-Weighted Essentially Non-Oscillatory Finite-Volume Interface-Capturing Schemes for Modeling Cavitation Induced by an Underwater Explosion

Ebenezer Mayowa Adebayo *, Panagiotis Tsoutsanis and Karl W. Jenkins

School of Aerospace, Transport and Manufacturing, Cranfield University, Cranfield MK43 0AL, UK; panagiotis.tsoutsanis@cranfield.ac.uk (P.T.); k.w.jenkins@cranfield.ac.uk (K.W.J.)
* Correspondence: e.adebayo@cranfield.ac.uk

Abstract: Cavitation resulting from underwater explosions in compressible multiphase or multicomponent flows presents significant challenges due to the dynamic nature of shock–cavitation–structure interactions, as well as the complex and discontinuous nature of the involved interfaces. Achieving accurate resolution of interfaces between different phases or components, in the presence of shocks, cavitating regions, and structural interactions, is crucial for modeling such problems. Furthermore, pressure convergence in simulations involving shock–cavitation–structure interactions requires accurate algorithms. In this research paper, we employ the diffuse interface method, also known as the interface-capturing scheme, to investigate cavitation in various underwater explosion test cases near different surfaces: a free surface and a rigid surface. The simulations are conducted using the unstructured compressible Navier–Stokes (UCNS3D) finite-volume framework employing central-weighted essentially non-oscillatory (CWENO) reconstruction schemes, utilizing the five-equation diffuse interface family of methods. Quantitative comparisons are made between the performance of both models. Additionally, we examine the effects of cavitation as a secondary loading source on structures, and evaluate the ability of the CWENO schemes to accurately capture and resolve material interfaces between fluids with minimal numerical dissipation or smearing. The results are compared with existing high-order methods and experimental data, where possible, to demonstrate the robustness of the CWENO schemes in simulating cavitation bubble dynamics, as well as their limitations within the current implementation of interface capturing.

Keywords: diffuse interface models; high-order methods; finite volume; cavitation; CWENO; underwater explosion

1. Introduction

Cavitation, which refers to the formation of vapour bubbles within a liquid due to high velocities and strong acceleration at low pressure, has diverse applications in various scientific and engineering fields. It can be beneficial in marine design engineering for reducing skin friction drag and achieving high speed for submerged projectiles, as well as in medicine for treating kidney stones and cancerous bodies. However, cavitation can also have adverse effects, such as material erosion, noise, and vibrations in marine propellers and turbomachinery, that can cause significant damage. Therefore, understanding the dynamics of cavitation is crucial, but it poses challenges in numerical simulations of compressible multiphase problems due to the complex physics involved in phase change, high gradients of flow variables, numerical discretization choices, and validation with experimental results. Numerical studies often focus on attached cavitations, as they are the primary subject of investigation in most research [1–3]. However, hull/bulk cavitation, which occurs during underwater explosions, is different as it evolves rapidly and unsteadily, changing dimensions before collapsing violently due to the surge in pressure, resulting in erratic and rapidly changing cavitation patterns. This makes numerical simulations of hull/bulk cavitation more challenging.

In hull cavitation, vapour bubbles are created on the surface of a submerged rigid surface, such as a projectile, ship hull, or submarine, due to rarefaction caused by high-pressure shock waves from an explosion. On the other hand, bulk cavitation occurs near a free surface between two interacting fluids due to the explosive shock waves, leading to widespread cavitation throughout the liquid. Hull cavitation is often more challenging to study using numerical methods due to its rapid and unsteady nature, while bulk cavitation is characterized by the formation of large-scale vapour bubbles throughout the liquid. Both forms of cavitation can have detrimental effects on marine structures and machinery.

In the context of hull or bulk cavitation, where regions experience extreme variations in pressure, negative pressure values are observed, which are physically unrealistic and can cause numerical codes to fail. To address this issue, two common approaches are used in diffuse interface models for simulating UNDEX (underwater explosion) phenomena. The first approach is to incorporate a single-fluid model, which has been used by several researchers [4–9] in similar UNDEX simulations. Various methods in this category, such as the cut-off model, vacuum model, isentropic model, Schmidt model, and modified Schmidt model, have been used to avoid negative pressure values during cavitation. The cut-off model assumes that cavitation bubbles do not form until the pressure drops below a critical value, known as the cut-off pressure, and is generally accurate in predicting peak pressure at the cavitation boundary but may not accurately predict cavitation collapse periods. The Schmidt model is accurate when the void fraction (ratio of vapour bubble density to total fluid density) is less than 10^{-5} [4] but has limitations in calculating the speed of sound, making it suitable for high-speed cavitating flows. The modified Schmidt model is an extension of the original Schmidt model and addresses its limitations, thereby expanding its range of applicability to larger vapour-to-liquid density ratios. The vacuum cavitation model assumes zero mass inside the cavitation bubble but is limited in capturing the complex shapes and dynamics of cavitation bubbles in multidimensional flows. The isentropic model, which has been used in the five-equation diffuse interface models [5,10], is robust as it considers the mixture as consisting of isentropic vapour and liquid phases, but it requires more computational resources and the predetermination of a model parameter k [4]. The cut-off model is preferred in this case as it is relatively simple to implement and does not alter peak pressure accuracy at the cavitation boundary, which is important for the UNDEX test cases considered in the studies. The one-fluid cavitation model aligns with the no-phase transition assumption of the five-equation interface-capturing models used in this study.

The accuracy of the one-fluid model in predicting the details of the cavitation process, including thermodynamic aspects, in compressible two-phase flows, is limited due to its inability to account for mass and heat transfer between the liquid and vapour phases. To overcome this limitation, researchers have extended simplified diffuse interface models such as the four-, five-, or six-equation models by incorporating source terms in the energy and mass conservation equations to represent mass and heat transfer. This approach, known as the "two-fluid cavitation model", aims to prevent non-physical pressure occurrences in the cavitation region. The source terms are implicitly expressed using an algebraic relaxation solver for phase transition modeling, as proposed by Pelanti and Shyue [11,12]. However, this method is more complex and computationally expensive, especially for seven-equation diffuse interface models such as the full non-equilibrium models of Baer–Nunziato [13] or Saurel–Abgrall [14]. A computational efficient and simple strategy in UNDEX is to employ a reduced diffuse interface model with phase transition, such as four- or five-equation models, which enforce mechanical, thermal, or both equilibrium. Jun et al. [15] extended the phase transition model used by Chiapolino et al. [16] for the four-equation model to study cavitation in underwater explosion test cases. Other existing diffuse interface models with phase transitions include those proposed by Pelanti and Shyue [9,11], Martelot et al. [17], Zein et al. [18], and others [14]. Ma et al. [19] utilised the Kapila [20] variant of the five-equation two-phase flow model with thermo-chemical relaxation terms to investigate liquid–vapour phase transition in cavitating flows for UNDEX test cases.

The present study focuses on hull and bulk cavitation in compressible multiphase or multicomponent flows, specifically in the context of underwater explosions. Two interface-capturing models, namely the ones proposed by Allaire et al. [21] and Kapila et al. [20], will be used to investigate the cavitation phenomenon in this study. The performance of these models will be evaluated using four underwater explosion (UNDEX) test cases in the open-source finite-volume UCNS3D solver [22].

To solve the interface-capturing models, high-order central-weighted essentially non-oscillatory (CWENO) schemes for multicomponent species within the finite-volume framework, as implemented by Tsoutsanis and Dumbser [23], will be employed. The CWENO schemes offer computational advantages such as lower cost and faster computation compared to classical WENO schemes, due to the use of low-order polynomials associated with directional stencils and their reduced size compared to the high-order polynomials related to the central stencil. These advantages will be fully utilised in capturing the dynamics of cavitation in the test cases considered herein.

The paper proceeds to Section 2, where it presents the governing equations that describe the diffuse interface models of Allaire et al. [21] and Kapila et al. [20], which consist of five equations. It includes details about the additional $K\nabla \cdot \mathbf{v}$ term that distinguishes these models, as well as the cut-off model used to prevent non-physical pressure values during cavitation. In Section 3, the implementation of the reconstruction process of the CWENO schemes is described, including the chosen fluxes and temporal discretization, within the finite-volume framework using the diffuse interface methods mentioned earlier. Section 4 is dedicated to the study of five test cases, where the proposed method is utilised to investigate cavitation occurrences in underwater explosions near different boundaries, such as a planar solid wall, a closed cylindrical container, and a free surface. The analysis focuses on parameters such as the influence of shock loading and cavitation collapse reloading on nearby structures like ship hulls or closed tanks. Additionally, the bubble dynamics of cavitating flows induced by underwater explosions near a free surface, as well as the impact of bubble separation distance on cavitation phenomena, are analyzed. The numerical results obtained are compared with analytical, experimental, and reference solutions whenever possible. A comparison between the five-equation models of Allaire et al. [21] and Kapila et al. [20] is also presented. Finally, the last section provides concluding remarks and a summary drawn from this study.

2. The Five-Equation Diffuse Interface Model

The five-equation model typically consists of two continuity equations: one momentum equation and one energy equation coupled with a non-conservative advection equation of the volume-fraction equation. The common five-equation models are the Allaire et al. [21], Kapila et al. [20], Murrone et al. [24], and Wackers et al. [25] models. All these are simpler formulations or reductions to lesser unknowns of the full seven (7) Baer–Nunziato [13] equations, which are often too complex to solve numerically. In this paper, we shall only be considering the Allaire and Kapila five-equation models. The difference between Allaire's and Kapila's models is only in the non-conservative advection equation in the volume-fraction equation. The equations are presented below:

$$\frac{\partial(\alpha_1\rho_1)}{\partial t} + \nabla \cdot (\alpha_1\rho_1\mathbf{v}) = 0, \tag{1}$$

$$\frac{\partial(\alpha_2\rho_2)}{\partial t} + \nabla \cdot (\alpha_2\rho_2\mathbf{v}) = 0, \tag{2}$$

$$\frac{\partial\rho\mathbf{v}}{\partial t} + \nabla \cdot (\rho\mathbf{v}\mathbf{v} + p\mathbf{I}) = 0, \tag{3}$$

$$\frac{\partial \rho E}{\partial t} + \nabla \cdot (\rho E + p)\mathbf{v} = 0, \tag{4}$$

$$\frac{\partial \alpha_1}{\partial t} + \mathbf{v} \cdot \nabla \alpha_1 = 0 \qquad \text{(Allaire et al.)}, \tag{5}$$

$$\frac{\partial \alpha_1}{\partial t} + \mathbf{v} \cdot \nabla \alpha_1 = K\nabla \cdot \mathbf{v} \qquad \text{(Kapila et al.)} \tag{6}$$

where the subscript $1, 2$ represents the fluid's component, α is the volume fraction of each component, ρ is the density, $\mathbf{v} = (u, v, w)$ is the velocity, p is the pressure, and E is the total energy.

This class of five-equation models (Allaire et al. [21] and Kapila et al. [20] model) is often referred to as mechanical equilibrium models. We assume that the momentum, energy, and mass transfer between the phases reach equilibrium due to the thermodynamic difference between each component. This reasonable flow assumption involves a free surface like the one treated in this paper. Note that this is a mechanical equilibrium of the fluid: each phase remains in thermal disequilibrium. All the models in this category can be expressed in the following general form of a non-linear system of partial differential equations in multiple space dimensions:

$$\frac{\partial \mathbf{Q}}{\partial t} + \nabla \cdot \mathbf{F}(\mathbf{Q}) + \mathbf{H}(\mathbf{Q})\nabla \cdot \mathbf{v} = \mathbf{S}(\mathbf{Q}) \tag{7}$$

where $\mathbf{Q}$ is the vector of evolution variables (conserved and not conserved), $\mathbf{F}$ is a flux function, $\mathbf{v}$ is the velocity field, and $\mathbf{H}$ and $\mathbf{S}$ are non-conservative quantities.

We then define $\mathbf{Q}$, $\mathbf{F}$, and $\mathbf{H}$ in the form as presented in Equation (8) (Allaire et al.'s model [21]) and Equation (9) (Kapila et al.'s model [21]) for a two-phase flow. The $K\nabla \cdot$ v term in Kapila's model [20] describes the compressibility and thermodynamic properties of the mixture region of the two-phase fluid. It is derived from the asymptotic reduction of Baer–Nunziato's seven-equation model [13,26,27]. From Equations (5) and (6), the $K\nabla \cdot \mathbf{v}$ term differentiates Kapila et al.'s model [20] from Allaire et al.'s model [21].

Allaire et al.'s five-equation model, presented in vector form:

$$Q = \begin{bmatrix} \alpha_1\rho_1 \\ \alpha_2\rho_2 \\ \rho\mathbf{v} \\ \rho E \\ \alpha_1 \end{bmatrix}, \quad F = \begin{bmatrix} \alpha_1\rho_1\mathbf{v} \\ \alpha_2\rho_2\mathbf{v} \\ \rho\mathbf{v}^2 + p \\ \mathbf{v}(\rho E + p) \\ \alpha_1\mathbf{v} \end{bmatrix} \quad H = \begin{bmatrix} 0 \\ 0 \\ 0 \\ 0 \\ -\alpha_1 \end{bmatrix}. \tag{8}$$

Kapila et al.'s five-equation model, presented in vector form:

$$Q = \begin{bmatrix} \alpha_1\rho_1 \\ \alpha_2\rho_2 \\ \rho\mathbf{v} \\ \rho E \\ \alpha_1 \end{bmatrix}, \quad F = \begin{bmatrix} \alpha_1\rho_1\mathbf{v} \\ \alpha_2\rho_2\mathbf{v} \\ \rho\mathbf{v}^2 + p \\ \mathbf{v}(\rho E + p) \\ \alpha_1\mathbf{v} \end{bmatrix} \quad H = \begin{bmatrix} 0 \\ 0 \\ 0 \\ 0 \\ -\alpha_1 - \mathbf{k} \end{bmatrix}. \tag{9}$$

For each fluid's component i, the mixture is assumed to be in mechanical equilibrium, as stated previously. The velocity and pressure of each component is $v = v_1 \equiv v_2$ and $p = p_1 \equiv p_2$ (i.e., we maintain a single velocity and single pressure for the mixture).

The total density, momentum, and kinetic and internal energy of the mixture are stated below:

$$\rho = (\alpha\rho)_1 + (\alpha\rho)_2 = \sum_{i=1}^{N} (\alpha\rho)_i : \text{ Mixture density.} \tag{10}$$

$$\rho\mathbf{v} = ((\alpha\rho)_1 + (\alpha\rho)_2)\mathbf{v} = \sum_{i=1}^{N} (\alpha\rho)_i\mathbf{v} : \text{ Mixture momentum.} \tag{11}$$

$$\rho e^{\mathrm{K}} = \frac{\rho_i \mathbf{v}^2}{2} : \text{ Mixture kinetic energy.} \tag{12}$$

$$\rho e^{\mathrm{I}} = (\alpha\rho)_1 e_1 + (\alpha\rho)_2 e_2 = \sum_{i=1}^{N} (\alpha\rho e)_i : \text{ Mixture internal energy.} \tag{13}$$

So, the total energy can be expressed as:

$$\rho E^{\mathrm{T}} = ((\alpha\rho)_1 e_1 + (\alpha\rho)_2 e_2 + \frac{\rho_i \mathbf{v}^2}{2}) \tag{14}$$

To determine the internal energy and put closure to the equations, it is convenient to use the stiffened equation of state (EOS) introduced by Harlow and Amsden [28] to describe the thermodynamic properties of both fluid components:

$$p_i = (\gamma_i - 1)\rho_i e_i - \gamma_i p_{c,i},\ i = 1,2 \tag{15}$$

Using the stiffened EOS, the internal energy can now be expressed as:

$$\rho e_i = \frac{p + \gamma_i p_c, i}{\gamma_i - 1} \tag{16}$$

and the total energy below:

$$\rho E = \sum_{i=1}^{N} \alpha_i (\frac{p + \gamma_i p_c, i}{\gamma_i - 1} + \frac{\rho_i . v^2}{2}) \tag{17}$$

From the five equations, as shown in Equations (5) and (6), the advection equation of the volume fraction is in a non-conservative form, which will present some difficulties when solving our five-equation model. We simplify it into a conservative term using Johnsen and Colonius's [29] approach.

First, let f represent a vector in three-dimensional form, defined as $f = \alpha_1 \mathbf{v} = [\alpha_1 u, \alpha_1 v, \alpha_1 w]^{\mathrm{T}}$.

The divergence of f then is:

$$\begin{aligned}\nabla \cdot f &= \frac{\partial}{\partial x}(\alpha_1 u) + \frac{\partial}{\partial y}(\alpha_1 v) + \frac{\partial}{\partial z}(\alpha_1 w) \\ &= \alpha_1 \left(\frac{\partial u}{\partial x} + \frac{\partial v}{\partial y} + \frac{\partial w}{\partial z}\right) + \left(u\frac{\partial \alpha_1}{\partial x} + v\frac{\partial \alpha_1}{\partial y} + w\frac{\partial \alpha_1}{\partial z}\right) \\ &= \alpha_1 \nabla \cdot \mathbf{v} + \mathbf{v} \cdot \nabla \alpha_1\end{aligned}$$

And then:

$$\mathbf{v} \cdot \nabla \alpha_1 = \nabla \cdot f - \alpha_1 \nabla \cdot \mathbf{v}.$$

Substituting $\mathbf{v} \cdot \nabla \alpha_1$ into the non-conservative advection equation of the volume fraction in Equations (5) and (6) to obtain the quasi-conservative form, the equation for the advection of the volume fraction can be written as Equations (18) and (19).

For Allaire et al.:

$$\frac{\partial \alpha_1}{\partial t} + \nabla \cdot f = (\alpha_1) \nabla \cdot v \tag{18}$$

For Kapila et al.:

$$\frac{\partial \alpha_1}{\partial t} + \nabla \cdot f = (\alpha_1 + K) \nabla \cdot v \tag{19}$$

The K function in Kapila et al.'s model can be determined from Equation (20), where the mixture's speed of sound is obtained using Wood's speed of sound, as seen in Equation (22)

$$K = \alpha_1 \alpha_2 \left(\frac{1}{\rho_1 c_1^2} - \frac{1}{\rho_2 c_2^2} \right) \rho c^2 \tag{20}$$

where c_1, c_2 and ρ are given as:

$$c_1 = \sqrt{\frac{\gamma_1}{\rho_1}(p + p_{c,1})}, c_2 = \sqrt{\frac{\gamma_2}{\rho_2}(p + p_{c,2})} \tag{21}$$

$$\frac{1}{\rho c^2} = \frac{\alpha_1}{\rho_1 c_1^2} + \frac{\alpha_2}{\rho_2 c_2^2} = \sum_{i=1}^{N} \frac{\alpha_i}{\rho_i c_i^2} \tag{22}$$

The "mixture-mixture" speed of sound derived from the stiffened-gas (SG)-EOS mixture is used for the Allaire et al. five equations [21] and is given as follows:

$$c = \sqrt{\frac{\gamma(p + p^{\infty})}{\rho}} \tag{23}$$

Cut-Off Methods

The stiffened-gas equation of state (SG-EOS) introduced by Harlow and Amsden [30] has been commonly used by various authors [29,31–33] to achieve thermodynamic closure for multiphase compressible flow, including flows involving cavitation. However, in cases where the pressure difference between the two-phase medium is considerably high, such as in underwater explosions, the use of stiffened-gas EOS alone may not be sufficient. This is because cavitation induced by the high-pressure surge is very fast, unsteady, and rapidly evolves into various dimensions before collapsing violently, resulting in non-physical negative pressures when using only the stiffened-gas EOS. Similar findings have been reported by other authors in their numerical simulations.

To improve the accuracy of simulations for cavitating flows, modifications are needed. One common approach is to incorporate an additional single-fluid model to circumvent the issue, such as the cut-off model, isentropic model, or Schmidt model, which have been used by authors in previous studies. Among these models, the cut-off model is often preferred due to its simplicity of implementation and preservation of peak pressure accuracy during cavitation collapse, which is important in underwater explosion (UNDEX) test cases. Additionally, these models are pure phase models that do not consider phase exchange, as noted in previous studies [4,34]. Further details of these models can be found in the references cited. In the cut-off model, once the pressure in the cavitation region drops below a certain fixed saturation pressure p_{sat}, the non-physical pressure computed assumes a new value, which is the value set for the p_{sat}.

$$p_i = \begin{cases} SG - EOS, & \text{if } p_i > p_{\text{sat}} \\ p_{\text{sat}}, & \text{if } p_i < p_{\text{sat}} \end{cases}$$

The physical saturation pressure (p_{sat}) is chosen to be a comparatively small value in relation to the significantly higher ambient pressure exerted by the fluid, and p_i refers to the static pressure at cell i. To avoid the negative physical pressure that occurs in the cavitating regions when the fluid experiences very steep and large variations in pressure, most authors [4,5,27,35] assume a value between 2000 and 5000 Pa.

The mixture's internal energy can be expressed as :

$$\rho_i e^{\text{I}}_{i\text{sat}} = \left(\frac{p_{\text{sat}} + \gamma_i p_{\text{c,i}}}{\gamma_i - 1}\right), \tag{24}$$

and the total energy is final given by:

$$\rho e_{\text{sat}} = \sum_{i=1}^{N} \alpha_i \left(\frac{p_{\text{sat}} + \gamma_i p_{c,i}}{\gamma_i - 1} + \frac{\rho_i . v^2}{2} \right). \tag{25}$$

3. Numerical Model

3.1. Finite-Volume Framework

Let us consider a 3D domain Ω consisting of conforming tetrahedral, hexahedral, prism, and pyramid, triangles, or quadrilateral cells, each one of them indexed by a unique mono-index i, and the governing equations of the five-equation model written in vector form as follows:

$$\frac{\partial}{\partial t} \int_{\Omega_i} \mathbf{Q} d\Omega + \oint_{\partial\Omega_i} \mathbf{F}_n((Q) dS = \int_{\Omega_i} \mathbf{G} d\Omega \tag{26}$$

where $\mathbf{Q} = \mathbf{Q}(\mathbf{x}, t)$ is the vector of conserved variables and the volume fraction of one species, and $\mathbf{F}_n$ is the non-linear flux in the direction normal to the cell interface, as given below:

$$\overset{Allaire}{\int_{\Omega_i}} \alpha_1 \nabla \cdot \mathbf{v} \, d\Omega \approx \int_{\Omega_i} \alpha_1 \, d\Omega \cdot \int_{\partial\Omega_i} (v_n)^{Riem.} \, dS. \tag{27}$$

$$\overset{Kapila}{\int_{\Omega_i}} (\alpha_1 + K) \nabla \cdot \mathbf{v} \, d\Omega \approx \int_{\Omega_i} (\alpha_1 + K) \, d\Omega \cdot \int_{\partial\Omega_i} (v_n)^{Riem.} \, dS. \tag{28}$$

and expressed as:

$$F_{\mathbf{n_{ij}}} = F_{\mathbf{n_{ij}}} \left(\mathbf{Q}^n_{ij,L}(\mathbf{x}_{ij,\alpha}, t), \mathbf{Q}^n_{ij,R}(\mathbf{x}_{ij,\alpha}, t) \right) \tag{29}$$

Integrating Equation (26) over the mesh element i using a high-order explicit finite-volume formulation, the following equation is obtained that incorporates the source term as previously defined:

$$\frac{d\mathbf{Q}_i}{dt} = \frac{1}{|V_i|} \sum_{j=1}^{N_f} \sum_{\alpha=1}^{N_{qp}} \left(F_{\mathbf{n_{ij}}} \left(\mathbf{Q}^n_{ij,L}(\mathbf{x}_{ij,\alpha}, t), \mathbf{Q}^n_{ij,R}(\mathbf{x}_{ij,\alpha}, t) \right) - \mathbf{a}^n_{i,1} \cdot v_n^{Riem}(\mathbf{x}_{ij,\alpha}, t) \right) \omega_\alpha |S_{ij}|, \tag{30}$$

where $\mathbf{Q}_i$ is the volume-averaged vector of variables

$$\mathbf{Q}_i = \frac{1}{|V_i|} \int_{V_i} \mathbf{Q}(x, y, z) \, dV, \tag{31}$$

and $F_{\mathbf{n_{ij}}}$ is a numerical flux function in the direction normal to the cell interface between a considered cell i and one of its neighboring cells j. N_f is the number of faces per element, N_{qp} is the number of quadrature points used for approximating the surface integrals, $|S_{ij}|$ is the surface area of the corresponding face, and $\mathbf{Q}^n_{ij,L}(\mathbf{x}_{ij,\alpha}, t)$ and $\mathbf{Q}^n_{ij,R}(\mathbf{x}_{ij,\alpha}, t)$ are the high-order approximations of the solutions for cell i and cell j, respectively. α corresponds to different Gaussian integration points $\mathbf{x}_\alpha$ and weights ω_α over each face. $\mathbf{a}^n_{i,1}$ corresponds to the volume-averaged volume fraction of cell i at time level n. A suitable Gauss–Legendre quadrature numerically approximates the volume, surface, and line integrals.

The reconstruction process adopted in `UCNS3D` [22] follows the approaches of Tsoutsanis et al. [23,36] and Titarev et al. [36] that have been previously applied to a wide range of smooth and discontinuous flow compressible flow problems, including laminar, transitional, and fully turbulent settings.

3.2. CWENO Scheme

Tsoutsanis's and Dumbser's [23] CWENO and CWENOZ schemes for mixed-element unstructured mesh schemes were used to test compressible multiphase components [33,37,38].

Their paper shows that the CWENO scheme can produce lesser oscillations near the interface at a cheaper computational cost than the WENO schemes on the same grid. This is made possible by the fact that in the presence of discontinuity, the CWENO scheme, unlike the traditional WENO schemes, uses the lower-order polynomials from the directional stencils that might contain a smooth solution for its reconstruction and, in the presence of smooth use, the optimal polynomial in the desired order of accuracy. This slight but significant difference gives the CWENO scheme advantages over the WENO schemes. The CWENO scheme will be discussed below briefly, and consequently, the same schemes will be used in capturing and resolving the material interface between the two-phase medium of UNDEX. More details of the CWENO schemes implemented by Tsoutsanis on an unstructured grid can be seen in Ref. [23,33].

The optimal polynomial is defined as:

$$p_{opt}(\xi,\eta,\zeta) = \sum_{s=1}^{s_t} \lambda_s p_s(\xi,\eta,\zeta), \tag{32}$$

where s is the stencil index, with $c = 1$ being the central stencil; $c = 2,3$, being the directional stencil; s_t being the total number of stencils; and λ_s being the linear coefficients for each stencil, whose sum is equal to 1. The p_1 polynomial is not computed directly but computed by subtracting the lower-order polynomials from the optimum polynomial as follows:

$$p_1(\xi,\eta,\zeta) = \frac{1}{\lambda_1}\left(p_{opt}(\xi,\eta,\zeta) - \sum_{s=2}^{s_t} \lambda_s p_s(\xi,\eta,\zeta)\right). \tag{33}$$

The CWENO scheme computes a reconstruction polynomial below as a non-linear combination of all the polynomials:

$$p(\xi,\eta,\zeta)^{\text{cweno}} = \sum_{s=1}^{s_t} \omega_s p_s(\xi,\eta,\zeta), \tag{34}$$

where ω_s corresponds to the non-linear weights assigned to each polynomial, and in regions with smooth data, $\omega_s \approx \lambda_s$, hence obtaining the high-order approximation from the central stencil. In regions of discontinuous solutions, the reconstructed solution will be mostly influenced from the lower-order polynomials of the directional stencils using the Type 3 directional stencils. The reader is referred to the work of Tsoutsanis [39] for the stencil selection algorithms. $\tilde{a}_k$ are the reconstructed degrees of freedom, and the non-linear weight ω_s is defined as such that the non-linear coefficients $\rightarrow_s$ are computed from the linear ones $\rightarrow_s$ as:

$$\omega_s = \frac{\tilde{\omega}_s}{\sum_{s=1}^{s_t} \tilde{\omega}_s} \quad \text{where} \quad \tilde{\omega}_s = \frac{\lambda_m}{(\epsilon + SI_s)^b}. \tag{35}$$

where ϵ is taken as 10^{-6} similarly to the WENO reconstructions, SI_s denotes the smoothness indicator, and $b = 4$. The readers are referrfed to [23] for a detailed description of the smoothness indicators.

3.3. *Fluxes and Time Advancement*

For the numerical fluxes, we employ the Harten–Lax–van Leer–Contact Riemann solver of Toro [40]. The solution is advanced in time by a third-order TVD Runge–Kutta method:

$$\mathbf{U}_i^{(n+1/3)} = \mathbf{U}_i^n + \Delta t R_i(\mathbf{U}), \tag{36}$$

$$\mathbf{U}_i^{n+2/3} = \frac{3}{4}\mathbf{U}_i^n + \frac{1}{4}\mathbf{U}_i^{(n+1/3)} + \frac{1}{4}\Delta t R_i(\mathbf{U}^{(n+1/3)}), \tag{37}$$

$$\mathbf{U}_i^{n+1} = \frac{1}{3}\mathbf{U}_i^n + \frac{2}{3}\mathbf{U}_i^{(n+2/3)} + \frac{2}{3}\Delta t R_i(\mathbf{U}^{(n+2/3)}), \tag{38}$$

and the time step Δt is selected according to:

$$\Delta t = \text{CFL}\frac{1}{d}\min_i(\frac{h_i}{S_i}), \quad (39)$$

where S_i is an estimate of the maximum in absolute value of the propagation speed in the cell i, d is the number of dimensions, h_i is a characteristic length of the element i, and CFL refers to the Courant–Friedrichs–Lewy condition. $CFL = 0.6$ is used for all tests unless otherwise specified. The reconstruction is performed with respect to the primitive variables since they were previously found [33] to be less oscillatory than the conserved variables.

All the volume/surface/line integrals are approximated by the Gaussian quadrature rule suitable for the order of polynomials employed. All the schemes developed are implemented in the UCNS3D CFD code [22], which is written in object-oriented Fortran 2008, employing the MPI message-passing interface (MPI) and the Open Multi-Processing (OpenMP) application programming interface (API), and the reader is referred to [41] for more details on implementation and performance benchmarks.

4. Test Cases

Using the high-order CWENO schemes in conjunction with the five-equation interface-capturing scheme models [20,21], five underwater test problems were selected (initial physical parameters of each fluid can be seen in Table 1) to study the cavitation effect of underwater shock near different boundaries:

1. 2D underwater explosion near a free surface.
2. 2D cylindrical underwater explosion near a planar rigid wall.
3. 3D underwater explosion in an enclosed, rigid cylindrical wall.
4. 2D underwater explosion of two bubbles placed horizontally near a free surface.
5. 3D underwater of two bubbles placed vertically in an enclosed region.

4.1. 2D Underwater Explosion Near a Free Surface

To study the dynamics of bubbles at the interface between air and water in an underwater explosion, a two-dimensional model of an underwater explosion near the free surface of water is considered. Previous studies by various authors have used numerical [5,15,42–47] and experimental [47,48] methods to investigate this phenomenon. For the experimental setup, a spherical charge of silver azide with known initial energy is detonated in an open cylinder half-filled with water, and the process involves charge detonation, bubble pulsation, and jet formation [47].

Table 1. Summary of the initial physical parameters of each fluid for the five (5) test cases.

Test Cases	Materials	ρ	u	v	p	γ	α
Test 4.1	Air	1.225	0	0	10^5	2.0	0
	Explosive	1250	0	0	10^9	2.0	1
	Water	1000	0	0	10^5	7.15	0
Test 4.2	Explosive	1270	0	0	8.29×10^8	2.0	1
	Water	1000	0	0	10^5	7.15	0
Test 4.3	Explosive	1770	0	0	2×10^9	2.0	1
	Water	1000	0	0	10^5	7.15	0
Test 4.4	Air	1.225	0	0	10^5	2.0	0
	Explosive	1250	0	0	10^9	2.0	1
	Water	1000	0	0	10^5	7.15	0
Test 4.5	Explosive	1770	0	0	2×10^9	2.0	1
	Water	1000	0	0	10^5	7.15	0

For all of our UNDEX numerical studies in this paper, a pressurized bubble will be used as a substitute for explosives in underwater explosions. The pressurized gas bubble with a diameter of 0.24 m is placed at [0, −0.3] m in the two-phase medium, using a similar setup as that seen in Shukla et al.'s work [46].

Our computational domain is a square area with $x \times y$ of dimensions $([-2,2] \times [-1.5,2.5])$ m (Figure 1) and discretized with a mixed-element unstructured mesh, incorporating quadrilateral and triangular elements in 2D, as shown in Figure 2, with the air–water interface positioned at the centre $[0,0]$.

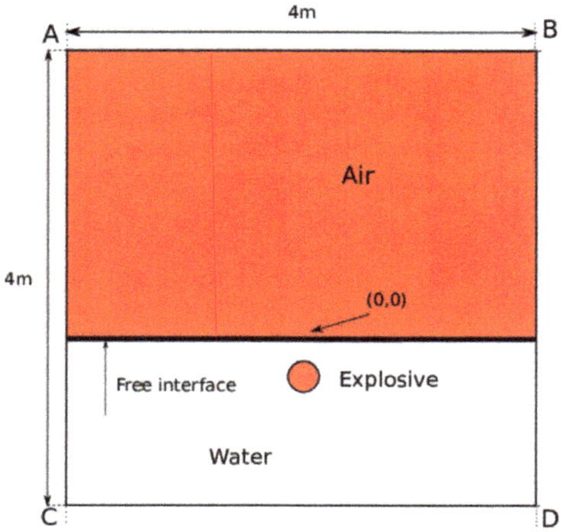

Figure 1. Computational domain setup for a two-dimensional underwater explosion near a free surface.

Transparent boundary conditions were set up at the top, left, and right surfaces, while the bottom was set to reflecting. The simulation was carried out until $t = 1.2$ ms as the final time, and the initial physical parameters of each fluid (density, velocities, pressure, specific heat ratio, and volume fraction) were set accordingly, as presented below:

$$(\rho, u, v, p, \gamma, \alpha) = \begin{cases} (1.225, 0, 0, 10^5, 2.0, 0), & \text{Air} \\ (1250, 0, 0, 10^9, 2.0, 1), & \text{Explosive} \\ (1000, 0, 0, 10^5, 7.15, 0), & \text{Water.} \end{cases} \tag{40}$$

We employ the high-order CWENO4 finite-volume method developed by Tsoutsanis et al. [23,33], on three grid resolutions (coarse, medium, and fine), in our study to investigate the interaction between gas bubbles and the free surface of an air–water medium. Numerical Schlieren visualisations were generated to capture density and pressure gradients at various instants, as depicted in Figures 3 and 4, respectively. The high-pressure air bubble undergoes an explosion, resulting in a shock wave that propagates outward towards the interface, reflects into the water upon hitting the air–water interface, and leaves behind cavitation bubbles, as evident from the pressure gradients in Figure 4. Subsequently, the bubble transforms from a spherical shape to an oval shape, as illustrated in the volume-fraction contour in Figure 5.

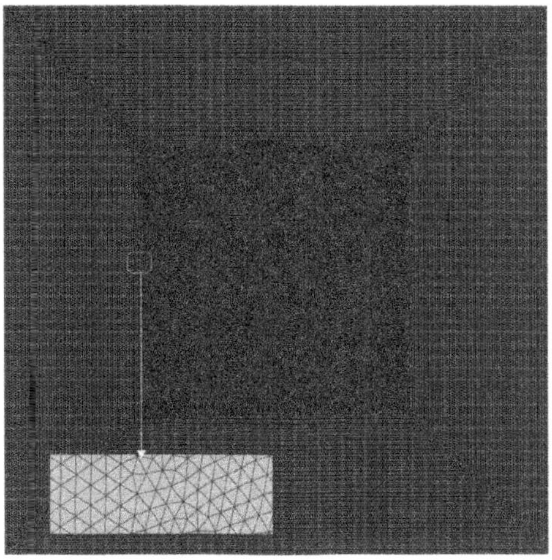

Figure 2. Underwater explosion near a free surface; 2D unstructured finest mesh with triangular elements used for this test case. The density of mesh elements in specific areas of the domain is increased to better capture the physical phenomena occurring in those regions. The zoomed-in area on the left provides a more detailed view of the mesh's structure.

We conducted a study to investigate the impact of grid resolution on the evolution of the interfacial region between the exploding gas bubble and the air–water free surface using the CWENO4 scheme, employing different grid sizes (400 × 400, 800 × 800, and 1600 × 1600). Changes in the CWENO4 scheme's behavior and its ability to capture the interfacial region between the gas bubble and the air–water free surface can be seen in Figure 6 as the grid size is refined. These include reductions in the smearing or numerical diffusion between the free surface and the expanding air bubble, resulting in a sharper description of the interface. The results are free from oscillations, which also demonstrates the robustness of the numerical framework employed. However, interface-sharpening or anti-diffusion techniques could prove beneficial, especially for up to third-order methods at coarse grid resolutions since they are the ones that can benefit most from this strategy [33,35,46].

We performed a quantitative investigation on the medium grid for the obtained density and pressure profiles along the horizontal axis (x = 0) and compared them with published results that utilize sharpening techniques and anti-diffusive techniques in conjunction with high-order schemes [9,49]. The results, as shown in Figures 7 and 8, are consistent with previously published findings [9,49,50].

Furthermore, we applied both the five-equation models proposed by Allaire et al. and Kapila et al. to investigate the influence of the additional term $K\nabla \cdot u$ in Kapila's model, thereby differentiating between the two models. The density and pressure plots, depicted in Figure 9, demonstrate that both the five-equation diffuse interface multicomponent models of Allaire et al. and Kapila et al. yield comparable results and are in good agreement with the published works of [49,51]. This shows that for cavitation test cases such as the one under consideration in this paper, we can affirm that the five-equation model of Allaire et al. and Kapila et al. accurately portrays both cavitation formation and collapse. However, this accuracy does not extend to spherical bubble collapse in a liquid, as highlighted in Ref. [26], where the five-equation model of Allaire et al. falls short in correctly representing the solution of spherical bubble collapse. In Figure 10, we observe that the maximum peak pressure on the finest mesh is approximately 85 MPa at x = -1.0 m, which is the location where the shock wave reflects upon hitting the expanding cavity from below. This finding is consistent with the observation made in [51] that traditional WENO schemes, although computationally more expensive than CWENO, fail to prevent the smearing of the air–water interfaces. Our results demonstrate that CWENO is capable of resolving material interfaces with reduced numerical smearing while maintaining their non-oscillatory nature, as reported in [33]. This makes CWENO schemes a favourable choice in terms of computational cost for simulating compressible multiphase or multicomponent flows involving cavitation.

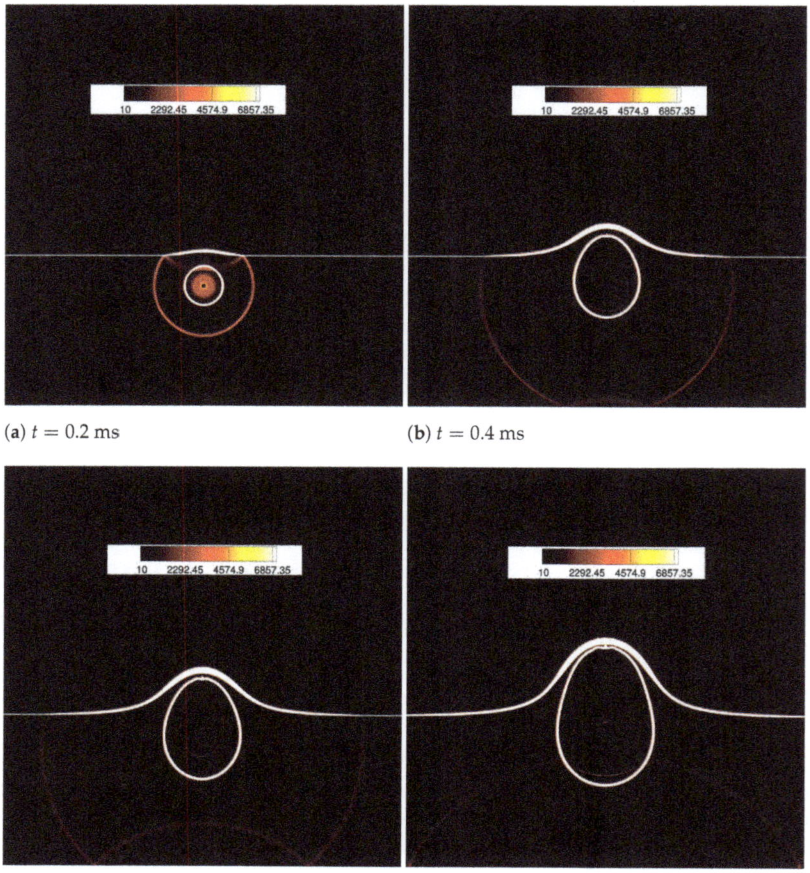

(**a**) $t = 0.2$ ms (**b**) $t = 0.4$ ms

(**c**) $t = 0.8$ ms (**d**) $t = 1.2$ ms

Figure 3. Underwater explosion near a free surface; Schileren visualisation of density gradient using a CWENO4 scheme at a resolution of 1600 × 1600 cells (finest mesh) at $t = (0.2, 0.4, 0.8, 1.2)$ ms. The colour legend represents density gradient, measured in units of kg/m^4. Energy release by the exploding bubble causes rapid pressure and temperature rise, heating and expanding the water, resulting in decreased density. Circular pressure waves are generated; a shock wave propagating through the water hits the free surface above it and a rarefaction wave is reflected and traverses towards the bubble ($t = 0.2$ ms). A wave diffraction occurs due to this interaction leading to a change in the shape of the gas bubble, from a circular to an oval form. Finally, the shock wave hits the wall at the bottom ($t = 0.8$ ms) and reflects back into the water hitting the bubble eventually at $t = 1.2$ ms.

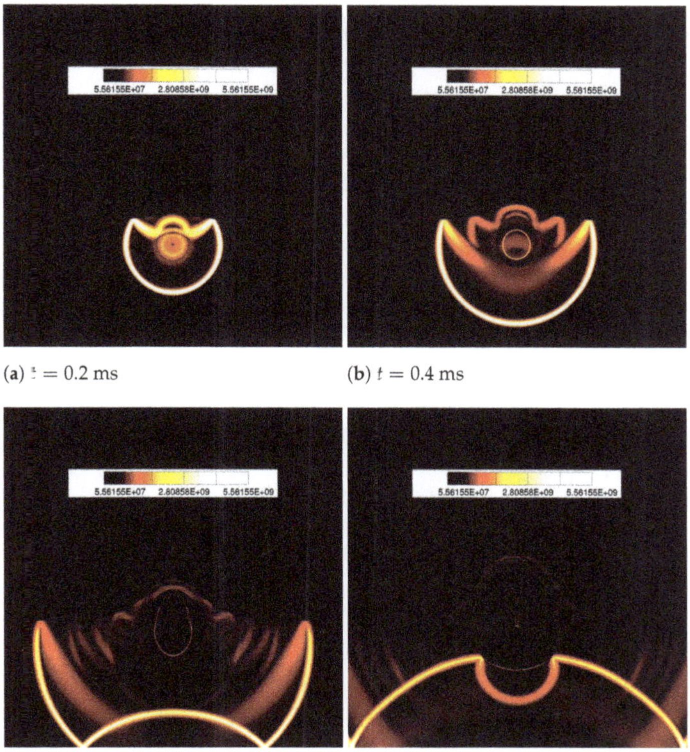

(**a**) $t = 0.2$ ms (**b**) $t = 0.4$ ms

(**c**) $t = 0.8$ ms (**d**) $t = 1.2$ ms

Figure 4. Underwater explosion near a free surface; Schileren visualisation of pressure gradient using a CWENO4 scheme on the finest mesh at $t = (0.2, 0.4, 0.8, 1.2)$ ms. The colour legend represents pressure gradient, measured in units of Pa/m. The visualisation displays the intricate wave patterns in the flow. At $t = 0.2$ ms, a shock wave expands outward through the water and causes rarefaction near the free surface. The rarefaction leads to a drastic drop in pressure in the fluid region below the free surface ($t = 0.4$ ms). The shock wave reaches the bottom wall at $t = 0.8$ ms, and then, at $t = 1.2$ ms, it reflects back into the water, striking the bubble from below.

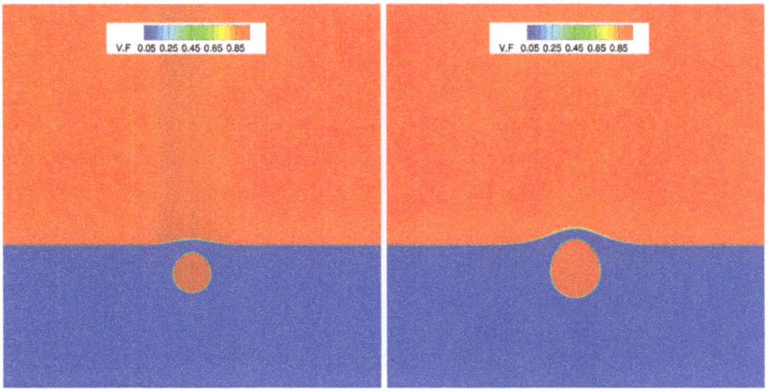

(**a**) $t = 0.2$ ms (**b**) $t = 0.4$ ms

Figure 5. *Cont.*

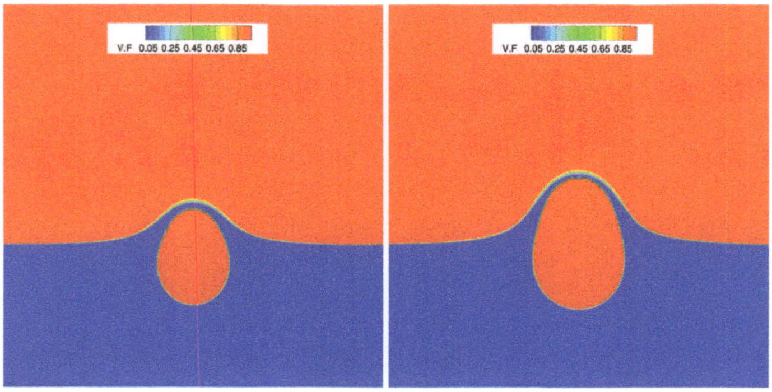

(**c**) $t = 0.8$ ms (**d**) $t = 1.2$ ms

Figure 5. Underwater explosion near a free surface; contours of volume fraction for underwater explosion near a free surface using a CWENO4 scheme on finest grid at $t = (0.2, 0.4, 0.8, 1.2)$ ms. As the bubble rises with the increasing pressure, it stretches and change shape from its spherical shape to an oval shape. The green-yellow areas depict the interface where water meets air as the bubble expands.

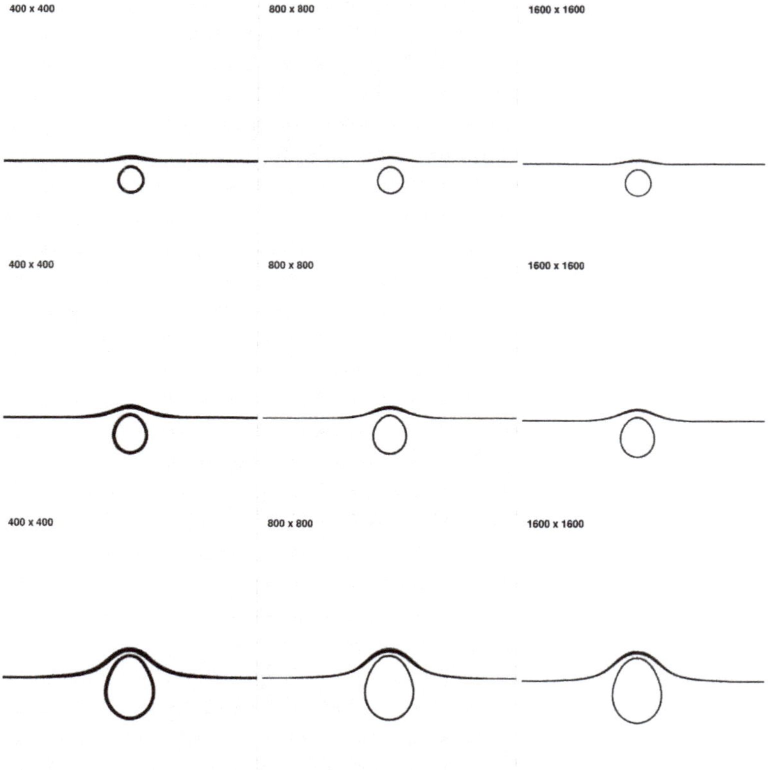

Figure 6. *Cont.*

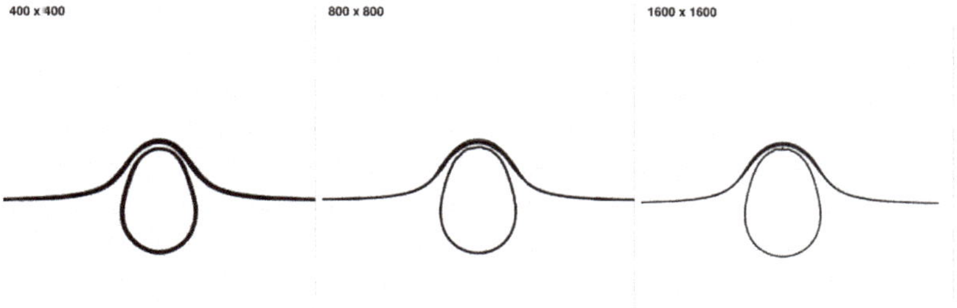

Figure 6. Underwater explosion near a free surface; density gradient contours at $t = (0.2, 0.4, 0.8, 1.2)$ ms. Captured using CWENO4 schemes with different grids resolution. More inter-facial structures are sharper as the grid resolution increases.

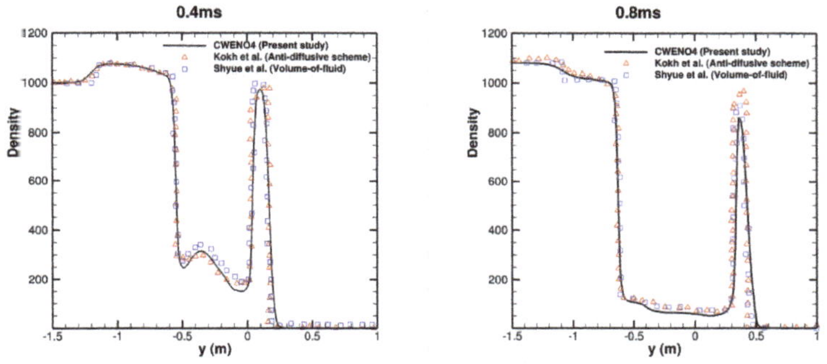

Figure 7. Underwater explosion near a free surface; plot of the density distribution along the horizontal axis x = 0 at two instants, specifically t = 0.4 ms and t = 0.8 ms. The obtained results are compared to those reported in previous studies of Kokh (2010) [49] and Shyue (2006) [52].

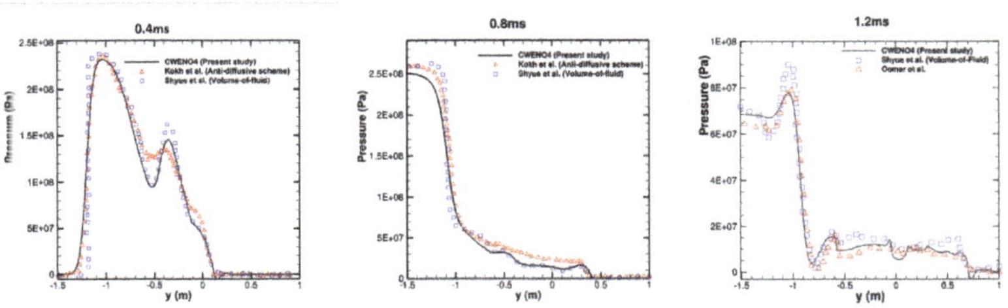

Figure 8. Underwater explosion near a free surface; pressure plot along horizontal axis x = 0 at instants t = (0.4; 0.8) ms. Comparison are made with the results of [49,50,52]

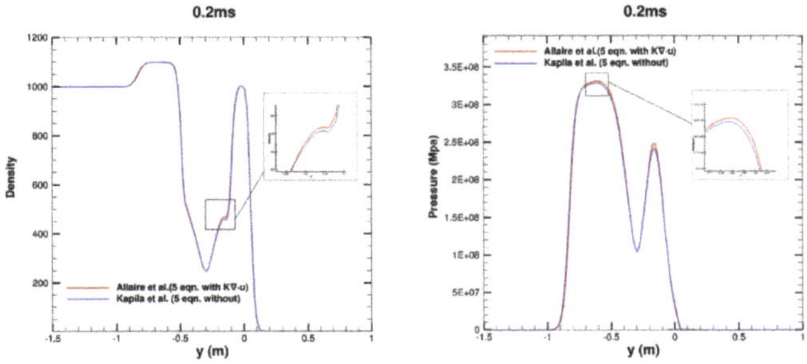

Figure 9. Underwater explosionnear a free surface; density and pressure plot along horizontal axis x = 0 at instants t = 0.2 ms using Allaire et al. [21] and Kapila et al. [20] models.

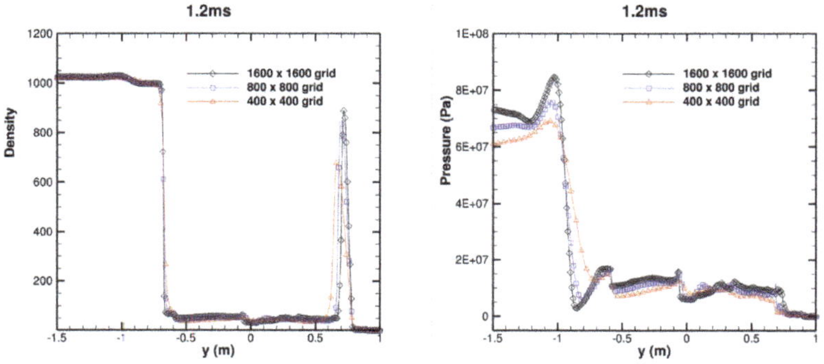

Figure 10. Underwater explosion near a free surface; density and pressure plot along horizontal axis x = 0 at instants t = 1.2 ms using different grid resolutions.

4.2. 2D Cylindrical Underwater Explosion Near a Planar Rigid Wall

This test case was carried out to examine the impact of shock loading and cavitation loading or reloading, caused by an underwater explosion, on a rigid structure (e.g., a ship's hull). Numerous studies, both numerical and experimental, have been conducted on this test case by various researchers [34,43,47,53,54]. The experiment involves detonating an explosive submerged in water underneath a plate.

The physical properties (density, velocities, pressure, specific heat ratio, and volume fraction) of each fluid involved in these test cases are presented below:

$$(\rho, u, v, p, \gamma, \alpha) = \begin{cases} (1270, 0, 0, 8.29 \times 10^8, 2, 1), & \text{Explosive} \\ (1000, 0, 0, 10^5, 7.15, 0), & \text{Water.} \end{cases} \quad (41)$$

The rectangular computational domain in two dimensions (as shown in Figure 11) has dimensions of $([-6,6] \times [-6,3])$ m. Similar to the previous test case, the explosive is modeled as a high-pressure air bubble with a spherical gas bubble of 2 m diameter placed in water at the origin $[0,0]$. A rigid wall is located at $y = 3$ m above the bubble, and a numerical sensor is placed at the centre point of the wall (denoted as P) to measure the load on the wall. The boundary conditions for the upper surface are set as reflecting, while the left, right, and bottom surfaces are set as non-reflecting. A coarse grid is used to discretize the domain, and the simulation is run for a final time of $t = 100$ ms.

The fourth-order CWENO scheme is used with the five-equation diffuse interface model for this test problem. To address the issue of non-physical negative pressure values (i.e., $p < 0$) that can occur in cavitation regions during the explosion, a cut-off model, described in Section 2, is introduced to maintain the pressure at 5000 Pa.

Contour plots of pressure obtained using the fourth-order CWENO scheme at different times during the explosion are shown in Figure 12. The plots show that the primary loading occurs when the pressurized air bubble explodes, generating shock waves that hit the wall above it at t = 1.5 ms and immediately reflect into the water, causing a rapid drop in density. Rarefaction can be observed in Figure 12 at t = 2 ms, as the pressure decreases around the wall, forming a low-pressure region at t = 3 ms. This pressure drop intensifies, resulting in a cavitation region around the wall at t = 4 ms and t = 5.3 ms. The cavitation collapses and generates a reload impact on the wall, known as cavitation collapse reloading (secondary reloading).

The pressure history of the wall along the horizontal axis is shown in Figure 13. The first peak pressure created by the shockwaves is 6200 MPa for the Allaire et al. model and 6400 MPa for the Kapila et al. model, indicating good agreement with the results of other authors [43,53]. As the bubbles continue to expand, the cavitation reloading impact on the wall occurs at a second peak pressure of 720 MPa for Allaire et al.'s model and

760 MPa for Kapila et al.'s model. It should be noted that although the impact of the shock loading is higher than that of the cavitation reloading, the duration of the cavitation reloading is longer, resulting in significant loading on the wall.

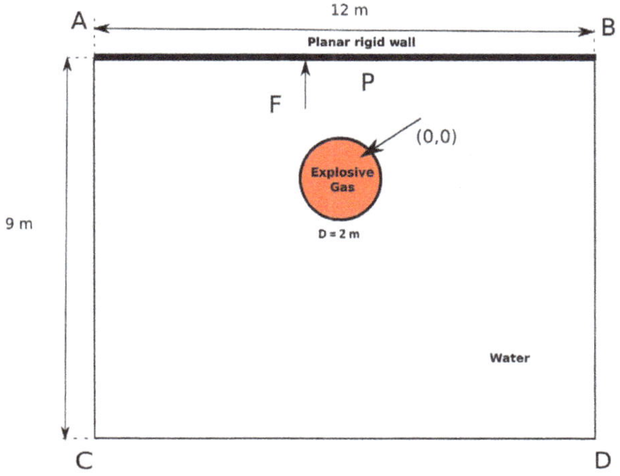

Figure 11. Computational domain setup for a two-dimensional underwater explosion near a planar rigid wall.

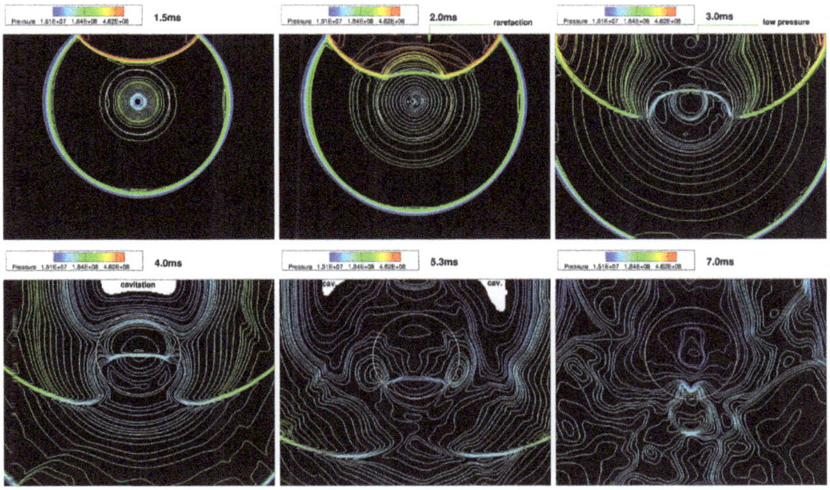

Figure 12. Underwater explosion near a rigid wall; the pressure contours at different instants. Captured using CWENO3 scheme with medium grid resolution. At very low pressure, cavitation can be noticed at times $t = 4$ ms and 5.3 ms near the rigid walls. The colour legend represents pressure gradient, measured in units of Pa/m.

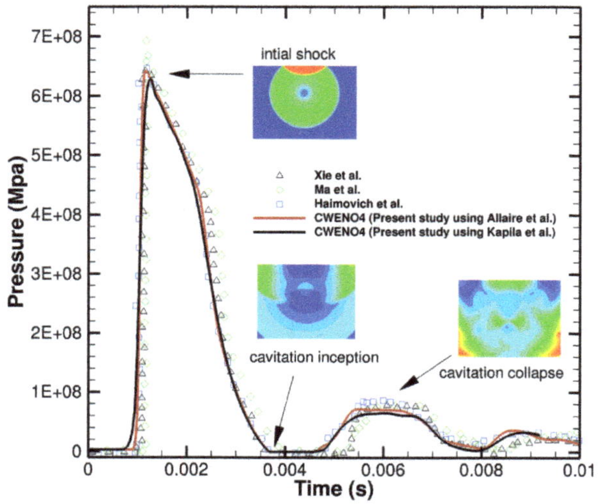

Figure 13. Underwater near a rigid wall; plots of pressure history for the underwater explosion near a planar rigid wall at the centre location of the upper wall obtained with CWENO4 schemes and compared with other published results [19,43,55].

4.3. 3D Underwater Explosion in an Enclosed, Rigid Cylindrical Wall

This test case has been investigated numerically by various authors [9,34,53,56–58]. It serves as an ideal problem for studying shock and cavitation responses in enclosed structures such as a cylindrical tank. The initial conditions for this simulation are based on Xie et al.'s work [4] and are presented below:

$$(\rho, u, v, p, \gamma, \alpha) = \begin{cases} (1770, 0, 0, 2 \times 10^9, 2, 1), & \text{Explosive} \\ (1000, 0, 0, 10^5, 7.15, 0), & \text{Water.} \end{cases} \quad (42)$$

The computational setup is illustrated in Figure 14, following that of [4,53].

The explosive, modeled as a gas bubble, has a diameter of 0.03 m and is placed within a rectangular cylindrical wall with dimensions of 0.2286 m in height and 0.0889 m in diameter. The water in the tank is partially aerated with 1.5% air. The test case can be simulated in two-dimensional axisymmetric form using cylindrical coordinates or in three-dimensional form. In our study, the test is conducted in a three-dimensional setup.

Reflecting boundary conditions are imposed on all surfaces, and a hybrid hexa-dominant mesh is used to discretize the computational domain with approximately 0.5 million cells as shown in Figure 15. A CWENO4 scheme is utilised for numerical simulations, and the simulation is carried out until $t = 120$ µs. The time evolution of the gas bubble's pressure is shown in the pressure contour plot in Figure 16. It can be observed that as the bubble explodes and expands, shock waves are generated and impact the side walls of the cylinder at t = 15 µs. These waves are immediately reflected back into the water, striking the expanding bubble at t = 30 µs. This results in rapid changes in density and rarefaction. At t = 45 µs, the shape of the bubble transforms into an oval. Over time, a low-pressure region is observed around the bubble and close to the centre of the lateral walls at t = 30 µs, leading to cavitation at t = 45 µs, 60 µs, and 70 µs, as evident in Figure 16. The cavitation pocket is most prominent on the wall at t = 60 µs. The violent collapse of cavitation generates reloading pressure on the walls of the cylindrical structure, as depicted in Figure 17. This phenomenon of cavitation-induced reloading can have a significant impact on enclosed structures, with effects comparable to those of the primary shock-wave loading.

The pressure history plot in Figure 17 reveals that the initial peak pressure obtained in our study using the diffuse interface methods proposed by Allaire et al. is 6600 MPa.

This value is in close agreement with the experimental result of Wardlaw et al. [59], where P = 6300 MPa. Other studies by Shyue et al. [9] and Ying et al. [60], which utilised the Allaire et al. diffuse interface methods in conjunction with cut-off cavitation models, reported numerical values in the range of 7000–7200 MPa, respectively.

The second impact of cavitation collapse occurs at a time of 90 µs, with a pressure value of P = 2100 MPa, as evident in our results. This value is lower than the numerical results reported by Shyue et al. [9] and Ying et al. [60] but closer to the experimental value than the latter. Notably, the study conducted by Chen [61] using smoothed particle hydrodynamics (SPH) models exhibited a delay in the cavitation collapse and reload, as the Schmidt model [4,55,62] was employed instead of the cut-off model, as utilised in the present study.

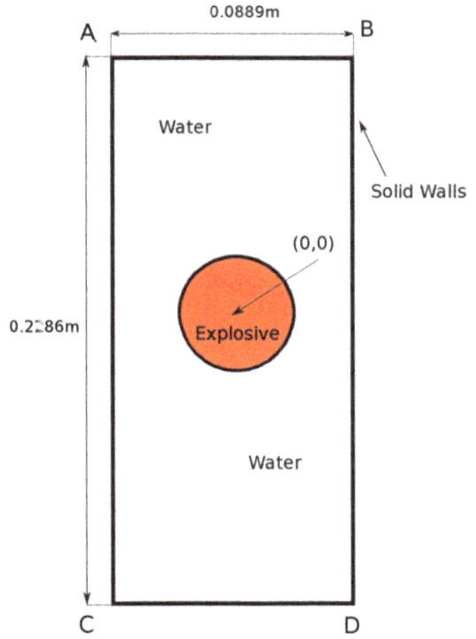

Figure 14. Computational domain setup for three-dimensional underwater explosion in a rigid cylinder.

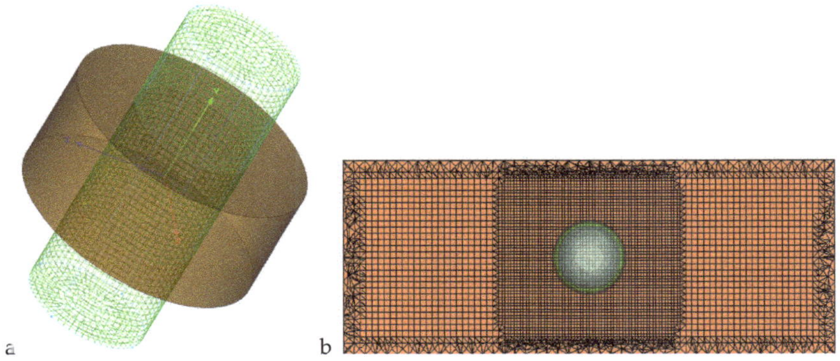

Figure 15. (**a**) Mesh used for three-dimensional cylindrical underwater explosions; (**b**) corresponding mesh refinement for simulation zone.

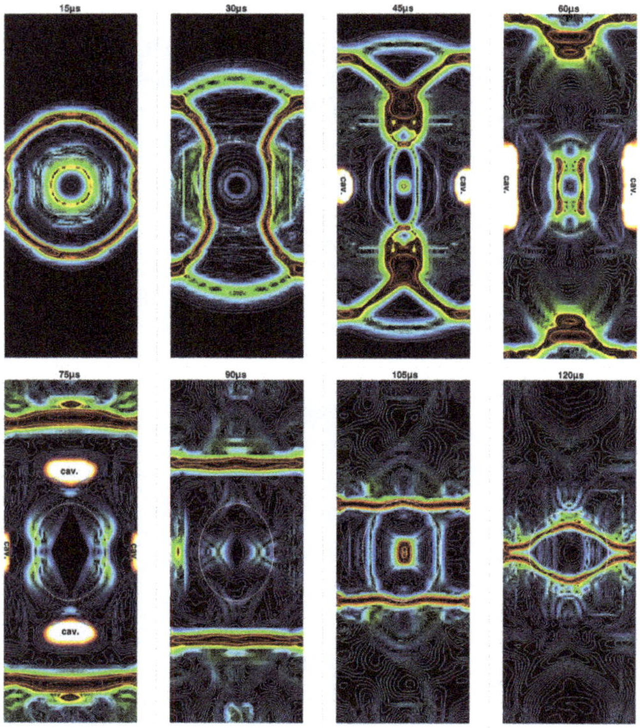

Figure 16. Underwater explosion in an enclosed cylindrical container; pressure gradient contours for the 3D underwater explosion in an enclosed, rigid cylindrical wall.

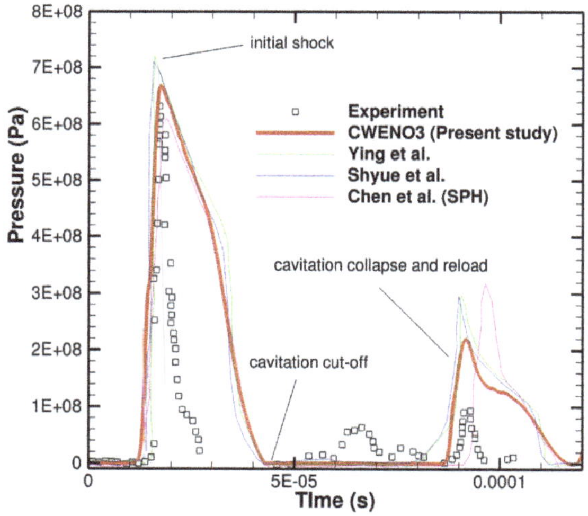

Figure 17. Underwater explosion in an enclosed cylindrical container; plots of pressure history (at the centre location of the left wall) for the underwater explosion in an enclosed cylindrical wall with CWENO4 schemes, and compared with other published results [9,59,61].

4.4. 2D Underwater Explosion of Two Bubbles Placed Horizontally Near a Free Surface

When two bubbles are positioned close to one another, the local pressure field may vary due to the bubble–bubble interactions. The growth, collapse, and oscillation periods of the bubbles, i.e., the time it takes for a bubble to complete one cycle of expansion and collapse can be affected by these changes in pressure. In a two-dimensional setup Figure 18, two bubbles are positioned horizontally in close proximity to each other near a free surface as an attempt to investigate the influence of the separation distance between the bubbles on cavitation. The separation distance, denoted as L_o, is varied at four different values: $(0.3, 0.5, 0.7, 0.9)$ m, between two identical bubbles located 0.3 m below the free surface. This configuration closely resembles the one utilised by Yu et al. [63]. Additionally, a single bubble with similar computational parameters and dimensions is also simulated for comparative analysis. The domain is discretized using a quadrilateral coarse mesh comprising approximately 0.025 million cells, with all surfaces imposing outflow boundary conditions.

The physical characteristics of the fluid involved in these test cases are described below, which includes parameters such as density, velocity, pressure, specific heat ratio (or specific heat capacity ratio), and volume fraction:

$$(\rho, u, v, p, \gamma, \alpha) = \begin{cases} (1.225, 0, 0, 10^5, 2.0, 0), & \text{Air} \\ (1250, 0, 0, 10^9, 2.0, 1), & \text{Explosive} \\ (1000, 0, 0, 10^5, 7.15, 0), & \text{Water.} \end{cases} \tag{43}$$

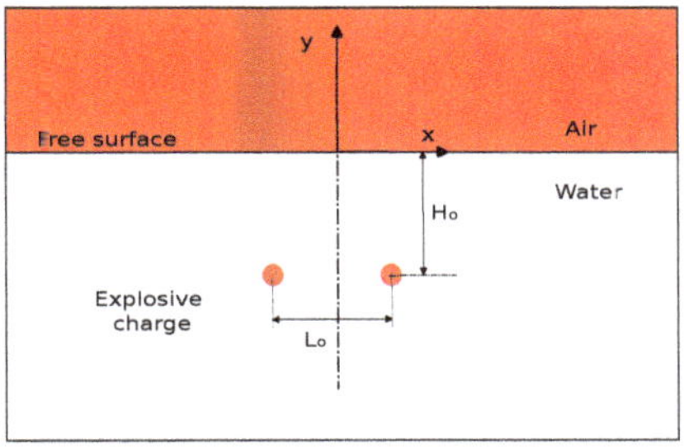

Figure 18. Computational domain setup for the two-dimensional underwater explosion of two bubbles placed horizontally near a free surface.

As observed in previous UNDEX (underwater explosion) test cases in this paper, the two bubbles undergo rapid expansion, generating a shock wave that propagates through the water and interacts with the surrounding bubbles. The pressure-time history at the centre of the explosion, depicted in the pressure contour plot Figure 19, reveals that the peak pressure caused by the initial shock wave hitting the free surface for a single bubble is 170 MPa, while for the two-bubble case with the largest inter-bubble distance of $L_o = 0.9$ m, it reaches 290 MPa, which is 1.7 times higher than that of the single bubble. Furthermore, as evident in Figure 19, the initial peak pressure, which is a result of the intensity of the shock wave on the free surface increases with decreasing inter-bubble distance, from the maximum value at $L_o = 0.9$ m to the smallest value at $L_o = 0.3$ m. The close proximity of the two bubbles allows for stronger interaction, leading to a higher shock intensity on the free surface.

The occurrence of cavitation beneath the free surface during a two-bubble explosion is influenced by the size and spacing of the bubbles. In same plot Figure 19, as the inter-distance bubbles increase, the cavitation bubble collapse that results in the cavitation reloading on the free surface is further delayed as seen in the second pressure peak. The second peak pressure associated with cavitation reloading is also more pronounced for the two charged bubbles compared to a single bubble. This observation further supports the conception that two charged bubbles have a higher potential for intensifying structural damage due to cavitation reloading compared to a single bubble. The ratio of the initial peak pressure on the free surface caused by the shock wave to that of the cavitation reloading is about four times for the smaller inter-bubble $L_o = (0.3, 0.5)$ m, and twice for the larger inter-bubble distances $L_o = (0.7, 0.9)$ m. This shows that for larger inter-bubble distances, the effect of structural damage caused by cavitation is more comparable to that of the primary shock-wave loading than for smaller inter-bubble distances.

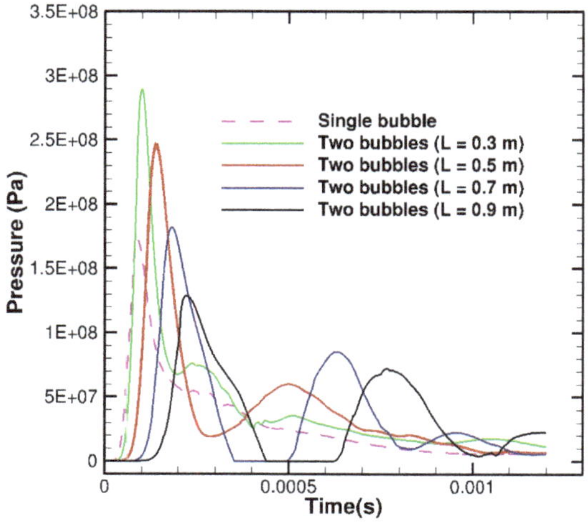

Figure 19. Underwater explosion of a single and two bubbles configurations placed horizontally near a free surface; plots of pressure history (at the centre location of the left wall) for the underwater explosion in an enclosed cylindrical wall with CWENO4 scheme, different inter-bubble distances $(0.3, 0.5, 0.7, 0.9)$ m are considered [9,64,65].

The volume-fraction contours presented in Figure 20 effectively illustrate the temporal evolution of the bubbles with the distances L_o ranging from 0.3 m to 0.9 m. As the separation distance L_o, increases from 0.3 m to 0.9 m, the height of the free surface centre consistently decreases at each time instant. In the same figure, it is observed that for the smallest inter-bubble distances $L_o = 0.3$ m, the bubbles expand and eventually coalesce or merge.

In Figure 21, the pressure gradient contour is depicted for that of a single charge and the two bubbles with varying inter-bubble distances. The contour illustrates a region of higher pressure near bubble(s), while the pressure diminishes further away as the bubbles expand. The magnitude of the pressure gradients varies based on the distance between the bubbles.

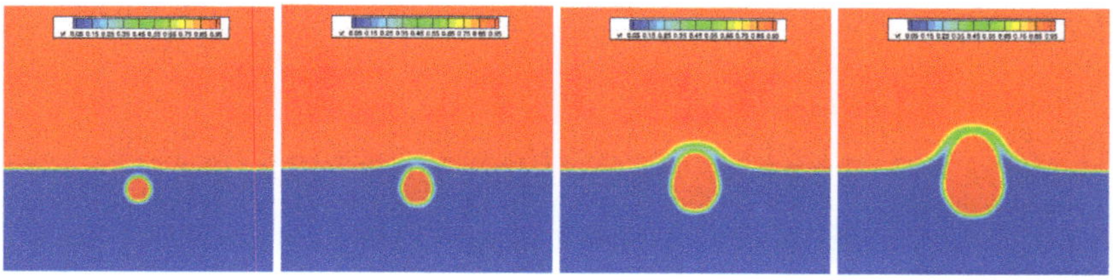

Figure 20. *Cont.*

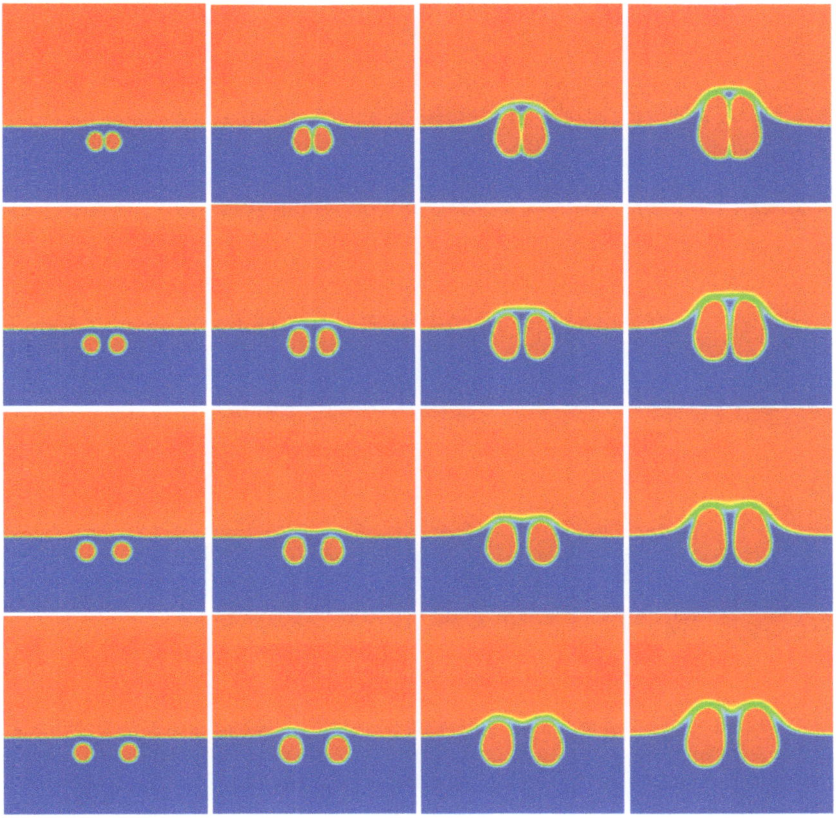

Figure 20. Underwater explosion of single bubble (top row) and two bubbles configurations placed horizontally near a free surface; contour of volume fraction using a CWENO4 scheme on a medium mesh at t = (0.2, 0.4, 0.8, and 1.2) ms (from left to right) and for inter-bubble lengths of 0.3, 0.5, 0.7, and 0.9 m from top to bottom, respectively.

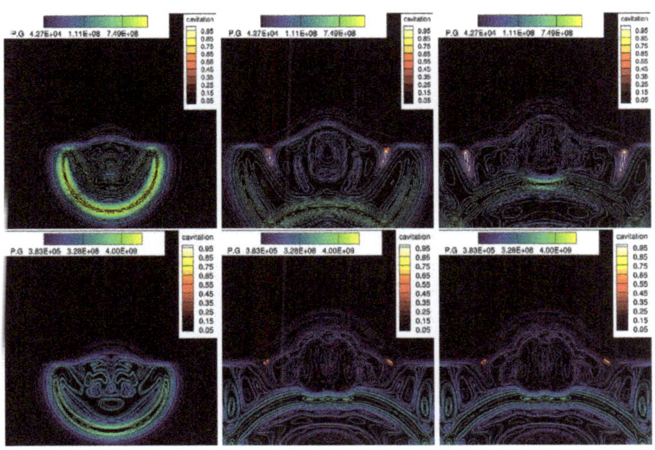

Figure 21. *Cont.*

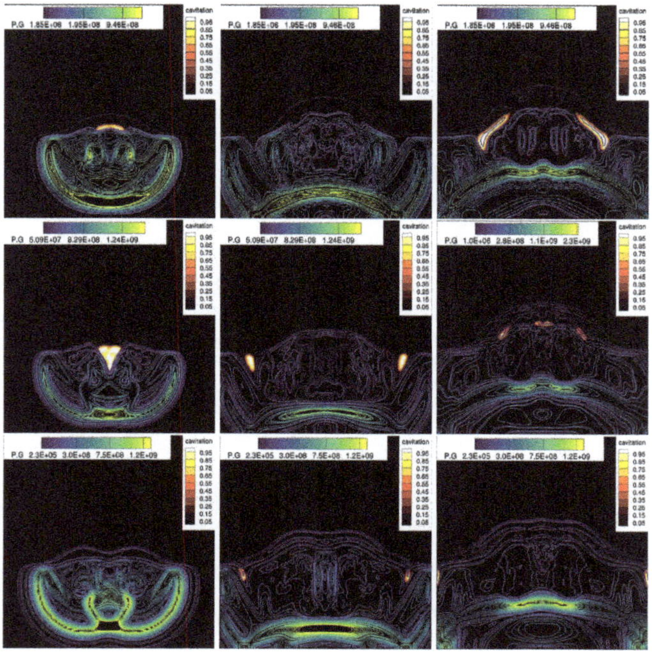

Figure 21. Underwater explosion of a single and two bubbles configurations placed horizontally near a free surface; contour of pressure gradient for the single bubble (Top) and the two-bubble cases subsequently below for inter-bubble lengths of 0.3, 0.5, 0.7, and 0.9 m, respectively, using a CWENO4 scheme on a medium mesh at t = (0.4, 0.8, and 1.06) ms. The red-yellow areas are cavitation zones. The colour legend represents pressure gradient, measured in units of Pa/m.

4.5. 3D Underwater of Two Bubbles Placed Vertically in an Enclosed Region

To investigate the interaction between two vertically aligned bubbles in water, enclosed in a region, it is best to consider this test case in either an axis-symmetric or three-dimensional setup, taking into account gravitational effects. The bubbles used in this test case are of smaller and identical diameter (d = 5.3 mm). They are initially positioned one above the other, with the first bubble placed at the bottom of the enclosed region and the second bubble above it.

Our numerical setup (Figure 22) uses similar computational dimensions as the Test Case section (Section 4.3) but with smaller bubble radii. The distance between the first and second bubbles is denoted by H_b, and chosen to be 15, 30, and 60, all of which are given in millimeters. Reflecting boundary conditions are applied to all surfaces, and a quadrilateral mesh discretizes the domain with approximately 0.4 million cells. The initial physical properties of both fluid is given below:

$$(\rho, u, v, p, \gamma, \alpha) = \begin{cases} (1770, 0, 0, 2 \times 10^9, 2.0, 1), & \text{Air bubble} \\ (1000, 0, 0, 10^5, 7.15, 0), & \text{Water.} \end{cases} \quad (44)$$

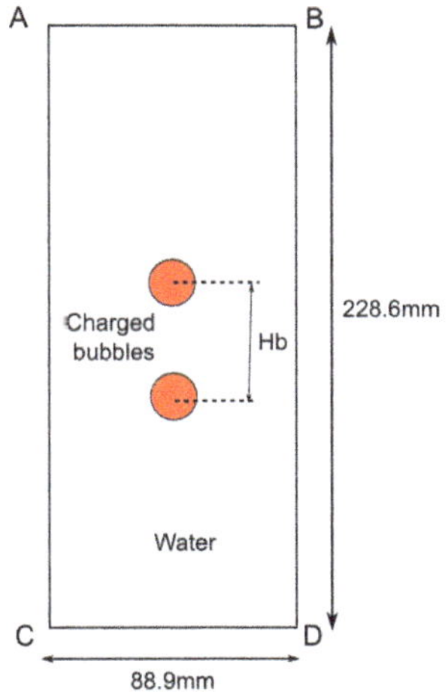

Figure 22. Computational domain setup for the three-dimensional underwater of two bubbles placed vertically in an enclosed region.

The volume-fraction contour in Figure 23 displays the shortest inter-bubble distance $H_b = 15$ mm, where the bubbles are in close proximity. As the explosion evolves with time, the water film between the interface of the two interacting gaseous bubbles thins out. This causes a surface tension force to pull the bubbles together until the thin film ruptures and the two bubbles merge together. The two bubbles combine to form a single, larger bubble after coalescing.

The pressure plot history at the centre of the left wall is displayed in Figure 24 for the three inter-bubble instances (H_b = 15, 30, and 60 mm). The pressure gradient contours in Figure 25 shows the cavitation regions are more prominent for the shortest inter-bubble distance $H_b = 15$ mm for varying time frames t = (1.54, 6, 30, 60, and 90) µs. The cavitation regions in Figure 25 can be observed to be majorly concentrated at the walls of the cylinder and the surrounding areas close to the bubbles.

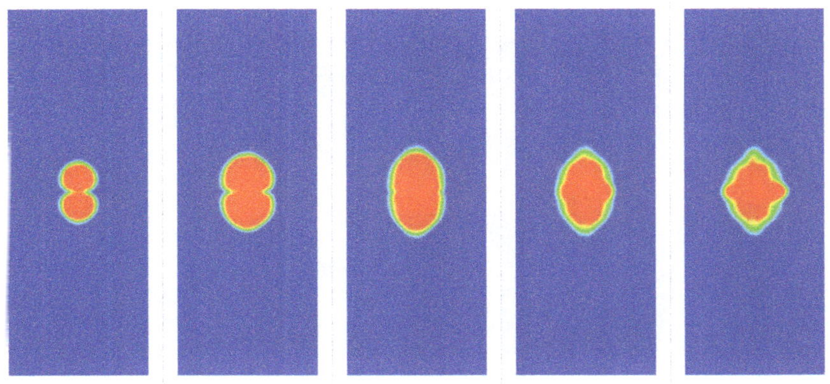

(**a**) $t = 10$ µs (**b**) $t = 30$ µs (**c**) $t = 60$ µs (**d**) $t = 90$ µs (**e**) $t = 120$ µs

Figure 23. Underwater explosion of two bubbles vertically placed in an enclosed region; volume-fraction contours for the 3D underwater explosion in an enclosed for $H_b = 30$ mm, rigid cylindrical wall. Captured using CWENO4. The two bubbles interact after explosion and at approximately $t = 60$ µs after initiation, the two bubbles eventually merge into a single larger bubble. Subsequently, the merged bubble continues to expand ($t = 90$ µs, $t = 120$ µs), reaching its maximum volume. The green-yellow areas show the interface between water and explosive.

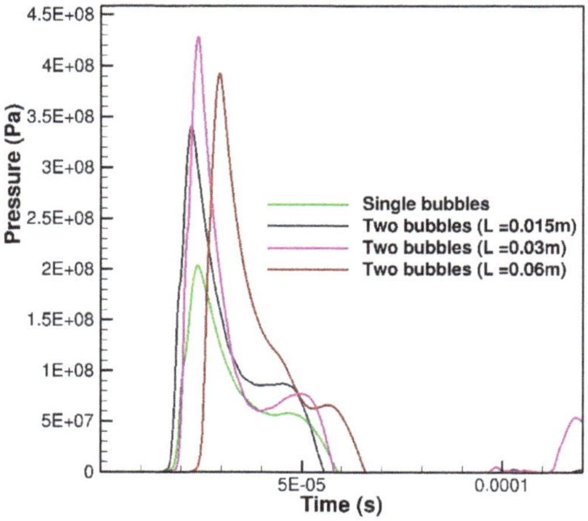

Figure 24. Underwater explosion of two bubbles vertically placed in an enclosed region; plots of pressure history (at the centre location of the left wall) for the underwater explosion in an enclosed cylindrical wall with CWENO4 schemes for inter-bubble lengths of 15, 30, and 60 mm.

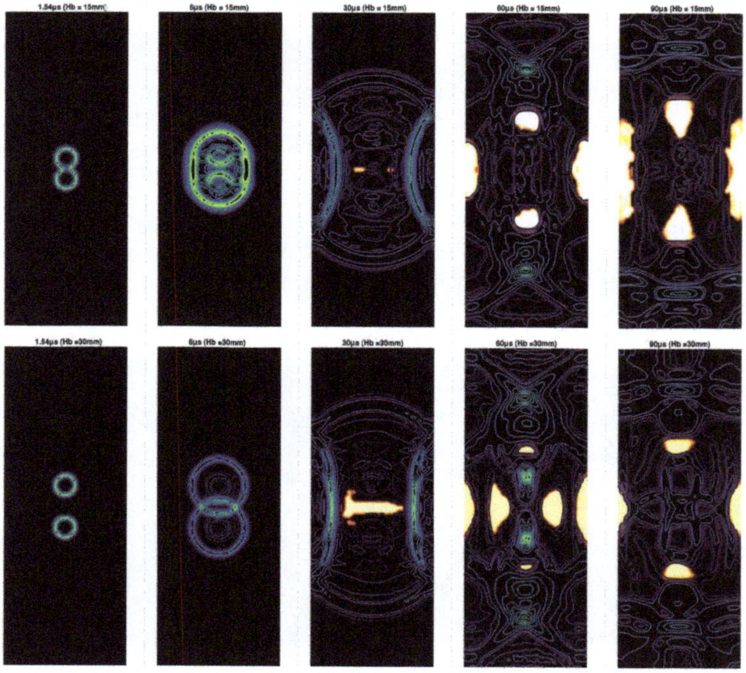

Figure 25. *Cont.*

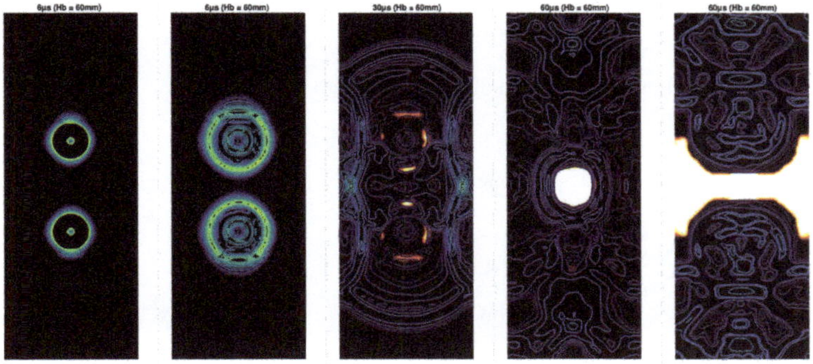

Figure 25. Underwater explosion of two bubbles vertically placed in an enclosed region; pressure gradient contours for the 3D underwater explosion in an enclosed rigid cylindrical wall with three different inter-bubble distances: H_b = 15 mm (**top**), 30 mm (**middle**), and 60 mm (**bottom**) for varying time frames t = (1.54, 6, 30, 60, and 90) µs. Captured using CWENO4. The white-red-yellow areas are cavitation zones, and the red regions are indicative of potential high cavitation activity.

5. Conclusions

This research extends the CWENO high-order finite-volume numerical framework employed for simulating multicomponent and multiphase flows on unstructured mesh, as detailed in [33], to specifically investigate cavitation phenomena in underwater explosions (UNDEX). The application of the CWENO schemes proved successful across a range of intricate two- and three-dimensional test cases related to underwater explosions, yielding noteworthy observations.

Notably, the CWENO schemes showcase a remarkable capability to provide high-resolution and sharply defined representations of shock–cavitation structures, as well as the infinitesimal interfacial region. This is achieved by resolving material interfaces with minimized numerical smearing, eliminating the necessity for additional interface sharpening [46] or compression [49]. This takes away the increase in computational cost associated with incorporating interface sharpening or compression into the diffuse interface models.

The compressibility effect introduced by the $K\nabla \cdot$ u term in the Kapila et al. five-equation models [20], which distinguishes it from the Allaire five-equation models [21], from our results shows that this additional term does not significantly impact cavitation. The similarity in results between the two models underscores their agreement with other published papers, affirming their reliability in simulating cavitation phenomena.

The results also show that the proximity of the bubbles played a crucial role in intensifying the shock wave impacts and the occurrence of cavitation around free surfaces and rigid boundaries. All these findings contribute valuable insights into the dynamics of cavitation within the context of underwater explosions.

Author Contributions: Conceptualization, E.M.A. and P.T.; methodology, E.M.A. and P.T.; software, E.M.A. and P.T.; validation, E.M.A. and P.T.; formal analysis, E.M.A. and P.T.; investigation, E.M.A. and P.T.; resources, E.M.A., P.T. and K.W.J.; data curation, E.M.A.; writing—original draft preparation, E.M.A. and P.T.; writing—review and editing, E.M.A. and P.T.; visualisation, E.M.A. and P.T.; supervision, K.W.J. and P.T.; project administration, K.W.J. and P.T.; funding acquisition, E.M.A. All authors have read and agreed to the published version of the manuscript

Funding: This research is partly financially supported by the The Petroleum Trust Development Fund (PTDF) Nigeria through grant number PTDF/ED/OSS/PHD/EMA/1591/19, accessed on 30 September 2019 (https://ptdf.gov.ng) for the first author's Ph.D. studies.

Data Availability Statement: The data that support the findings of this study are available from the corresponding author upon reasonable request.

Acknowledgments: The authors acknowledge the computing time at Cranfield University Delta2 HPC facility.

Conflicts of Interest: The authors declare that they have no conflicts of interest.

Abbreviations

The following abbreviations are used in this manuscript:

UNDEX	Underwater Explosion
DIM	Diffuse Interface Models
UCNS3D	Unstructured Compressible Navier–Stokes 3D
CWENO	Central-Weighted Essentially Non-Oscillatory
FV	Finite Volume

References

1. Petitpas, F.; Massoni, J.; Saurel, R.; Lapebie, E.; Munier, L. Diffuse interface model for high speed cavitating underwater systems. *Int. J. Multiph. Flow* **2009**, *35*, 747–759. [CrossRef]
2. Petitpas, F.; Saurel, R.; Ahn, B.K.; Ko, S. Modelling cavitating flow around underwater missiles. *Int. J. Nav. Archit. Ocean. Eng.* **2011**, *3*, 263–273. [CrossRef]
3. Kwack, Y.K.; Ko, S.H. Numerical analysis for supercavitating flows around axisymmetric cavitators. *Int. J. Nav. Archit. Ocean. Eng.* **2013**, *5*, 325–332. [CrossRef]
4. Wenfeng, X. A Numerical Simulation of Underwater Shock-Cavitation-Structure Interaction. Ph.D. Thesis, National University of Singapore, Singapore, 2005; pp. 1–207.
5. Daramizadeh, A.; Ansari, M.R. Numerical simulation of underwater explosion near air-water free surface using a five-equation reduced model. *Ocean. Eng.* **2015**, *110*, 25–35. [CrossRef]
6. Xie, W.F.; Liu, T.G.; Khoo, B.C. The simulation of cavitating flows induced by underwater shock and free surface interaction. *Appl. Numer. Math.* **2007**, *57*, 734–745. [CrossRef]
7. Saurel, R.; Lemetayer, O. A multiphase model for compressible flows with interfaces, shocks, detonation waves and cavitation. *J. Fluid Mech.* **2001**, *431*, 239–271. [CrossRef]
8. Ghidaglia, J.M.; Mrabet, A.A. A regularized stiffened-gas equation of state. *J. Appl. Anal. Comput.* **2018**, *8*, 675–689. [CrossRef]
9. Shyue, K.M. An Anti-Diffusion based Eulerian Interface-Sharpening Algorithm for Compressible Two-Phase Flow with Cavitation. In Proceedings of the 8th International Symposium on Cavitation, Singapore, 13–16 August 2012; Volume 268, pp. 7–12. [CrossRef]
10. Maltsev, V.; Skote, M.; Tsoutsanis, P. High-order methods for diffuse-interface models in compressible multi-medium flows: A review. *Phys. Fluids* **2022**, *34*, 021301. [CrossRef]
11. Pelanti, M.; Shyue, K.M. A mixture-energy-consistent six-equation two-phase numerical model for fluids with interfaces, cavitation and evaporation waves. *J. Comput. Phys.* **2014**, *259*, 331–357. [CrossRef]
12. Pelanti, M.; Shyue, K.M. A numerical model for multiphase liquid–vapour–gas flows with interfaces and cavitation. *Int. J. Multiph. Flow* **2019**, *113*, 208–230. [CrossRef]
13. Baer, M.; Nunziato, J. A two-phase mixture theory for the deflagration-to-detonation transition (ddt) in reactive granular materials. *Int. J. Multiph. Flow* **1986**, *12*, 861–889. [CrossRef]
14. Saurel, R.; Petitpas, F.; Abgrall, R. Modelling phase transition in metastable liquids: Application to cavitating and flashing flows. *J. Fluid Mech.* **2008**, *607*, 313–350. [CrossRef]
15. Jun, Y.; Jian-qiang, P.; Hai-kun, W.; Hai-bin, M. Application of Compressible Multi-component Flow in Underwater Explosion Problems. *Procedia Eng.* **2015**, *126*, 339–343. [CrossRef]
16. Chiapolino, A.; Saurel, R.; Nkonga, B. Sharpening diffuse interfaces with compressible fluids on unstructured meshes. *J. Comput. Phys.* **2017**, *340*, 389–417. [CrossRef]
17. LeMartelot, S.; Nkonga, B.; Saurel, R. Liquid and liquid–gas flows at all speeds. *J. Comput. Phys.* **2013**, *255*, 53–82. [CrossRef]
18. Zein, A. Numerical Methods for Multiphase Mixture Conservation Laws with Phase Transition. Ph.D. Thesis, Otto von Guericke University Magdeburg, Germany, 2010.
19. Ma, Z.H.; Causon, D.M.; Qian, L.; Gu, H.B.; Mingham, C.G.; Martínez Ferrer, P. A GPU based compressible multiphase hydrocode for modelling violent hydrodynamic impact problems. *Comput. Fluids* **2015**, *120*, 1–23. [CrossRef]
20. Kapila, A.K.; Menikoff, R.; Bdzil, J.B.; Son, S.F.; Stewart, D.S. Two-phase modeling of deflagration-to-detonation transition in granular materials: Reduced equations. *Phys. Fluids* **2001**, *13*, 3002–3024. [CrossRef]
21. Allaire, G.; Clerc, S.; Kokh, S. A five-equation model for the simulation of interfaces between compressible fluids. *J. Comput. Phys.* **2002**, *181*, 577–616. [CrossRef]
22. Antoniadis, A.; Drikakis, D.; Farmakis, P.; Fu, L.; Kokkinakis, I.; Nogueira, X.; Silva, P.; Skote, M.; Titarev, V.; Tsoutsanis, P. UCNS3D: An open-source high-order finite-volume unstructured CFD solver. *Comput. Phys. Commun.* **2022**, *279*, 108453. [CrossRef]

3. Tsoutsanis, P.; Dumbser, M. Arbitrary high order central non-oscillatory schemes on mixed-element unstructured meshes. *Comput. Fluids* **2021**, *225*, 104961. [CrossRef]
4. Murrone, A.; Guillard, H. A five equation reduced model for compressible two phase flow problems. *J. Comput. Phys.* **2005**, *202*, 664–698. [CrossRef]
5. Wackers, J.; Koren, B. A fully conservative model for compressible two-fluid flow. *Int. J. Numer. Methods Fluids* **2005**, *47*, 1337–1343. [CrossRef]
6. Schmidmayer, K.; Bryngelson, S.H.; Colonius, T. An assessment of multicomponent flow models and interface capturing schemes for spherical bubble dynamics. *J. Comput. Phys.* **2020**, *402*, 109080. [CrossRef]
7. Zhang, J. A simple and effective five-equation two-phase numerical model for liquid-vapour phase transition in cavitating flows. *Int. J. Multiph. Flow* **2020**, *132*, 103417. [CrossRef]
8. Harlow, F.H.; Amsden, A.A. Numerical calculation of almost incompressible flow. *J. Comput. Phys.* **1968**, *3*, 80–93. [CrossRef]
9. Johnsen, E.; Colonius, T. Implementation of WENO schemes in compressible multicomponent flow problems. *J. Comput. Phys.* **2006**, *219*, 715–732. [CrossRef]
0. Harlow, F.H.; Amsden, A.A. A numerical fluid dynamics calculation method for all flow speeds. *J. Comput. Phys.* **1971**, *8*, 197–213. [CrossRef]
1. Saurel, R.; Abgrall, R. Simple method for compressible multifluid flows. *Siam J. Sci. Comput.* **1999**, *21*, 1115–1145. [CrossRef]
2. Quirk, J.J.; Karni, S. On the dynamics of a shock-bubble interaction. *J. Fluid Mech.* **1996**, *318*, 129–163. [CrossRef]
3. Tsoutsanis, P.; Adebayo, E.; Merino, A.; Arjona, A.; Skote, M. CWENO Finite-Volume Interface Capturing Schemes for Multicomponent Flows Using Unstructured Meshes. *J. Sci. Comput.* **2021**, *89*, 64,. [CrossRef]
4. Liu, T.; Khoo, B.; Xie, W. Isentropic one-fluid modelling of unsteady cavitating flow. *J. Comput. Phys.* **2004**, *201*, 80–108. [CrossRef]
5. Ansari, M.; Daramizadeh, A. Numerical simulation of compressible two-phase flow using a diffuse interface method. *Int. J. Heat Fluid Flow* **2013**, *42*, 209–223. [CrossRef]
6. Tsoutsanis, P.; Titarev, V.; Drikakis, D. WENO schemes on arbitrary mixed-element unstructured meshes in three space dimensions. *J. Comput. Phys.* **2011**, *230*, 1585–1601. [CrossRef]
7. Tsoutsanis, P.; Pavan Kumar, M.; Farmakis, P. A relaxed a posteriori MOOD algorithm for multicomponent compressible flows using high-order finite-volume methods on unstructured meshes. *Appl. Math. Comput.* **2023**, *437*, 127544.. [CrossRef]
8. Tsoutsanis, P.; Farmakis, P. A posteriori mood limiting approach for multicomponent flows on unstructured meshes. In Proceedings of the World Congress in Computational Mechanics and ECCOMAS Congress, Oslo, Norway, 5–9 June 2022. [CrossRef]
9. Tsoutsanis, P. Stencil selection algorithms for WENO schemes on unstructured meshes. *J. Comput. Phys.* **2023**, *475*, 100037. [CrossRef]
0. Toro, E.; Spruce, M.; Speares, W. Restoration of the contact surface in the HLL-Riemann solver. *Shock Waves* **1994**, *4*, 25–34. [CrossRef]
1. Tsoutsanis, P. Antoniadis, A.; Jenkins, K. Improvement of the computational performance of a parallel unstructured WENO finite volume CFD code for Implicit Large Eddy Simulation. *Comput. Fluids* **2018**, *173*, 157–170. [CrossRef]
2. Petrov, N.V.; Schmidt, A.A. Multiphase phenomena in underwater explosion. *Exp. Therm. Fluid Sci.* **2015**, *60*, 367–373. [CrossRef]
3. Haimovich, O.; Frankel, S.H. Numerical simulations of compressible multicomponent and multiphase flow using a high-order targeted ENO (TENO) finite-volume method. *Comput. Fluids* **2017**, *146*, 105–116. [CrossRef]
4. Yeom, G.S.; Chang, K.S. A modified HLLC-type Riemann solver for the compressible six-equation two-fluid model. *Comput. Fluids* **2013**, *76*, 86–104. [CrossRef]
5. Shukla, R.K.; Pantano, C.; Freund, J.B. An interface capturing method for the simulation of multi-phase compressible flows. *J. Comput. Phys* **2010**, *229*, 7411–7439. [CrossRef]
6. Shukla, R.K. Nonlinear preconditioning for efficient and accurate interface capturing in simulation of multicomponent compressible flows. *J. Comput. Phys.* **2014**, *276*, 508–540. [CrossRef]
7. Zhang, Z.F.; Wang, C.; Zhang, A.M.; Silberschmidt, V.V.; Wang, L.K. SPH-BEM simulation of underwater explosion and bubble dynamics near rigid wall. *Sci. China Technol. Sci.* **2019**, *62*, 1082–1093. [CrossRef]
8. Kleine, H.; Tepper, S.; Takehara, K.; Etoh, T.; Hiraki, K. Cavitation induced by low-speed underwater impact. *Shockwaves* **2009**, *2*, 895–900. [CrossRef]
9. Kokh, S.; Lagoutière, F. An anti-diffusive numerical scheme for the simulation of interfaces between compressible fluids by means of a five-equation model. *J. Comput. Phys.* **2010**, *229*, 2773–2809. [CrossRef]
0. Oomar, M.; Malan, A.; Jones, B.; Horwitz, R.; Langdon, G. An all-Mach number HLLC based scheme for Multi-phase Flow with Surface Tension. *Appl. Sci.* **2021**, *11*, 3413. [CrossRef]
1. Deng, X.; Inaba, S.; Xie, B.; Shyue, K.M.; Xiao, F. High fidelity discontinuity-resolving reconstruction for compressible multiphase flows with moving interfaces. *J. Comput. Phys.* **2018**, *371*, 945–966. [CrossRef]
2. Shyue, K.M. A wave-propagation based volume tracking method for compressible multicomponent flow in two space dimensions. *J. Comput. Phys.* **2006**, *215*, 219–244. [CrossRef]
3. Xie, W.F.; Young, Y.L.; Liu, T.G.; Khoo, B.C. Dynamic response of deformable structures subjected to shock load and cavitation reload. *Comput. Mech.* **2007**, *40*, 667–681. [CrossRef]

54. Qin, Y.; Wang, Y.; Wang, Z.; Yao, X. The Influence of Various Structure Surface Boundary Conditions on Pressure Characteristics of Underwater Explosion. *Comput. Model. Eng. Sci.* **2021**, *126*, 1093–1123. [CrossRef]
55. Xie, W.; Liu, T.; Khoo, B. Application of a one-fluid model for large scale homogeneous unsteady cavitation: The modified Schmidt model. *Comput. Fluids* **2006**, *35*, 1177–1192. [CrossRef]
56. Zhang, Z.F.; Wang, C.; Wang, L.K.; Zhang, A.M.; Silberschmidt, V.V. Underwater explosion of cylindrical charge near plates: Analysis of pressure characteristics and cavitation effects. *Int. J. Impact Eng.* **2018**, *121*, 91–105. [CrossRef]
57. Hong, Y.; Wang, B.; Liu, H. Numerical simulation of compressible multiphase hydrodynamic problems using reduced five-equation model on body-fitted grids. *Comput. Fluids* **2022**, *233*, 105243. [CrossRef]
58. Jafarian, A.; Pishevar, A. An exact multiphase Riemann solver for compressible cavitating flows. *Int. J. Multiph. Flow* **2017**, *88*, 152–166. [CrossRef]
59. Wardlaw, A.B.; Luton, J.A. Fluid-structure interaction mechanisms for close-in explosions. *Shock Vib.* **2000**, *7*, 265–275. [CrossRef]
60. Wang, Y.; Yao, X.; Qin, Y. Investigation on influence factors about damage characteristics of ice sheet subjected to explosion loads: Underwater explosion and air contact explosion. *Ocean. Eng.* **2022**, *260*, 111828. [CrossRef]
61. Chen, J.Y.; Peng, C.; Lien, F.S.; Yee, E.; Zhao, X.H. Simulations for the explosion in a water-filled tube including cavitation using the SPH method. *Comput. Part. Mech.* **2019**, *6*, 515–527. [CrossRef]
62. Schmidt, D.; Rutland, C.; Corradini, M. A fully compressible, two-dimensional model of small, high-speed, cavitating nozzles *At. Sprays* **1999**, *9*, 255–276. [CrossRef]
63. Yu, J.; Liu, J.h.; Wang, H.k.; Wang, J.; Zhou, Z.t.; Mao, H.b. Application of two-phase transition model in underwater explosion cavitation based on compressible multiphase flows. *Aip Adv.* **2022**, *12*, 025209. [CrossRef]
64. Cheng, J.; Shu, C.W. High Order Schemes for CFD: A Review. *Chin. J. Comput. Phys.* **2009**, *26*, 633. Available online: http://www.cjcp.org.cn/EN/abstract/article_825.shtml (accessed on 23 January 2024).
65. Yingtang, D.; Lanhao, Z.; Jia, M.; Eldad, A. A resolved CFD-DEM-IBM algorithm for water entry problems. *Ocean. Eng.* **2021**, *240*, 110014. [CrossRef]

fluids

Article

Underwater Shock Wave-Enhanced Cavitation to Induce Morphological Changes and Cell Permeabilization in Microscopic Fungi

Miguel A. Martínez-Maldonado [1], Blanca E. Millán-Chiu [2,*], Francisco Fernández [1], Daniel Larrañaga [3], Miguel A. Gómez-Lim [4] and Achim M. Loske [1,*]

1 Centro de Física Aplicada y Tecnología Avanzada, Universidad Nacional Autónoma de México, Blvd. Juriquilla 3001, Querétaro 76230, Mexico; miguel.mtz.m@outlook.com (M.A.M.-M.); fcofdez@unam.mx (F.F.)
2 CONACyT, Centro de Física Aplicada y Tecnología Avanzada, Universidad Nacional Autónoma de México, Blvd. Juriquilla 3001, Querétaro 76230, Mexico
3 Posgrado en Ciencia e Ingeniería de los Materiales, Centro de Física Aplicada y Tecnología Avanzada, Universidad Nacional Autónoma de México, Blvd. Juriquilla 3001, Querétaro 76230, Mexico; daniel.larranaga@outlook.es
4 Centro de Investigación y Estudios Avanzados del Instituto Politécnico Nacional, Unidad Irapuato, Km 9.6 Libramiento Norte Carretera Irapuato-León, Irapuato 36824, Mexico; miguel.gomez@cinvestav.mx
* Correspondence: blanca.millan@fata.unam.mx (B.E.M.-C.); loske@fata.unam.mx (A.M.L.); Tel.: +52-442-1926126 (B.E.M.-C.); +52-442-2381164 (A.M.L.)

Abstract: Since the discovery of extracorporeal lithotripsy, there has been an increased interest in studying shock wave-induced cavitation, both to improve this technique and to explore novel biotechnological applications. As shock waves propagate through fluids, pre-existing microbubbles undergo expansion and collapse, emitting high-speed microjets. These microjets play a crucial role in the pulverization of urinary stones during lithotripsy and have been utilized in the delivery of drugs and genetic materials into cells. Their intensity can be amplified using tandem shock waves, generated so that the second wave reaches the bubbles, expanded by the first wave, during their collapse. Nevertheless, there is little information regarding the control of microjet emissions. This study aimed to demonstrate that specific effects can be obtained by tuning the delay between the first and second shock waves. Suspensions containing *Aspergillus niger*, a microscopic fungus that produces metabolites with high commercial value, were exposed to single-pulse and tandem shock waves. Morphological changes were analyzed by scanning and transmission electron microscopy. Proteins released into the medium after shock wave exposure were also studied. Our findings suggest that, with enhanced control over cavitation, the detachment of proteins using conventional methods could be significantly optimized in future studies.

Keywords: acoustic cavitation; microjets; tandem shock waves; biotechnological applications; *Aspergillus niger*; morphological changes; cell permeabilization; genetic transformation; cell surface shaving; protein detachment

1. Introduction

Acoustic cavitation, initially studied in the early 20th century due to its role in liquid-impact erosion [1], also made a significant breakthrough in medicine in the 1980s with the advent of extracorporeal shock wave lithotripsy (SWL) [2,3]. For over 40 years, SWL has been successfully used for the non-invasive disintegration of urinary stones. Shock waves are generated outside the body in water and focused on the stone, leading to its pulverization after several hundred shock fronts. The theory of underwater shock wave-induced cavitation for medical and biotechnological applications is well-established. Research on this topic has been extensively conducted and documented by several authors since the early 1980s [4–10].

Cavitation plays a critical role in SWL [11–13]. When a shock front reaches a urinary stone and the surrounding fluid, pre-existing microbubbles undergo a forced collapse, followed by a rapid rebound that causes them to grow to volumes much larger than the initial ones. Within hundreds of microseconds, each bubble suffers an asymmetric inertial collapse, partly invaginating and generating a high-speed microjet of liquid, which pierces through the bubble and exits it at the opposite side. According to studies published by Philipp and colleagues [4], the speed of the microjets exceeds 700 m/s. Despite their limited range, microjets emitted by bubbles near the stone significantly contribute to its pulverization. The phenomenon is similar to the erosion experienced by propeller blades or turbines in liquids. The impact between the microjet and the collapsing bubble's wall is so intense that a secondary shock wave is generated, which can contribute to the observed effects and interact with neighboring bubbles.

During bubble collapse, the local energy density increases enormously. This can generate extremely high temperatures as well as free radicals capable of inactivating enzymes and causing cellular apoptosis [8]. Because the temperature increases are highly localized, they virtually do not contribute to a rise in temperature within the vial.

The two most popular equations for studying the dynamics of bubbles exposed to shock waves of the type mentioned here are the Rayleigh–Plesset equation and the Gilmore equation [9]. An advantage of the Gilmore equation is that it accounts for the compressibility of the fluid, which proves useful for describing inertial collapses.

From the late 20th century to the beginning of the current one, significant efforts were made to enhance the efficiency of extracorporeal lithotripters by adjusting the power of microjets, aiming to shorten treatment times and minimize tissue damage [14–16]. A notable innovation in this field is the tandem system [17], which basically involves emitting pairs of successive shock fronts, at a controlled, adjustable delay ranging from 50 to 950 μs. Each emitted pair, known as an "event", can be repeated at frequencies adjustable from 1/3 to 3 Hz, depending on the application.

In the past, we used different pressure profiles with the aim of controlling bubble collapse and microjet formation [17–19] and hypothesized that enhancing bubble collapse and microjet emission is possible when the second shock front in each event arrives during the inertial collapse of the bubbles generated by the first shock wave. This has been demonstrated in both in vitro [17,18] and in vivo [18,19] studies. Cavitation, being a multi-bubble phenomenon with uncontrollable variables like the number, size, and distribution of pre-existing microbubbles, makes it challenging to determine an optimal delay applicable to all treatments. Nevertheless, delays between 250 and 350 μs have been shown to improve stone-phantom fragmentation efficiency, supporting the described hypothesis.

The knowledge that the diameter of a microjet emitted during shock wave-induced bubble collapse is approximately one-tenth of the bubble´s diameter [20–22] suggests the possibility of using tandem shock waves to selectively favor the formation of microjets of a specific diameter. These high-energy, short-range microjets may function like microscopic syringes [20], injecting liquid into nearby cells.

The phenomenon has been successfully utilized in biotechnology for the genetic transformation of bacteria and fungi [23,24], as well as for extracting phenolic acids and flavonoids from plants [25]. Bacteria and fungi were subjected to shock waves in a vial containing an aqueous solution and the transformant. The shock wave-induced fluid microjets temporally increased cell membrane permeability, facilitating the internalization of DNA into the cells. For phytochemical extraction, plant samples in powdered form were placed in a vial with water or ethanol as a solvent and exposed to shock waves. The resulting microjets broke down the plant particles, releasing their contents.

The complexity of the described processes arises from the multitude of variables involved. To date, there are no precise methods for controlling microjet emissions. Thus, it is not surprising that, despite careful estimation of the parameters before an experiment, in most cases, replicates are necessary to adjust the number of shock waves, the pressure profile and the delay to continue exploring the biotechnological applications of induced cavitation. Moreover, the dynamics of a bubble exposed to the action of a shock wave in water depend on other factors, such as the content of dissolved gases, the pressure profile, of the impinging shock wave, the vapor pressure, and the presence of cavitation nuclei [26]. Determining an optimal delay is further complicated by the large number of bubbles present in the suspension, all of which are collapsing and interacting with each other.

In cases related to genetic transformation or the extraction of compounds, it is worth highlighting the size of the microjet in relation to the target (on a statistical level). Thus, microjets with a size comparable to the object can disintegrate it, while much smaller microjets have facilitated the genetic transformation of bacteria and microscopic fungi, raising the question about the effects that microjets of more extreme sizes could cause. Some could detach the so-called "ornamentation" of fungal conidia without damaging the overall structure. This is of interest for the extraction of proteins relevant to biotechnology, which are in the outer layers.

Following the aforementioned discussion about the possibility of using tandem shock waves to selectively favor the formation of microjets of a specific diameter, the aim of this study was to analyze cell wall proteins from *A. niger* conidia (see next Section) using single-pulse (SP) shock waves and tandem shock waves with two different delays (SDT: short-delay tandem and LDT: long-delay tandem), recording the cell wall morphological changes by scanning electron microscopy (SEM) and transmission electron microscopy (TEM), and analyzing the proteins released into the medium after shock wave application by sodium dodecyl sulfate-polyacrylamide gel electrophoresis (SDS-PAGE). Our findings suggest that protein detachment can be optimized by selecting an adequate delay between the first and second shock waves.

2. Materials and Methods

2.1. Biological Subject

In this study, we used the fungus *Aspergillus niger* as a model microorganism due to its significant relevance in research and industrial applications [27]. It is a filamentous fungus that has been harnessed, modified, and used in the biotechnological industry for over a century. Among the principal features of *A. niger* is its capability to produce primary and secondary metabolites with high commercial value, such as organic acids, proteins, and enzymes [28]. In addition, its capacity to perform posttranslational modifications and its secretory systems make it a valuable microorganism for use as an expression platform for producing recombinant molecules [29]. *A. niger* has been called a "cell factory" and classified as "generally regarded as safe" (GRAS) [30,31]. A recent use of this fungus is the postfermentation harvesting of cell wall products to obtain polysaccharides such as chitin and chitosan, which have proven to be useful in medicine, biotechnology, and agriculture [32]. *A. niger*, like other fungal species, has specific molecular biosynthesis needs, such as the conservation of its osmotic balance, its cellular structure, the morphogenesis process, and protection against environmental factors. These requirements are achieved to a large extent by the cell wall, which undergoes constant and diverse structural changes [32,33]. The cell wall is characterized by a layered structure composed of matrix polymer components such as carbohydrates (α-glucans, β-glucans, chitin, and galactomannan), surface proteins, glycoproteins, hydrophobins, and pigments such as melanin [34,35]. These components, principally proteins, play an essential role in the interaction between cells and their environment, participating in crucial processes such as signal transduction, cell adhesion, cell–cell interactions, molecular transport, and pathogenicity [35]. The content and composition of the cell wall differ among strains and depend on environmental factors, such as nutrients, growth culture medium, mycelial age, and stress [36]. Enhanced entry of plasmids into the cell via shock wave-induced transient pores has been possible despite the barrier of the cell wall [37].

2.2. Fungal Growth and Conidia Harvesting

The *A. niger* CBS 513.88 strain was used. Fungal inoculation was performed in Petri dishes with minimal medium agar, and the plates were incubated for six days at 30 °C. After that, conidia were collected by adding 5 mL of minimal medium broth over the colony and lightly scratching the colony to harvest it. Three layers of sterile Miracloth (EMD Millipore, MA, USA, catalog number 475855) were used as filters for removing conidiophores in the suspension. Conidia were counted with a Neubauer chamber and adjusted with minimal medium to a final concentration of 2×10^6 conidia/mL.

2.3. The Shock Wave Generator

Shock waves were produced with a modified Piezolith 2501 shock wave source (Richard Wolf GmbH, Knittlingen, Germany) consisting of about 3000 piezoelectric elements mounted on the concave side of a hemispherical aluminum backing (radius 345 mm) embedded in an insulating polymer (Figure 1). The shock wave source is placed at the bottom of a 675 × 675 mm Lucite water tank (height 450 mm). To generate a shock wave, a high-voltage (2–6 kV) discharge is applied to the piezoelectric array. Each element produces a pressure pulse that propagates through the water toward the center (focus *F*) of the system. The superposition of all pulses and nonlinear effects steepen the pressure pulse into a shock wave as it approaches *F*. Two independent capacitor charging units are used to generate tandem shock waves with a delay of 10 to 100 µs (±2 µs) and 100 to 900 µs (±5 µs) between pulses. The vials to be treated were positioned inside the water tank with an *XYZ* positioning system (UniSlide Assemblies Velmex, Inc., Bloomfield, NY, USA). The error in positioning each sample was approximately ±0.5 mm.

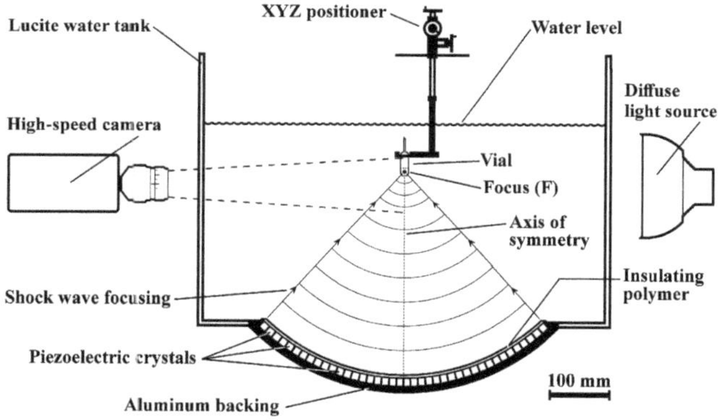

Figure 1. Sketch of the experimental setup showing a test vial containing conidial suspension centered at the axis of symmetry of the hemispherical shock wave source inside the water tank. The field of view of the camera is shown with dotted lines.

2.4. Shock Wave Exposure

Sterile 4 mL transfer pipettes (Thermo Fisher Scientific, Waltham, MA, USA) were filled with 1.5 mL conidial suspension (see Section 2.2), heat-sealed, and cut 20 mm above the bulb. For shock wave exposure, each vial was fixed vertically along the axis of symmetry of the shock wave source so that focus F coincided with the center of curvature of the bottom of the pipette (Figure 1). Shock waves were generated using a discharge voltage of 5.5 ± 0.125 kV throughout the whole experiment. The pressure profile is similar to that of several extracorporeal shock wave lithotripters and shock wave sources used for biotechnological applications. At this voltage, the peak positive pressure, recorded with a polyvinylidene difluoride needle hydrophone (Imotec GmbH, Würselen, Germany), had an amplitude of 73.62 ± 1.44 MPa. The measured full width at half maximum (FWHM) of the pressure pulse was 194 ± 8 ns. The −6 dB focal volume had the shape of an ellipsoid with minor and major axes of approximately 2 and 8 mm, respectively. Either single-pulse (SP) or tandem shock waves were emitted at a rate of 0.5 Hz. Due to the similarity in acoustic impedances, the suspension–polyethylene–suspension interfaces did not significantly modify the pressure profile. The tub was filled with 24 °C tap water up to 80 mm above F. A total of 30 vials were divided into 10 groups to receive either sham- or shock-wave-treatment, as indicated in Table 1. Each group consisted of three vials; that is, the experiment was repeated three times. In the tandem mode, an "event" was defined as a pair of shock waves; that is, to equalize the energy delivered to each sample, the number of tandem "events" was halved compared to the number of SP shock waves. Two delays (40 and 160 μs) between the first and second shock waves were used in the tandem mode and defined as the SDT and LDT modes. As explained in the next section, high-speed imaging of acoustic cavitation was used as a tool to select these delays.

Table 1. Experimental design.

Group	Mode	Tandem Events	Number of SWs [1]
Control	-	-	0
1	SP [2]	-	50
2	SP	-	100
3	SP	-	200
4	SDT [3]	25	50
5	SDT	50	100
6	SDT	100	200
7	LDT [4]	25	50
8	LDT	50	100
9	LDT	100	200

[1] Shock waves. [2] Single pulse. [3] Short-delay tandem (delay = 40 μs). [4] Long-delay tandem (delay = 160 μs).

2.5. High-Speed Imaging

Images of the bubble dynamics were captured with a high-speed MotionPro X4 (Integrated Design Tools, Inc., Pasadena, CA, USA) camera equipped with a Mamiya/Sekor sx 1:1.4, f = 55 mm lens (made in Japan) in the SP mode and both tandem modes at a speed of 5×10^4 frames per second (fps). At this speed, each image is composed of 44 × 508 pixels. A vial containing 1.5 mL of a suspension of distilled water and A. niger conidia was placed inside the water tank and illuminated with back-illumination as shown in Figure 1. The camera was triggered 170 μs after the shock wave generator. All recordings were performed in duplicate. Bubbles were seen as dark discs; however, some of them overlapped and were difficult to distinguish from one another.

Figure 2 shows sequences of images recorded after the passage of SP, SDT, and LDT shock waves. The time T between one image and the next was 20 μs. t = 0 was set at the instant when bubbles could be seen through the vial with the naked eye for the first time. The markers labeled 1–8 were added to describe some features (see the Figure caption). In general, it can be noticed that bubbles underwent fast growth, some of them up to a diameter of more than 1 mm, followed by inertial collapse. Images (b) and (c) (tandem mode) show increased bubble density, shortly after the second shock wave passed, both in the SDT (t = 2T) and the LDT (t = 8T) modes.

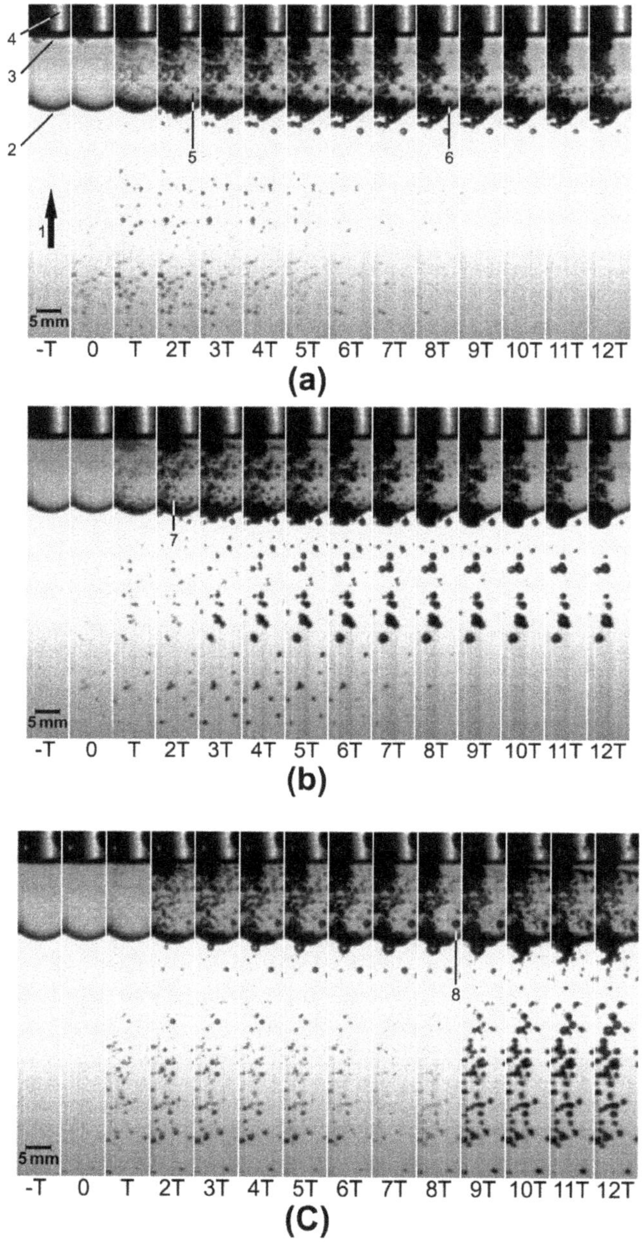

Figure 2. Sequence of images (T = 20 μs) showing acoustic cavitation outside and inside a sealed vial containing distilled water and *A. niger* conidia, exposed to (**a**) an SP, (**b**) an SDT, and (**c**) an LDT shock wave. In the SDT mode, the second shock wave arrived at t = 2T (delay = 40 μs), and in the LDT mode, the second shock wave reached the vial at t = 8T (delay = 160 μs). Markers refer to: 1. Shock wave direction as it propagates toward the vial. 2. Vial. 3. Suspension–air interface inside the vial. 4. Undesirable diffraction of light (not relevant for this study) from the back side of the vial. 5. and 7. Examples of the smallest visible bubbles at the instant of their maximum expansion. 6. and 8. Examples of the largest bubbles at the instant of their maximum expansion.

As already mentioned, the diameter of a microjet emitted during bubble collapse is approximately one-tenth of the bubble´s original diameter [20–22], meaning that larger bubbles generate microjets with a larger diameter than smaller bubbles. Normally, in a suspension, there is a distribution of bubbles of many sizes. In this study, we proposed a method by which the diameter of the emitted microjets can be influenced to some extent. This is achieved by generating two shock waves. The first wave produces bubbles of all sizes. If the second wave is sent such that its peak of positive pressure arrives when a bubble of a specific size begins to collapse, this collapse and the microjet emission are significantly reinforced as the pressure from the second wave compresses the collapsing bubble. The collapse of bubbles with sizes different from those selected is not reinforced. In other words, if the arrival of the second shock wave coincides with the instant when small (large) bubbles begin their collapse, the generation of microjets with a small (large) diameter is enhanced. Considering this, the criterion for selecting the short (long) delay in this study was based on the time needed for the smallest (largest) bubbles to begin collapsing after the passage of an SP shock wave. The instant when the smallest bubbles begin their collapse, following the passage of an SP shock wave, was estimated using high-speed recordings. The same was done for the largest bubbles observed in the suspension. If the effects on the cell wall of *A. niger* conidia depend on the diameter of the microjets, exposing the conidial suspension to tandem shock waves emitted at two significantly different delays, will produce different degrees and types of damage.

As mentioned above, Figure 2a shows a sequence of images obtained after exposing the conidial suspension to an SP shock wave. At t = 2T, that is, 40 µs after the arrival of the shock wave, the smallest bubbles observable with the naked eye acquired their maximum diameter, and their size began to decrease at t = 3T. An example is the bubble marked with the number 5. At t = 8T, 160 µs after the arrival of the shock wave, the largest bubbles observed inside the suspension started to collapse. In this case, a representative bubble is number 6. This bubble appears smaller on each image after t = 8T. Figure 2b shows a sequence recorded after the passage of a tandem shock wave with a delay of 40 µs (SDT shock wave). The second shock wave arrived at t = 2T. Because of this, in the following images, more bubbles are seen both outside and inside the vial. Number 7 is one of the smallest bubbles at its maximum size. Analogously, Figure 2c shows a sequence obtained after the passage of an LDT shock wave. At t = 8T, 160 µs after the passage of the first shock wave, the second shock wave enhanced the collapse of the largest bubbles (see number 8) observed inside the vial. Increased bubble activity appeared after t = 8T.

2.6. *Electron Microscopy Sample Preparation*

Samples destined for SEM and TEM observation were processed equally until dehydration. Briefly, 1 mL of conidial suspension was fixed for 1 h by adding 1 mL of glutaraldehyde (3%) dissolved in a sodium cacodylate buffer (0.1 M at pH 7.4). After this time, the conidia were washed twice for 10 min with 1 mL of cacodylate buffer, and they were recovered by centrifugation at 7500× *g*. The last wash was performed for 24 h. Then, the conidia were incubated with osmium tetroxide (1% in 0.1 M sodium cacodylate buffer at pH 7.4) for 1 h. Then, 3 washes in cacodylate buffer were performed: 2 for 10 min and the last for 24 h. Finally, the samples were dehydrated by submerging them in increasing percentages of ethanol (10, 30, 50, 70, 96, and 100%) for 10 min each.

For SEM, samples immersed in 100% ethanol were treated to replace ethanol with liquid CO_2 and then desiccated at critical point temperature using a critical point dryer CPD2 (Ted Pella, Inc., Redding, CA, USA), according to the manufacturer's instructions. Once desiccated, conidia were recovered and coated with a 20 nm gold nanoparticle layer using a sputter coater EMS 550 (Electron Microscopy Sciences, Hatfield, PA, USA) and observed with a JSM-6060LV electron microscope (JEOL, Tokyo, Japan).

For TEM, ethanol (concentration 100%) was withdrawn from the samples by centrifugation at 7500× *g*; 1 mL of propylene oxide was added to the pellet and incubated for 30 min at 20 °C; the last step was repeated. Then, to infiltrate the cells, 1 mL of resin and propylene oxide (rate 1:1) was added and left to evaporate for 24 h at RT; this step was repeated twice, increasing the resin portion each time (2:1 and 3:1) until it reached honey consistency after evaporation. After infiltration, the samples were placed in pure resin and polymerized at 60 °C for 36 h. Microtome sections with a thickness between 60 and 90 nm were obtained using an ultramicrotome MTX-RMC (Boeckeler Instruments Inc., Tucson, AZ, USA) and collected on 300 mesh copper grids. Then, the samples were counterstained with a 2% uranyl acetate and 2% lead citrate solution. Observations were made at 80 kV with a JM100 electron transmission microscope (JEOL). The images were recorded with a CCD camera (Orious, Gatan, Inc., Pittsburgh, PA, USA) coupled to the microscope.

2.7. Protein Release Analysis

Immediately after shock wave exposure, samples in minimal medium broth were centrifuged at 12,000× *g*; 1.2 mL of the supernatant with the released proteins was carefully separated and collected in a clean tube without disturbing the conidial pellet. Next, the proteins were precipitated by adding 4 volumes of ice-cold acetone and incubation for 15 min on ice; then, samples were centrifuged at 16,600× *g*, the acetone was dismissed, and the protein pellet was dried at 20 °C. Finally, protein-enriched fractions were resuspended in 200 μL of 7 M urea and analyzed.

A bicinchoninic acid (BCA) protein quantification assay in a 96-well microplate (ABP Biosciences, catalog P011, Beltsville, MD, USA) was employed to quantify proteins released into the medium after shock wave treatments, according to the instructions of the manufacturer. Briefly, 20 μL of each sample was added to 200 μL of working reagent and homogenized. Then, the plate was incubated at 30 °C for 30 min and cooled at 20 °C. Measurements were taken at 562 nm using a Varioskan microplate reader (Thermo Fisher Scientific Inc., Waltham, MA, USA). Bovine serum albumin was used as a standard protein. Each sample was measured in triplicate.

SDS-PAGE was used to analyze the proteins released into the medium by the action of shock wave treatments. For electrophoresis, acrylamide 15% separating gel, 4% stacking gel, 1× Tris/Glycine/SDS buffer, and a molecular weight marker were used (Sigma-Aldrich, St. Louis, MO, USA). Electrophoretic separation was performed for 1.5 h at 100 V. Protein bands were revealed with silver staining. For this, gels were fixed in a solution of acetic acid:methanol:water (5:50:45) for 1 h; subsequently, two 20 min washes with 50% methanol and a 20 min wash with 30% methanol were performed. The gels were immersed in a 0.02% sodium thiosulfate (Na_2SO_3) solution for 1 min, with two subsequent 20 min washes with water, followed by incubation in a 1% silver nitrate solution for 20 min, after which they were washed 3 times with water. Finally, they were immersed in the developer solution (2% $NaCO_3$; 0.04% formaldehyde) for up to a maximum of 10 min. The gels were visualized using a gel photo documenter system in gel Doc EZ (Bio-Rad Laboratories, Hercules, CA, USA).

2.8. Electrophoretic Light Scattering

Immediately after shock wave application, the conidia were recovered and stored at −20 °C in the culture medium. Subsequently, they were fixed with formaldehyde (2%) in phosphate-buffered saline (PBS pH 7.4) for 15 min. Next, two washes with PBS and three with sterile distilled water were performed, and finally, 1.5×10^4 conidia were suspended in 1 mL of injectable-grade sterile distilled water. Surface charge data were obtained from 1000 conidia using a particle size analyzer (Litesizer 500, Anton Paar México S.A. de C.V., Mexico City). The software (Kalliope version 2.4, 2019) uses the Helmotz-Smoluchowski equation to calculate the Z potential from their electrophoretic mobility values.

3. Results

3.1. Morphological Changes in A. niger Observed through SEM

The shape of all observed control conidia through SEM (see Section 2.6) was quasi-spherical, with an average diameter of approximately 3–5 μm. Figure 3a shows the typical ornamentation of *A. niger* conidia on the surface of control cells [38]. Its cell wall is rough with folds and grooves due to the presence of structures made up of chitin, glucans, proteins, and pigments [35]. The filter paper that served as support is seen in the background. Shock wave-exposed conidia revealed a kind of shaving on the cell wall surface in addition to the presence of pores. Cells treated with SP shock waves (Figure 3b) showed a rough cell wall but with less ornamentation. Conidia exposed to SDT shock waves also exhibited a smoother cell wall surface (Figure 3c) and evident loss of cell wall material. Most images of conidia subjected to LDT shock waves showed erosion of their cell surface (Figure 3d). Collapsed cell walls, such as the one in Figure 3d, were only seen after using one of the three shock wave modes (SP, SDT, or LDT), not in the control group. Figure 3 is a representative example of a larger group of obtained images. Comparing all SEM images of shock wave-exposed conidia, the effect over the cellular surface was found to be different with every applied treatment, and LDT treatment was the most efficient procedure to shave surfaces. Images at a lower magnification can bee seen in Figure S1.

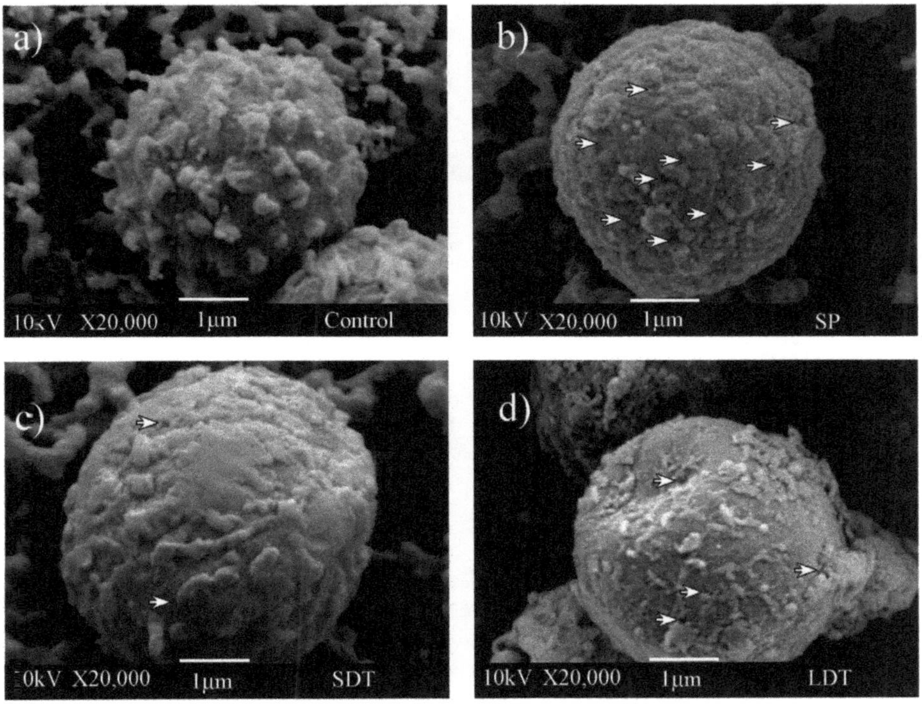

Figure 3. SEM images of (**a**) nontreated conidia (control) and conidia exposed to (**b**) 200 SP shock waves, (**c**) 200 SDT shock waves, and (**d**) 200 LDT shock waves. White arrowheads indicate foreground cell pores.

3.2. Internal Changes of A. niger Observed through TEM

The internal structures of the cells in the control samples were the same as typically reported; they were covered with a cell wall of approximately 200 nm thickness, and an electron-dense outer layer consistent with melanin was embedded in the cell wall components [39,40]. A representative image is shown in Figure 4a. As with SEM, shock-wave-treated conidia revealed different degrees of shaving effect on their outer layer. SP shock waves tended to detach the outer layer of the conidium (Figure 4b). SDT shock waves induced damage to the external structure of the conidia, apparently thickening the cell wall (Figure 4c). Additionally, the formation of vesicle-like structures can be observed, probably attributed to shear forces exerted by the pressure changes after the passage of the shock wave. As for the conidia that received LDT shock waves (Figure 4d), a partial wasting effect on the wall surface was observed. Figure 4d shows a conidium with erosion areas where there is detachment of cell wall components, including melanin. Figure 4e–h are magnifications of Figure 4a–d, respectively. In most of the images obtained, the changes in the cell wall morphologies were more evident when SDT shock waves were applied, while greater wear was observed with the LDT shock wave treatment.

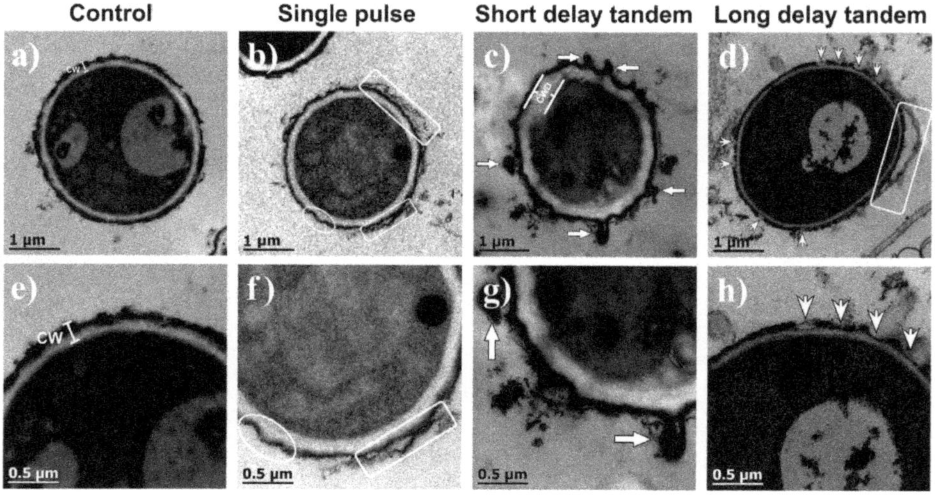

Figure 4. TEM images of (**a**) a nontreated conidium (control) and a conidium exposed to (**b**) 200 SP shock waves, (**c**) 200 SDT shock waves, and (**d**) 200 LDT shock waves. (**e–h**) are magnifications of the images shown above. CW means cell wall. Rectangles enclose evidence of a detached outer layer. The ellipse in (**f**) shows cell wall morphology changes, such as depressions. Arrows show the formation of diverticula-like structures. Arrowheads indicate detachments resembling shavings of the outer cell wall.

3.3. Analysis of Released Protein

The quantity of protein released from the outer layer of the conidial wall into the medium after shock wave exposure was measured. As shown in Figure 5, the protein concentration ranged from 0.23 to 0.35 mg/mL. The shock wave mode (SP, SDT, or LDT) influenced the quantity of released protein. Compared with nontreated conidia, a significant increase in protein was obtained in most shock wave-treated groups. Exposure to 200 SP shock waves resulted in a 1.44-fold higher value (0.33 mg/mL). Applying 200 LDT shock waves increased the protein concentrations up to 1.52 times (0.35 mg/mL) the value recorded for the control group. The lowest protein concentrations were obtained from samples exposed to 200 SDT shock waves, similar to the control group.

SDS-PAGE was performed to confirm the shock wave-shaving effect on the cell surface, allowing the obtention of non-degraded proteins. In this analysis, the resulting electrophoresis gel allowed for the detection of protein bands with molecular weights between 6.5 and 120 kDa (Figure 6). Lane 1 (MM), corresponding to the protein content in the culture medium, shows bands at 45, 55, and 66 kDa. The control sample (Lane 2) revealed protein bands of 17, 40, and 97 kDa in addition to the bands obtained from the culture medium. Samples treated with 50 and 100 SP shock waves (Lanes 3 and 4) showed a similar banding pattern as the control group, although the bands were slightly more intense. Furthermore, a band corresponding to high-molecular-weight proteins (116 kDa) appeared. This was probably generated by the enrichment of proteins due to shock wave action. Interestingly, 200 SP shock wave treatments seem to have a damaging and possibly degrading effect (Lane 5). In the three SDT shock wave groups (Lanes 6 to 8), only samples treated with 50 shock waves revealed a banding pattern with bands of 23, 30, and 37 kDa.

Finally, when analyzing the proteins obtained with LDT shock waves (Lanes 9 to 11), various molecular weights, from 6.5 kDa to fragments or proteins with molecular weights greater than 100 kDa, were obtained. The effect was more evident as the number of LDT shock waves increased.

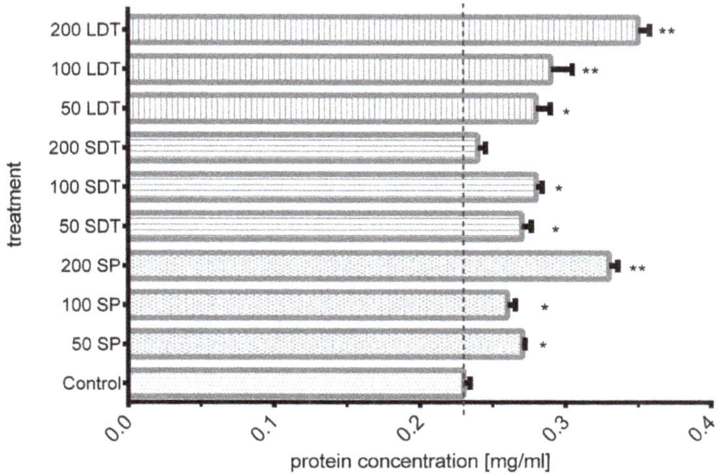

Figure 5. BCA protein quantification of control and shock wave-treated samples. The length of each bar is the average of three results. * Significant differences between shock wave groups and the control group. ** Significant differences between groups. ($p > 0.001$).

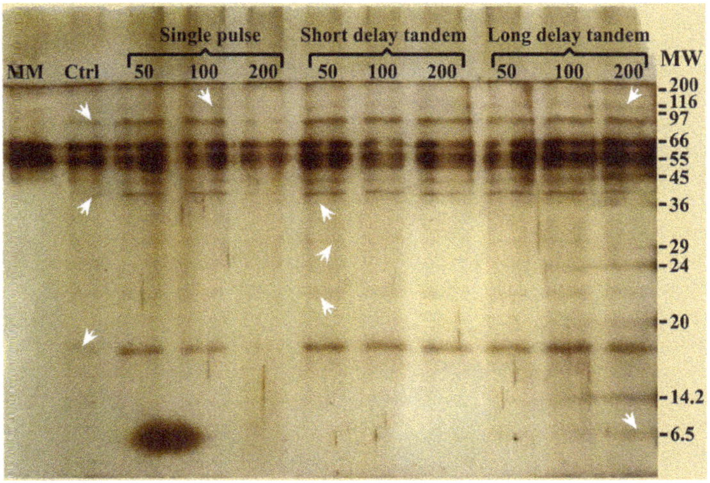

Figure 6. SDS-PAGE gel electrophoresis of different shock wave treatments and the control group. Arrowheads point to some of the new, more intense bands observed in the lanes of treated samples. Minimal medium (MM); control sample (Ctrl); molecular weight marker (MW) in kDa.

The surface electrostatic charge of conidia depends on the molecular interaction between different components of the cell wall. This was assessed by microelectrophoresis, electrophoretic mobility, and the Z potential. As shown in Table 2, both the control and shock wave-treated samples only expressed negative values and increased electrophoretic mobility. All conidia that released proteins and other cell wall compounds because of shock wave exposure, presented significantly more negative values (from −26.98 ± 0.59 mV to −33.34 ± 0.35 mV) than those recorded in the control samples (−20.37 ± 0.94 mV).

Table 2. Zeta potential and electrophoretic mobility of *Aspergillus niger*.

Sample	Mean Z Potential (mV) ± SD [1]	Electrophoretic Mobility (μm × cm/Vs)
Control without SWs [2]	−20.37 ± 0.94	−1.44
SP [3] 50 SWs	−28.85 ± 0.82	−2.04
SP 100 SWs	−26.98 ± 0.59	−1.91
SP 200 SWs	−30.09 ± 0.74	−2.13
SDT [4] 50 SWs	−33.34 ±0.35	−2.36
SDT 100 SWs	−29.56 ± 1.33	−2.09
SDT 200 SWs	−32.66 ± 0.81	−2.31
LDT [5] 50 SWs	−27.3 ± 1.0	−1.93
LDT 100 SWs	−29.57 ± 1.22	−2.09
LDT 200 SWs	−30.36 ± 0.04	−2.15

[1] Standard deviation. [2] Shock waves. [3] Single pulse. [4] Short-delay tandem (delay = 40 μs). [5] Long-delay tandem (delay = 160 μs).

4. Discussion

Our SEM and TEM images revealed that conidia exposed to different shock wave conditions suffered a variety of modifications to their cell structure, from pore and vesicle formation to slight wear of the cell wall and deformation to cell wall surface shaving. Representative examples are shown in Figures 3 and 4. Because it is not possible to observe the interaction between a shock wave and the cell surface in real time, other authors [40–42] performed simulations in silico, and they concluded that after passage of the positive pressure pulse, the change in membrane permeability is caused by a disturbance of the phospholipid fatty acid chains. It has also been reported that shock waves with a smaller positive pressure amplitude but longer FWHM increased the membrane permeability more than shock waves with higher pressure and a shorter FWHM [42]. According to these authors, shock wave-induced shear stress, tangential to the cell surface, contributed to cell permeability. As mentioned above, shock waves can induce permeabilization in several kinds of cells [42–44]. Compression, as well as tensile and shearing stress, may be partially responsible for the changes to the cell wall, principally pore formation, as evidenced in Figure 3b–d; however, the accepted hypothesis is that acoustic cavitation is the main phenomenon responsible for this effect.

Bubble growth and collapse depend not only on the applied shock front but also on the number of suspended microbubbles and impurities, such as solid kernels. The cell wall of *A. niger* conidia is rough (see Figure 3a) and hydrophobic, owing to the presence of hydrophobins, melanin, α-1,3-glucan, and chitin [31,32]. These physicochemical properties of the cell wall components generate low wettability in *A. niger* conidia compared with other fungal species [45]. This reduces the adhesion force between the surface of the wall and the surrounding liquid, enhancing the nucleation of microbubbles [46–48].

A shortcoming of the methodology described here is that cell shaving and sonoporation are difficult to control. Using tandem shock waves with specific delays can exert a certain influence on the impact on the cells, such as the pore size, cell deformation, or modification of the cell wall. Nevertheless, because cavitation is a multibubble phenomenon and depends on several factors, such as the properties of the fluid, the bubble size distribution, the existence of cavitation nuclei, and the pressure profile [49], the most appropriate delay for a specific application is difficult to determine. Moreover, several phenomena act synergistically, complicating the scenario.

A useful tool to estimate the size of bubbles immersed in a liquid versus the time after their interaction with a shock wave is the Gilmore–Akulichev equation [50]. By obtaining the maximum diameter of shock wave-exposed bubbles using high-speed imaging, it is possible to calculate their initial size, the approximate diameter of the microjets emitted after collapse, and the delay to enhance the collapse energy of these specific bubbles. This method could allow relative control over acoustic cavitation and microjet formation, influencing possible effects on the cells.

Deformation of the cell wall and cell membrane can also be produced by direct shock wave-induced stress; this is without the presence of cavitation and solely dependent on the pressure profile of the shock wave [51]. On the other hand, if a microbubble smaller than a cell is compressed close enough to its surface, it causes a vacuum effect, pulling part of the outer layer of the cell surface. During expansion, the bubble hits the cell surface [51–53]. Both effects (suctioning and hitting) alternate at high speed, resulting in cell surface deformation. These effects are caused by harmonic waves that reach megahertz frequencies, such as those generated by ultrasound devices [52–54]. Furthermore, shear stress produced by cavitation after sonoporation of conidia has been reported to cause changes in cell permeability. Consequently, if cells are in the vicinity of cavitating bubbles, transient pores derived from the viscoelastic shear forces of these microstreams may appear on their surface [55–57]. During shock wave treatment, bubbles normally do not oscillate; however, both bubble expansion and compression have much higher energy, and strong shear stress can be expected.

In this study, SDT shock waves generated significantly more cell wall deformations than SP shock waves, but with less protein release. This may occur because the SDT mode favors small microjets but inhibits larger ones, as generated in the SP mode. Inferences can be drawn from the number of cellular pores appearing in the foreground of our SEM images. Even though TEM revealed greater changes in the cell wall with SDT, the formation of diverticulum-like agglomerates was also evident, which perhaps accumulated the proteins instead of releasing them. Undoubtedly, more tests are required to confirm or correct such a hypothesis.

LDT shock waves were more effective in shaving the outer layer with diminished cell deformation, releasing proteins of different sizes (Figure 6). Moreover, the quantity of protein released into the medium was higher after exposure to LDT shock waves. SDS-PAGE revealed that the recovery of proteins of variable size was achieved with this treatment. Other authors reported detaching effects during the sonication of *A. nidulans* conidia suspended in sterile deionized water. That method generated streams that peeled off the hydrophobin layers without deforming the morphology of the cells [58].

In recent years, the study of surface proteins has stood out for its importance in understanding many cellular interaction processes with the environment, as well as its importance in plant, animal, and human pathogenicity. Few methods to perform the shaving of superficial proteins have been reported [59,60]. Some of them use enzymes or chemicals for the treatment of samples to obtain proteins and proceed to analysis and characterization [61–63]. As demonstrated in this study, shock waves can be considered a new approach for detaching proteins from the conidial surface. With standardized protocols for fungi, shock waves could constitute an alternative method to study cell wall proteins, avoiding subcellular fractionation. The advantages of shock wave treatments are that the procedure is relatively fast and that there is no need for special sample preparations. Moreover, using tandem shock waves, the same number of shock waves can be applied in half the time as with SP treatments.

The Z potential value is associated with the cell surface charge, which is dictated by its physicochemical characteristics. Fungal cell walls are negatively charged because of molecules such as β-1,3-glucans and melanin. As reported by Pihet et al. [40] and Wargenau et al. [39], the presence of melanin forms a negatively charged shell. In the control group, the Z potential was −20 mV. This agrees with the values reported by Wang et al. [64] for fungal cells dispersed in deionized water. For conidia exposed to any of the three shock wave treatment modes (SP, SDT, and LDT), the Z potential values decreased, indicating the detachment of proteins, such as hydrophobins, from the conidial outer layers.

5. Conclusions

Single-pulse and both SDT and LDT shock waves induced various modifications on the cell wall surface of *A. niger* conidia, from a superficial shaving effect releasing proteins attached to the cell wall to the disruption of the fungal cell wall layers. Our results support the hypothesis that acoustic cavitation is one of the most important phenomena of shock wave treatments, not only for cell permeabilization but also for modifying or extracting components from the surface of cells covered by a cell wall. This shaving effect may be relevant in biomedicine as a method to obtain antigenic biomolecules from the surface of pathogenic fungi, such as *Cryptococcus neoformans*, *Candida* spp., and *Sporothrix schenckii*. To the best of our knowledge, this is the first article reporting shock wave-induced changes to the conidial cell wall. Furthermore, our results may be useful for establishing the physical parameters and experimental conditions of shock waves in new fields of application.

Supplementary Materials: The following supporting information can be downloaded at: https://www.mdpi.com/article/10.3390/fluids9040081/s1, Figure S1: SEM images at a lower magnification (×10,000) compared to Figure 3, providing a wider field of view. The shock wave treatment type (Control, SP, SDT, LDT) is indicated in the upper right corner of each image. Samples for SP, SDT, and LDT were subjected to 50 shock waves.

Author Contributions: Conceptualization, B.E.M.-C., F.F. and A.M.L.; methodology, M.A.M.-M., B.E.M.-C., F.F., D.L. and A.M.L.; validation, M.A.G.-L.; formal analysis, B.E.M.-C., F.F., M.A.G.-L. and A.M.L.; investigation, M.A.M.-M., B.E.M.-C., F.F., D.L. and A.M.L.; resources, B.E.M.-C., M.A.G.-L. and A.M.L.; writing—original draft preparation, M.A.M.-M., B.E.M.-C., D.L. and A.M.L.; writing—review and editing, B.E.M.-C., F.F. and A.M.L.; visualization, B.E.M.-C., F.F., M.A.G.-L. and A.M.L.; supervision, B.E.M.-C., F.F. and A.M.L.; project administration, B.E.M.-C., F.F. and A.M.L.; funding acquisition, M.A.G.-L. and A.M.L. All authors have read and agreed to the published version of the manuscript.

Funding: This research was funded by the "Consejo Nacional de Ciencia y Tecnología" (CONACYT), grants number 372 and CB-2017-2018-A1-S-21130, and a scholarship from the Postgraduate Studies Support Program (PAEP) of the "Universidad Nacional Autónoma de México" (UNAM).

Data Availability Statement: Data are contained within the article.

Acknowledgments: The authors would like to thank Manuel Aguilar Franco, Paula Bernardino, Gerardo Fonseca, Lourdes Palma Tirado, René Preza, Ana Lucía Tovar and Guillermo Vázquez for technical assistance.

Conflicts of Interest: The authors declare no conflicts of interest.

References

1. Field, J.E. The physics of liquid impact, shock wave interactions with cavities, and the implications to shock wave lithotripsy. *Phys. Med. Biol.* **1991**, *36*, 1475–1484. [CrossRef]
2. Chaussy, C.; Brendel, W.; Schmiedt, E. Extracorporeally induced destruction of kidney stones by shock waves. *Lancet* **1980**, *316*, 1265–1268. [CrossRef]
3. Chaussy, C.; Schmiedt, E.; Jocham, B.; Brendel, W.; Forssmann, B.; Walther, V. First clinical experience with extracorporeally induced destruction of kidney stones by shock waves. *J. Urol.* **1982**, *127*, 417–420. [CrossRef]
4. Philipp, A.; Delius, M.; Scheffczyk, C.; Vogel, A.; Lauterborn, W. Interaction of lithotripter-generated shock waves with air bubbles. *J. Acoust. Soc. Am.* **1993**, *93*, 2496–2509. [CrossRef]
5. Lauterborn, W.; Ohl, C.D. The peculiar dynamics of cavitation bubbles. *Appl. Sci. Res.* **1998**, *58*, 63–76. [CrossRef]
6. Brujan, E.A.; Ikeda, T.; Matsumoto, Y. On the pressure of cavitation bubbles. *Exp. Thermal Fluid Sci.* **2008**, *32*, 1188–1191. [CrossRef]
7. Klaseboer, E.; Fong, S.W.; Turangan, C.K.; Khoo, B.C.; Szeri, A.J.; Calvisi, M.L.; Sankin, G.N.; Zhong, P. Interaction of lithotripter shockwaves with single inertial cavitation bubbles. *J. Fluid Mech.* **2007**, *593*, 33–56. [CrossRef]
8. Choi, M.J.; Coleman, A.J.; Saunders, J.E. The influence of fluid properties and pulse amplitude on bubble dynamics in the field of a shock wave lithotripter. *J. Phys. Med. Biol.* **1993**, *38*, 1561–1573. [CrossRef]
9. Prosperetti, A.; Lezzi, A. Bubble dynamics in a compressible liquid. Part 1. First-order theory. *J. Fluid Mech.* **1986**, *168*, 457–478. [CrossRef]
10. Zhong, P.; Lin, H.; Xi, X.; Zhu, S.; Bhogte, E.S. Shock wave-inertial microbubble interaction: Methodology, physical characterization, and bioeffect study. *J. Acoust. Soc. Am.* **1999**, *105*, 1997–2009. [CrossRef]
11. Coleman, A.J.; Saunders, J.E.; Crum, L.A.; Dyson, M. Acoustic cavitation generated by an extracorporeal shock wave lithotripter. *Ultrasound Med. Biol.* **1987**, *13*, 69–76. [CrossRef]

12. Crum, L.A. Cavitation microjets as a contributory mechanism for renal calculi disintegration in ESWL. *J. Urol.* **1988**, *140*, 1587–1590. [CrossRef]
13. Delacrétaz, G.; Rink, K.; Pittomvils, G.; Lafaut, J.P.; Vandeursen, H.; Boving, R. Importance of the implosion of ESWL-induced cavitation bubbles. *Ultrasound Med. Biol.* **1995**, *21*, 97–103. [CrossRef]
14. Zhong, P.; Cocks, F.H.; Cioanta, I.; Preminger, G.M. Controlled, forced collapse of cavitation bubbles for improved stone fragmentation during shock wave lithotripsy. *J. Urol.* **1997**, *158*, 2323–2328. [CrossRef]
15. Cathignol, E.; Tavakkoli, J.; Birer, A.; Arefiev, A. Comparison between the effects of cavitation induced by two different pressure-time shock waveform pulses. *IEEE Trans. Ultrason. Ferroelectr. Freq. Control* **1998**, *45*, 788–799. [CrossRef]
16. Xi, X.; Zhong, P. Improvement of stone fragmentation during shock-wave lithotripsy using a combined EH/PEAA shock-wave generator-in vitro experiments. *Ultrasound Med. Biol.* **2000**, *26*, 457–467. [CrossRef]
17. Loske, A.M.; Prieto, F.E.; Fernández, F.; van Cauwelaert, J. Tandem shock wave cavitation enhancement for extracorporeal lithotripsy. *Phys. Med. Biol.* **2002**, *47*, 3945–3957. [CrossRef]
18. Loske, A.M.; Fernández, F.; Zendejas, H.; Paredes, M.; Castaño-Tostado, E. Dual pulse shock wave lithotripsy: In vitro and in vivo study. *J. Urol.* **2005**, *174*, 2388–2392. [CrossRef]
19. Fernández, F.; Fernández, G.; Loske, A.M. Treatment time reduction using tandem shockwaves for lithotripsy: An in vivo study. *J. Endourol.* **2009**, *23*, 1247–1253. [CrossRef]
20. Ohl, C.D.; Ikink, R. Shock-wave-induced jetting of micron-size bubble. *Phys. Rev. Lett.* **2003**, *90*, 214502. [CrossRef]
21. Kodama, T.; Takayama, K. Dynamic behavior of bubbles during extracorporeal shock-wave lithotripsy. *Ultrasound Med. Biol.* **1998**, *24*, 723–738. [CrossRef]
22. Tzanakis, I.; Eskin, D.G.; Georgoulas, A.; Fytanidis, D.K. Incubation pit analysis and calculation of the hydrodynamic impact pressure from the implosion of an acoustic cavitation bubble. *Ultrason. Sonochem.* **2014**, *21*, 866–878. [CrossRef]
23. Loske, A.M.; Campos-Guillen, J.; Fernández, F.; Castaño-Tostado, E. Enhanced shock wave-assisted transformation of *Escherichia coli*. *Ultrasound Med. Biol.* **2011**, *37*, 502–510. [CrossRef]
24. Magaña-Ortíz, D.; Coconi-Linares, N.; Ortiz-Vazquez, E.; Fernández, F.; Loske, A.M.; Gómez-Lim, M.A. A novel and highly efficient method for genetic transformation of fungi employing shock waves. *Fungal Genet. Biol.* **2013**, *56*, 9–16. [CrossRef]
25. Molina, G.A.; González-Fuentes, F.; Loske, A.M.; Fernández, F.; Estevez, M. Shock wave-assisted extraction of phenolic acids and flavonoids from *Eysenhardtia polystachya* heartwood: A novel method and its comparison with conventional methodologies. *Ultrason. Sonochem.* **2020**, *61*, 104809. [CrossRef]
26. Church, C.C. A Theoretical study of cavitation generated by an extracorporeal shock wave lithotripter. *J. Acoust. Soc. Am.* **1989**, *86*, 215–227. [CrossRef]
27. Cairns, T.C.; Barthel, L.; Meyer, V. Something old, something new: Challenges and developments in *Aspergillus niger* biotechnology. *Essays Biochem.* **2021**, *65*, 213–224. [CrossRef]
28. Meyer, V.; Wu, B.; Ram, A.F.J. *Aspergillus* as a multi-purpose cell factory: Current status and perspectives. *Biotechnol. Lett.* **2011**, *33*, 469–476. [CrossRef]
29. Rose, S.H.; van Zyl, W.H. Exploitation of *Aspergillus niger* for the heterologous production of cellulases and hemicellulases. *Open Biotechnol. J.* **2008**, *2*, 167–175. [CrossRef]
30. Fleißner, A.; Dersch, P. Expression and export: Recombinant protein production systems for *Aspergillus*. *Appl. Microbiol. Biot.* **2010**, *87*, 1255–1270. [CrossRef]
31. Sun, X.; Wu, H.; Zhao, G.; Li, Z.; Wu, X.; Liu, H.; Zheng, Z. Morphological regulation of *Aspergillus niger* to improve citric acid production by *chsC* gene silencing. *Bioproc. Biosyst. Eng.* **2018**, *41*, 1029–1038. [CrossRef]
32. van Leeuwe, T.M.; Gerritsen, A.; Arentshorst, M.; Punt, P.J.; Ram, A.F.J. Rab GDP-dissociation inhibitor gdiA is an essential gene required for cell wall chitin deposition in *Aspergillus niger*. *Fungal Genet. Biol.* **2020**, *136*, 103319. [CrossRef]
33. de Groot, P.W.J.; Brandt, B.W.; Horiuchi, H.; Ram, A.F.J.; de Koster, C.G.; Klis, F.M. Comprehensive genomic analysis of cell wall genes in *Aspergillus nidulans*. *Fungal Genet. Biol.* **2009**, *46*, S72–S81. [CrossRef]
34. Gow, N.A.R.; Latge, J.P.; Munro, C.A. The fungal cell wall: Structure, biosynthesis, and function. *Microbiol. Spectr.* **2017**, *5*, 1–25. [CrossRef]
35. Ruiz-Herrera, J.; Ortiz-Castellanos, L. Cell wall glucans of fungi. A review. *Cell Surface* **2019**, *5*, 100022. [CrossRef]
36. Olaya-Abril, A.; Jiménez-Munguía, I.; Gómez-Gascón, L.; Rodríguez-Ortega, M.J. Surfomics: Shaving live organisms for a fast proteomic identification of surface proteins. *J. Proteom.* **2014**, *97*, 164–176. [CrossRef]
37. Lichius, A.; Ruiz, D.M.; Zeilinger, S. Genetic Transformation of Filamentous Fungi: Achievements and Challenges. In *Grand Challenges in Biology and Biotechnology*; Nevalainen, H., Ed.; Springer: Cham, Switzerland, 2020; Volume Grand Challenges in Fungal Biotechnology, pp. 123–164. [CrossRef]
38. Fang, T.H.; Kang, S.H.; Hong, Z.H.; Wu, C.D. Elasticity and nanomechanical response of *Aspergillus niger* spores using atomic force microscopy. *Micron* **2012**, *43*, 407–411. [CrossRef]
39. Wargenau, A.; Fleißner, A.; Bolten, C.J.; Rohde, M.; Kampen, I.; Kwade, A. On the origin of the electrostatic surface potential of *Aspergillus niger* spores in acidic environments. *Res. Microbiol.* **2011**, *162*, 1011–1017. [CrossRef]
40. Pihet, M.; Vandeputte, P.; Tronchin, G.; Renier, G.; Saulnier, P.; Georgeault, S.; Bouchara, J.P. Melanin is an essential component for the integrity of the cell wall of *Aspergillus fumigatus* conidia. *BMC Microbiol.* **2009**, *9*, 177. [CrossRef]

41. Koshiyama, K.; Kodama, T.; Yano, T.; Fujikawa, S. Structural change in lipid bilayers and water penetration induced by shock waves: Molecular dynamics simulations. *Biophys. J.* **2006**, *91*, 2198–2205. [CrossRef]
42. Kodama, T.; Hamblin, M.R.; Doukas, A.G. Cytoplasmic molecular delivery with shock waves: Importance of impulse. *Biophys. J* **2000**, *79*, 1821–1832. [CrossRef]
43. Kuraya, E.; Miyafuji, Y.; Takemoto, A.; Itoh, S. The effect of underwater shock waves on steam distillation of *Alpinia zerumbet* leaves. *Trans. Mater. Res. Soc. Jpn.* **2014**, *39*, 447–449. [CrossRef]
44. López-Marín, L.M.; Millán-Chiu, B.E.; Castaño-González, K.; Aceves, C.; Fernández, F.; Varela-Echavarría, A.; Loske, A.M. Shock wave-induced damage and poration in eukaryotic cell membranes. *J. Membr. Biol.* **2017**, *250*, 41–52. [CrossRef]
45. Liauw, C.M.; Slate, A.J.; Butler, J.A.; Wilson-Nieuwenhuis, J.S.; Deisenroth, T.; Preuss, A.; Whitehead, K.A. The effect of surface hydrophobicity on the attachment of fungal conidia to substrates of polyvinyl acetate and polyvinyl alcohol. *J. Polym. Environ* **2020**, *28*, 1450–1464. [CrossRef]
46. Arora, M.; Ohl, C.D.; Mørch, K.A. Cavitation inception on microparticles: A self-propelled particle accelerator. *Phys. Rev. Lett.* **2004**, *92*, 174501. [CrossRef] [PubMed]
47. Newton Harvey, E.; Barnes, D.K.; McElroy, W.D.; Whiteley, A.H.; Pease, D.C.; Cooper, K.W. Bubble formation in animals. I Physical factors. *J. Cell. Comp. Physiol.* **1944**, *24*, 1–22. [CrossRef]
48. Newton Harvey, E.; McElroy, W.D.; Whiteley, A.H. On cavity formation in water. *J. Appl. Phys.* **1947**, *18*, 162–172. [CrossRef]
49. Kfoury, R.; Marzban, B.; Makki, E.; Greenfield, M.L.; Yuan, H. Effect of pressure profile of shock waves on lipid membrane deformation. *PLoS ONE* **2019**, *14*, e0212566. [CrossRef] [PubMed]
50. de Icaza-Herrera, M.; Fernández, F.; Loske, A.M. Combined short and long-delay tandem shock waves to improve shock wave lithotripsy according to the Gilmore–Akulichev theory. *Ultrasonics* **2015**, *58*, 53–59. [CrossRef] [PubMed]
51. Li, D.; Pellegrino, A.; Hallack, A.; Petrinic, N.; Jérusalem, A.; Cleveland, R.O. Response of single cells to shock waves and numerically optimized waveforms for cancer therapy. *Biophys. J.* **2018**, *114*, 1433–1439. [CrossRef]
52. Kodama, T.; Tomita, Y. Cavitation bubble behavior and bubble–shock wave interaction near a gelatin surface as a study of in vivo bubble dynamics. *Appl. Phys. B* **2000**, *70*, 139–149. [CrossRef]
53. Qin, P.; Han, T.; Alfred, C.; Xu, L. Mechanistic understanding the bioeffects of ultrasound-driven microbubbles to enhance macromolecule delivery. *J. Control. Release* **2018**, *272*, 169–181. [CrossRef] [PubMed]
54. van Wamel, A.; Kooiman, K.; Harteveld, M.; Emmer, M.; ten Cate, F.J.; Versluis, M.; de Jong, N. Vibrating microbubbles poking individual cells: Drug transfer into cells via sonoporation. *J. Control. Release* **2006**, *112*, 149–155. [CrossRef] [PubMed]
55. Li, W.; Juan, T.; Guo, X.S.; Di, X.; Dong, Z. Microstreaming velocity field and shear stress created by an oscillating encapsulated microbubble near a cell membrane. *Chin. Phys. B* **2014**, *23*, 124302. [CrossRef]
56. Le Gac, S.; Zwaan, E.; van den Berg, A.; Ohl, C.D. Sonoporation of suspension cells with a single cavitation bubble in a microfluidic confinement. *Lab Chip* **2007**, *7*, 1666–1672. [CrossRef]
57. Marmottant, P.; Hilgenfeldt, S. Controlled vesicle deformation and lysis by single oscillating bubbles. *Nature* **2003**, *423*, 153–156. [CrossRef]
58. Zhao, L.; Schaefer, D.; Marten, M.R. Assessment of elasticity and topography of *Aspergillus nidulans* spores via atomic force microscopy. *Appl. Environ. Microb.* **2005**, *71*, 955–960. [CrossRef]
59. Hernáez, M.L.; Ximénez-Embún, P.; Martínez-Gomariz, M.; Gutiérrez-Blázquez, M.D.; Nombela, C.; Gil, C. Identification of *Candida albicans* exposed surface proteins in vivo by a rapid proteomic approach. *J. Proteom.* **2010**, *73*, 1404–1409. [CrossRef]
60. Karkowska-Kuleta, J.; Satala, D.; Bochenska, O.; Rapala-Kozik, M.; Kozik, A. Moonlighting proteins are variably exposed at the cell surfaces of *Candida glabrata*, *Candida parapsilosis* and *Candida tropicalis* under certain growth conditions. *BMC Microbiol.* **2019**, *19*, 149. [CrossRef] [PubMed]
61. Gow, N.A.R.; Lenardon, M.D. Architecture of the dynamic fungal cell wall. *Nat. Rev. Microbiol.* **2023**, *21*, 248–259. [CrossRef] [PubMed]
62. Gil-Bona, A.; Parra-Giraldo, C.M.; Hernáez, M.L.; Reales-Calderon, J.A.; Solis, N.V.; Filler, S.G.; Gil, C. *Candida albicans* cell shaving uncovers new proteins involved in cell wall integrity, yeast to hypha transition, stress response and host–pathogen interaction. *J. Proteom.* **2015**, *127*, 340–351. [CrossRef] [PubMed]
63. Yin, Q.Y.; de Groot, P.W.; de Koster, C.G.; Klis, F.M. Mass spectrometry-based proteomics of fungal wall glycoproteins. *Trends Microbiol.* **2008**, *16*, 20–26. [CrossRef] [PubMed]
64. Wang, H.Y.; Hua, X.W.; Jia, H.R.; Li, C.; Lin, F.; Chen, Z.; Wu, F.G. Universal cell surface imaging for mammalian, fungal, and bacterial cells. *ACS Biomater. Sci. Eng.* **2016**, *2*, 987–997. [CrossRef] [PubMed]

fluids MDPI

Article

Ultrasonic Bubble Cleaner as a Sustainable Solution

Austin Howell [1], Emerson Ham [1] and Sunghwan Jung [2,*]

[1] Department of Biomedical Engineering and Mechanics, Virginia Tech, Blacksburg, VA 24061, USA
[2] Department of Biological and Environmental Engineering, Cornell University, Ithaca, NY 14853, USA
* Correspondence: sj737@cornell.edu

Abstract: We aim to develop a floor-cleaning design by exploiting oscillating bubbles combined with ambient pressure waves to clean various surfaces. Previous studies of this method in lab settings have proven its efficacy, but practical applications, especially concerning real-world conditions like dirt surfaces, remain largely unprobed. Our findings indicate that, excluding a configuration with a heavy mass bottom transducer, all tested configurations achieved approximately 60–70% cleaning performance. A slight improvement in cleaning performance was observed with the introduction of microbubbles, although it was within the error margin. Particularly noteworthy is the substantial reduction in water consumption in configurations with a water pocket, decreasing from 280 mL to a mere 3 mL, marking a significant step toward more environmentally sustainable cleaning practices, such as reduced water usage. This research provides implications for real-world cleaning applications, promising an eco-friendly and efficient cleaning alternative that reduces water usage and handles a variety of materials without causing damage.

Keywords: acoustic pressure; cleaning; soil dirt; bubbles; cavitation

1. Introduction

Pressure-wave cleaning typically involves transducer-induced pressure waves in a liquid [1]. Some pressure waves will induce the creation of bubbles in a liquid, a process referred to as cavitation [2]. Ultrasonic waves induce a number of unstable cavitation bubbles that will collapse and generate a very powerful jet toward solid boundaries [3,4]. Studies have explored temperature [5–8], geometry [9–14], and elasticity [15–21] as critical factors to manage the intensity and dynamics of cavitation. This strong jet and shock wave kills the bacteria [22,23] or further erodes the solid surfaces [24–26] rather than gently cleans/scrubs adhered particles. Oftentimes, the damage caused by cavitation is so pronounced that solid structures are significantly eroded or completely destroyed [24–27].

There are three different regimes of pressure waves in fluids. Acoustic waves are audible and range up to around 25 kHz. Above those, there are ultrasonic and megasonic waves. Typically, in industry, the frequencies used for ultrasonic cleaning are between 25 kHz and 300 kHz. For megasonic cleaning, much higher frequencies are used that typically range from 500 kHz up to 10 MHz. Typically, both ultrasonic and megasonic pressure waves are generated by piezo-transducers, with costs ranging from one to ten USD depending on the size and quality of the transducers.

Typically, ultrasonic cleaning is not preferred on brittle or otherwise vulnerable solids as they may suffer damage or scars in the process. When megasonic transducers are used, the higher frequency generates smaller wavelengths, resulting in fewer and smaller cavitation bubbles [28].

In terms of floor cleaning, both ultrasonic and megasonic waves can be used. Ultrasonic waves exhibit a powerful cleaning action, though surface damage can be significant. On the other hand, megasonic waves produce fewer and smaller cavitation bubbles, resulting in little to no surface damage. However, the effectiveness of cleaning is questionable due to the presence of fewer bubbles in a liquid. These observations support our main goal of combining megasonic waves with pre-existing bubbles.

Using bubbles to clean surfaces has proven to be useful in the food industry to clean fruits and vegetables [29–34]. Typically, fruits or vegetables are placed in a water bath, and bubbles are injected at the bottom of the tank. Bubbles will then rise, impact, and scrub the fruits by sliding along the surface. The main benefit of cleaning using injected bubbles is that bubbles are highly deformable, so the bubbles can go into small crevices or grooves to clean. The surface tension of the bubbles plays a crucial role, enabling them to adapt to the surface topography and ensuring thorough cleaning.

There are two problems to consider when applying this technique to floor cleaning applications: rising bubbles and limited cleaning power. First, due to buoyancy, bubbles rise up to the top surface, which makes it hard to keep them in contact with the surface to be scrubbed in a floor cleaning application. One way to rectify this rising-bubble problem is to utilize external pressure to push the bubbles to the targeted area. The second problem is that slowly rising bubbles do not produce strong cleaning forces (i.e., shear forces). To powerfully clean stained/dirty surfaces, external excitation is required to create strong shear forces and streaming circulations to efficiently scrub surfaces.

Since using pressure waves or bubbles alone shows limited capability when cleaning surfaces, what happens when we combine them? A method using pre-dispersed gas bubbles with acoustic waves has been used in the semiconductor industry [35]. This method has been proven an effective means to remove particulate contaminants from semiconductor surfaces. The acoustically activated bubbles will stay close to the liquid–surface interface and vibrate at a high frequency. This, in turn, generates small vortices (micro streaming) and strong shear forces [36–40], which cause the removal of particles from the surface without otherwise damaging it. Currently, this technology has been applied to the semiconductor industry to liberate particulate contaminants on small surface areas (silicon wafers or CPU chips). However, this combined method should work for cleaning large-scale floors and adhered strains/dirt, an application that has not been investigated extensively.

In this paper, we aim to test a method for cleaning soiled floors using ultrasonic piezo transducers. To integrate the cavitation method into surface cleaning applications effectively, we encounter several challenges. Firstly, the method traditionally requires a fully immersed liquid bath, which is impractical due to water usage constraints. Secondly, a comprehensive comparison of its cleaning performance relative to conventional methods, such as brushes, is yet to be conducted.

These transducers in our study oscillate back and forth with enough power to create tiny cavitation bubbles in water, which appear and collapse periodically. The collapsing of these bubbles can create very high shock-wave forces, which presumably can be used to shake loose embedded dirt and soil in flooring. The benefit of this system is that it could potentially be used in combination with other cleaning systems. Further benefits include that cavitation can work down into cracks and ridges and does not require cleaning soap solutions to operate. We test the cleaning performance of this system in combination with a metal brush, with no brush (direct transducer contact with water only), and with a water pocket. For each of these setups, we performed scrubbing tests with the transducer and micro-bubbler off, with the transducer on, and with the transducer and micro-bubbler on to help isolate the effects of each of these systems. Finally, we will present the results of different configurations.

2. Materials and Methods

Our experimental setup, as shown in Figure 1, is composed of five separate parts, many of which were modified at different points throughout the development of this system.

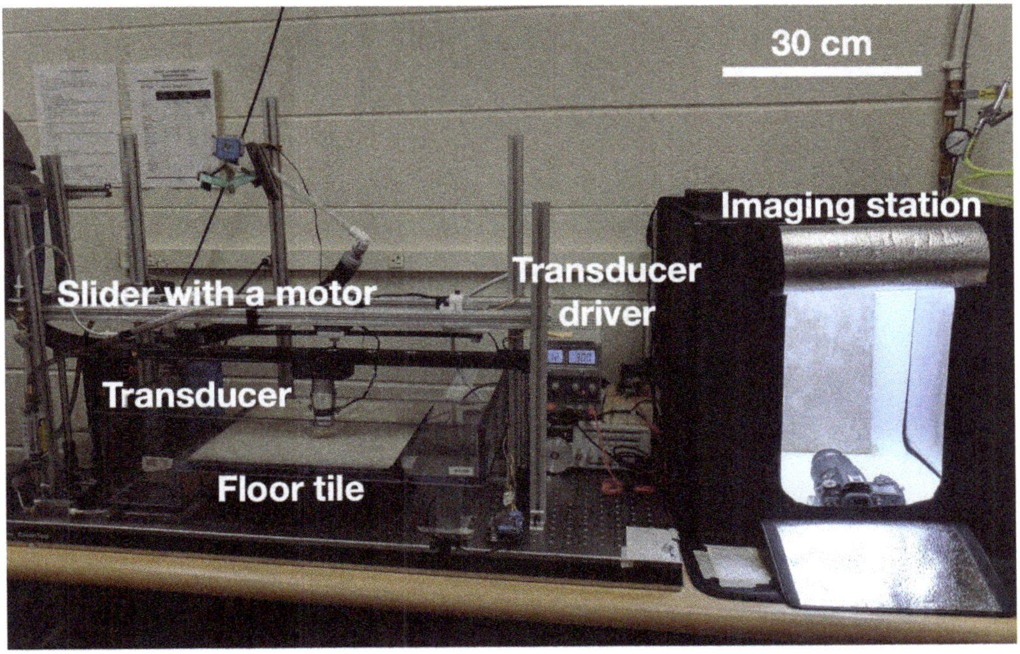

Figure 1. Experimental setup: a transducer, a driver, a transducer slider, and a test floor tile.

The first is the transducer driver system, which is composed of a signal generator, a power amplifier, a variable power supply, and the transducer itself. The signal generator is a GFG-8216A (GW Instek Co., New Taipei City, Taiwan), which generates a low-voltage frequency signal. This signal is amplified using a MX200 piezo driver (PiezoDrive Co., Shortland, Australia) to an output power of approximately 30 watts. This is connected directly to the 40 KHz Bolt Clamped Langevin Transducers (STEMiNC Inc., Davenport, FL, USA) we used in our experiments. The transducer manufacturer positioned a holding connection on the light mass side rather than on the heavy mass. Nevertheless, we examined two distinct configurations: one with a heavy mass at the bottom (Figure 2a) and another with a light mass at the bottom (Figure 2b). Vibration tests revealed that the configuration with a lighter mass at the bottom produces a more pronounced vibration than its heavy-mass counterpart. We prepared separate transducers for the plastic bristle brush, the brass bristle brush (Figure 2c), and the water pocket (Figure 2(d1,d2)) experiments. The ultimate design features a water pocket with a rubber enclosure integrated around a transducer. This enclosure serves dual functions; it acts as a squeegee and also confines water, minimizing water loss from the cylindrical structure. Due to the added mass of the brush heads, each transducer had to be tuned individually, and all were driven with an input power of 30 watts for our tests.

The second main part of the experimental setup is the scrubbing system based around a SLIDECAM V 800 camera slider (Varavon Inc., Gimpo, South Korea). This has a screw-on slider mount, which we used to hold the transducers. To automate the scrubbing process and to reduce variability between samples, we used a stepper motor and a belt drive to move the transducer/scrubber back and forth. The stepper motor is driven by an Arduino with a stepper motor shield for simplicity of use. The slider is programmed to move at a pace that completes one path in 20 s.

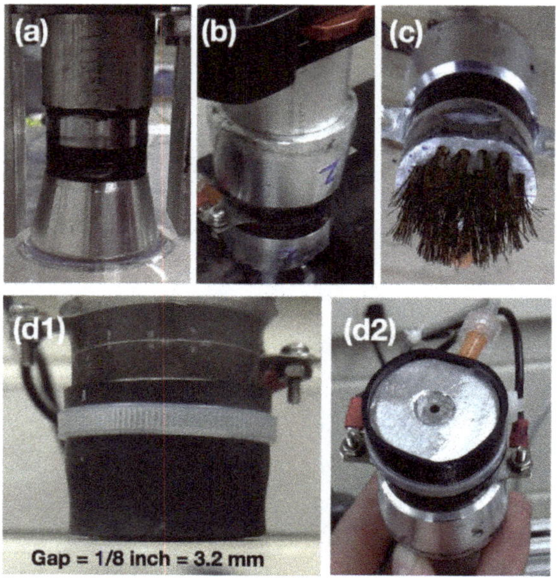

Figure 2. Four configurations of experiments: (**a**) a heavy mass on the bottom, (**b**) a light mass on the bottom, (**c**) metal brushes on the bottom, (**d**) a water pocket around the transducer (**d1**) from the side view and (**d2**) from the bottom view.

The third major component of the experimental setup is a custom micro-bubble generator, as shown in Figure 3. Commercially available micro-bubble generators were prohibitively expensive and did not provide the low flow rates we desired since most commercial systems are designed for industrial-scale use. In the end, we were able to create our own micro-bubble generator capable of making bubbles as small as 200 microns at total flow rates of about 110 mL per minute. Measurements from a camera, while a thin laser sheet illuminated a plane, may lack precision as some bubbles deform into non-spherical shapes. The final micro-bubble generator design consists of an adjustable water pump, an air pump, and a vortex mixing device. It creates bubbles at an air/water volume ratio of about 5% (Figure 4). A short and narrow hose runs from the device directly to the base of the transducer, where the micro-bubbles are injected.

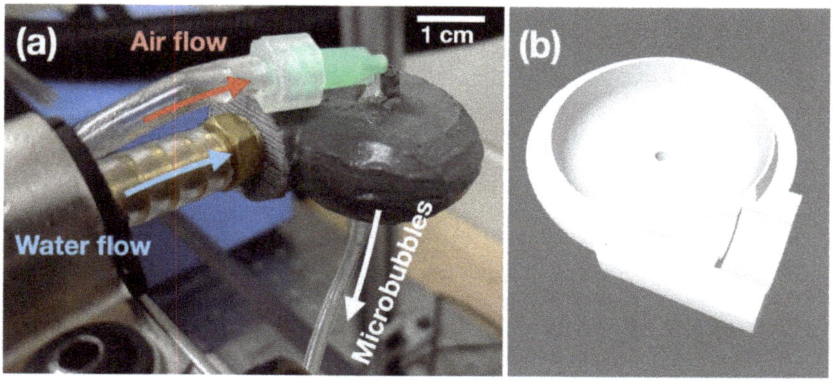

Figure 3. Micro-bubble generator. (**a**) As both air and water are fed into a vortex circulator, micro-bubbles are generated through the tube on the bottom. (**b**) CAD file to show the internal structure.

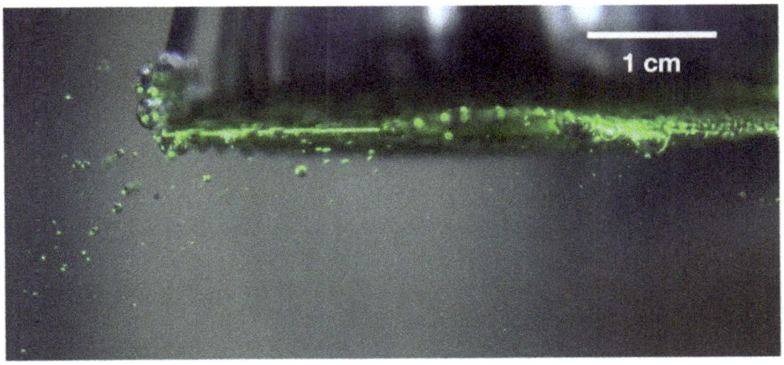

Figure 4. Micro-bubbles around a transducer.

An optional squeegee and vacuum system are the fourth main component of the experimental setup. There is one squeegee on each side of the transducer since it slides in both directions, and a servo moves the desired one into a position coordinated with the movement of the slider system. We decided that this system was another source of potential noise in our data, so we elected to remove it during our primary data collection; however, we did observe a few interesting behaviors with this system, which we will discuss in our results.

Additionally, we were interested in the effects of physically implanting the brush bristles into the end of the transducer. We tested two methods of securing the brush bristles. The first was a JB weld epoxy, and the second method used a small length of brass rod that would be punched into a drilled hole, as shown in Figure 5.

Figure 5. Process for mechanically securing brass bristles into a transducer head using brass.

The last overall system of our experimental setup was a method for reliably measuring the effectiveness of our cleaning mechanism. We used protein-based soil (2.5 g Nigrosin (A18147.22 from Thermo Scientific Chemicals, Waltham, MA, USA) + 190 g water + 220 g milk + 26 g white wheat flour). This is a very dark soil, so the European Cleaning Machines Association recommends measuring the darkness of the tile to measure how clean it is. We used a high-end digital camera on full manual settings to take photos of the floor tiles. Lighting conditions of our lab were held constant by using a purpose-built LED lighting box for all our pictures. We measured the darkness of each tile when it was completely clean, both after it was soiled and after it was scrubbed, for every test we performed. The darkness measuring system was automated using MATLAB 2017b, which corrects for any differences in the exposure brightness and averages the darkness of the entire scrubbed area to help minimize the variation in the cleaned area. To obtain a "percent cleaned" numerical result for each test, we mapped the measured darkness of the tile after scrubbing from a range of [when soil was applied, when the tile was fully clean] to a range of [0%, 100%] using this formula:

$$\% \text{ Clean} = \frac{(\text{cleaned} - \text{dirty})}{(\text{before} - \text{dirty})} \times 100\%$$

where 'cleaned' refers to the level of darkness of the soil-dirt tile after a cleaning test has been conducted, while 'dirty' denotes the darkness level of the soil-dirt tile before the cleaning test. The term 'before' is used to represent the darkness level of a clean tile surface. The output of this formula is as follows: 100% means the tile was completely cleaned, and 0% means the tile was just as visibly dirty as it was before cleaning. In this study, cleaning performance is quantified based on proficiency in contaminant removal from surfaces. We utilize images of cleaned surfaces and the cleaning percentage to measure the surface cleaning performance. While efficiency can also be delineated through the mechanical power input and output, we do not explore mechanical efficiency in this study.

3. Results

The amount of cavitation formed is dependent on the vibration amplitude of the transducer. Ultrasonic transducers are mass-spring-mass resonant systems and, therefore, operate at significantly higher output when tuned to resonance. To quantify the surface amplitude, we employed a surface vibrometer (OFV 5000/505, Polytec Inc., Irvine, CA, USA). This measurement was performed in air, not in water, due to the substantial mass of the transducer; we surmise that its amplitude remains consistent whether in air or water. We drove the transducer at a range of frequencies from 10 kHz to 40 kHz and found the resonant peaks. We ran the sweep at both 10 V and 100 V to observe the effects of nonlinear dynamics on the vibrating system, as shown in Figure 6.

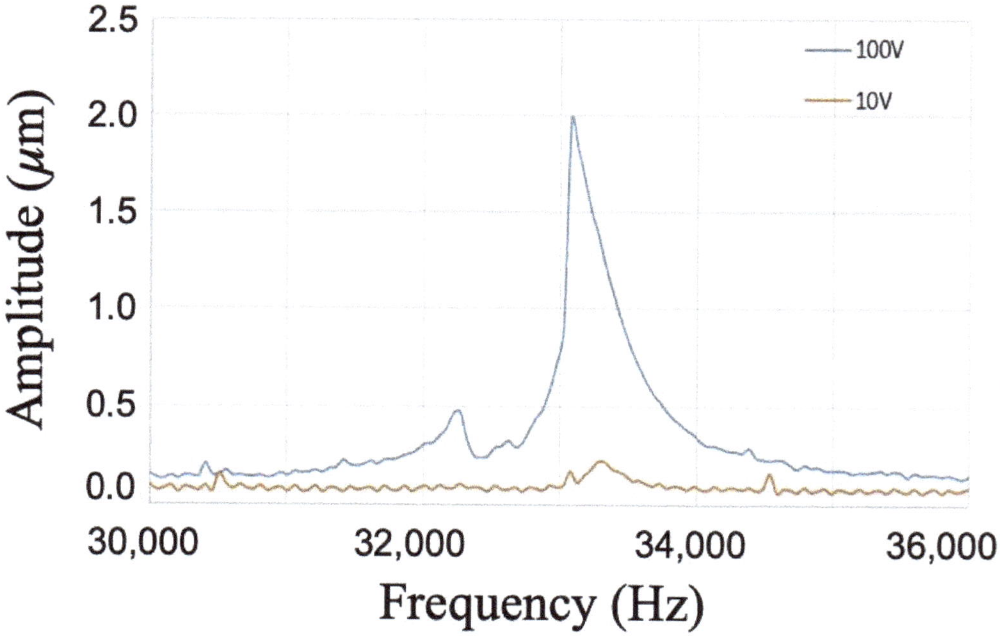

Figure 6. Amplitude of vibration vs. frequency with a 40 kHz transducer.

We found that the added mass of the slider mechanism did decrease the resonant frequency of the transducer from 40,000 Hz to about 33,200 Hz. The peak visibly tilts to the left, showing nonlinear softening behavior in the resonant system. This is important to note when tuning the system as the resonant frequency will decrease as the amplitude is increased, meaning a standard low amplitude analysis will not be correct in determining the resonant frequency at full power. Correctly tuning the system provided a 14× increase in amplitude versus non-resonant frequencies. Since this process takes a lot of time, we were only able to perform this detail of a resonance analysis for this 40 kHz transducer once with a heavy mass on the bottom, as shown in Figure 2a.

Figure 7 shows how the protein soil has been cleaned with various methods. The upper panel shows a tile with the protein soil before the cleaning test, whereas the lower panel shows the tile after the cleaning test. As described above, when the heavy mass is placed on the bottom, the transducer does not create strong vibrations. Hence, the tile does not get cleaned after the test, as shown in Figure 7a. On the other hand, when the light mass is placed on the bottom, the tile gets cleaned quite well (see Figure 7b). With the metal brush, it cleans the title relatively well. Figure 7c shows that the dirt gets dispersed around the tile. Lastly, a transducer with a water pocket cleans the surface at the center region but leaves some dirt on the side. Even though the overall cleaning with the water pocket is not improved much, the water use is significantly reduced.

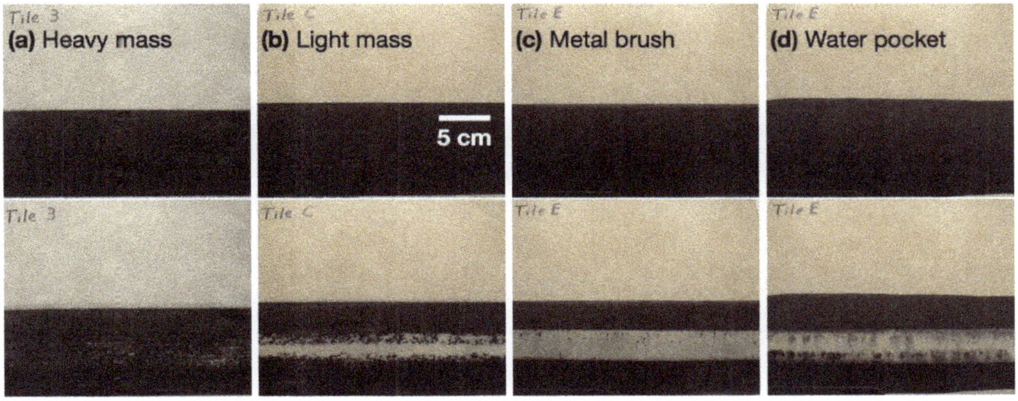

Figure 7. Cleaning test results. The upper panel is for a tile with a protein soil before the cleaning test, and the lower panel is after the test. Different transducers are used: (**a**) a heavy mass, (**b**) light mass, (**c**) metal brushes, and (**d**) a water pocket on the bottom.

Figure 8 presents a sequence of images captured from a side view, detailing the disintegration process of a dirt particle under the effect of ultrasonic waves. In the first two images, we observe the liberation of a small dirt particle from the surface. As time progresses from 3.7 ms to 20.3 ms, multiple cavitation processes act upon the particle, gradually breaking it down and forming a nebulous cloud of smaller dirt fragments. This complex transformation is not only a reduction in size but also the particle being pushed upwards into the bulk fluid. Ultimately, the small particle continues to decrease in size until it virtually disappears, leaving behind a clear surface at around 22.2 ms. This intriguing process is further illustrated in a high-speed video provided in the Supplementary Material.

One interesting finding is a crack pattern when a transducer with a light mass on the bottom passes by. Figure 9 shows the detailed structure/pattern of the removed protein soil. Such disjointed cleaned patterns are quite unique among all four configurations that we tested. We speculate that this pattern is a consequence of cavitation bubbles collapsing on the wall [41]. Since the cavitation nucleation and collapsing are a random process, the location of cavitation cleaning is stochastic.

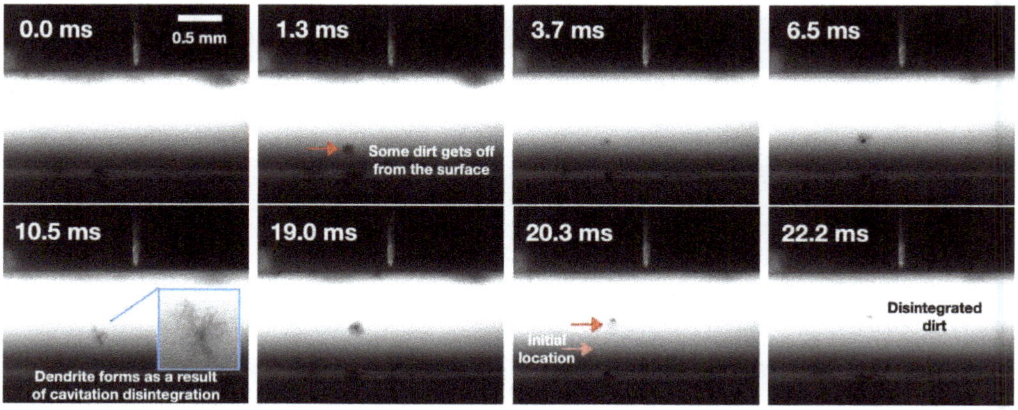

Figure 8. Sequence of images illustrating the disintegration process of a dirt particle under ultrasonic influence. The images capture the liberation, breakdown, and eventual disappearance of the particle as it is acted upon by cavitation processes over a timeframe from 3.7 ms to 20.3 ms. The process represents a complex interplay of forces that ultimately leads to a clear surface. A high-speed video detailing this phenomenon is available in the Supplementary Material.

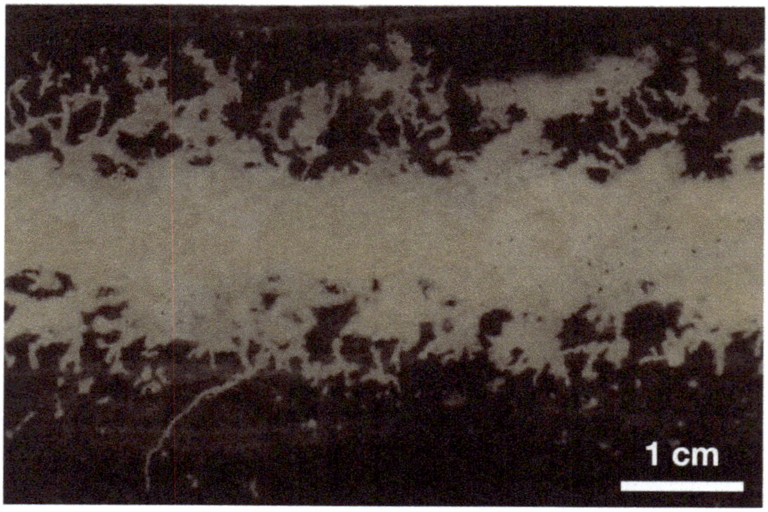

Figure 9. Crack pattern as an ultrasonic transducer with a light mass on the bottom passes by.

As indicated in Tables 1–5, the performance of a transducer equipped with a heavy mass on the bottom is markedly deficient, failing to effectively clean the surface even when the transducer is active and microbubbles are injected. In contrast, all other tested configurations yield comparable cleaning performance, achieving around 60–70% surface cleanliness within the measured error bars. With the injection of microbubbles, there is a slight increase in cleaning performance, observing a rise from 2% to 9% with a heavy mass bottom and from 52% to 55% with a light mass bottom. However, given the large error, we cannot conclude whether the microbubbles significantly enhance the cleaning process or not. Though the cleaning performance is similar across these other configurations, a noteworthy distinction lies in the water consumption. Specifically, the design incorporating a water pocket dramatically reduces water use. While the necessary volume for the water pocket is a mere 3 mL, other configurations demand a significantly higher water volume, approximately 280 mL. This underlines the value of the water pocket design not just in terms of cleaning effectiveness but also in environmental sustainability.

Table 1. Heavy mass on the bottom with two passes over title at 40.12 kHz and V_{max} 150 V.

Transducer Only	Avg. % Cleaned	Std.	Sample Number
Transducer Off	2%	3%	3
Transducer On	2%	7%	3
Microbubbles	9%	6%	3

Table 2. Light mass on the bottom with two passes over title at 40.12 kHz and V_{max} 150 V.

Transducer Only	Avg. % Cleaned	Std.	Sample Number
Transducer Off	−7%	1%	3
Transducer On	52%	5%	3
Microbubbles	55%	7%	3

Table 3. Metal brush with one pass over title at 40.12 kHz & V_{max} 150 V.

Transducer Only	Avg. % Cleaned	Std.	Sample Number
Transducer Off	25%	12%	3
Transducer On	69%	32%	3

Table 4. Water pocket with one pass over title at 40.12 kHz and V_{max} 150 V.

Transducer Only	Avg. % Cleaned	Std.	Sample Number
Transducer Off	0%	4%	3
Transducer On	60%	22%	3

Table 5. Central region of water pocket with one pass over title at 40.12 kHz and V_{max} 150 V.

Transducer Only	Avg. % Cleaned	Std.	Sample Number
Transducer Off	1%	15%	3
Transducer On	72%	15%	3

4. Discussion

This research has successfully explored a practical design in floor cleaning by combining ultrasonic waves with pre-existing bubbles in a confined water pocket. While extensive scientific studies have been conducted on the impact of single or multiple cavitation bubbles in terms of erosion and cleaning, there has been limited research dealing with real tiled surfaces laden with dirt, especially when compared with other cleaning methodologies. The cleaning performance we achieved, particularly in configurations involving water pockets, highlights the potential of our method to redefine conventional cleaning techniques. Our experiments reveal a cleaning performance ranging from 60 to 70%, presenting a significant advancement in the field. Perhaps the most striking finding is the reduction in water usage when utilizing a water pocket configuration, where the necessary volume decreased from 280 mL to a mere 3 mL. This not only supports cleaning performance but also represents a meaningful contribution to sustainability. Additionally, our method's ability to minimize surface damage in comparison with metal brushes offers a solution to a longstanding issue in ultrasonic cleaning, expanding the range of materials that can be safely cleaned. Moreover, the cost-effectiveness of the piezo-transducers used provides a potentially economical alternative for industrial cleaning applications.

Our study shows the promise of a method to clean surfaces through the combination of pressure waves with pre-existing bubbles, which offers promising results in terms of floor cleaning. This device does not require cleaning soap solutions to operate, which is an environmentally friendly option to clean surfaces. The study sheds light on an approach to conserve water during the cleaning process. The inception of a transducer embedded with a water pocket is quite novel, reducing water consumption significantly while maintaining optimum cleaning performance. It shows sustainable cleaning solutions, leading the way to eco-friendly advancements in cleaning technologies. The method's success in removing embedded dirt and soil from flooring, along with its applicability beyond the semiconductor industry, holds significant implications for large-scale floor cleaning and adhered strains/dirt removal.

Several tasks remain for future work. First, we plan to quantify and visualize the effect of injected bubbles. First, we need to develop a method to count the number of microbubbles, a task that presents substantial challenges due to two principal reasons. First, the bubbles are injected using a pump and quickly disperse in water. Many bubbles are pushed out from a thin gap between the transducer and the surface. Secondly, as the transducer slides along the surface, some bubbles manage to enter the small spaces, while others are forced out due to the motion. As of now, we have not observed any notable differences in the structures of cavitation.

Furthermore, it is crucial to identify resonance on different transducers. At present, we have determined the resonant frequency on a bare 40 kHz transducer and have applied this to other transducers, assuming that minor modifications will not alter its resonant frequency. However, confirming the resonant frequency of all the different transducers is essential to ensure the accuracy and reliability of our findings. Moreover, a thorough study should be conducted on acoustic impedance matching related to epoxy, drilled holes, and affixed metal brushes, as these modifications, depending on their magnitude and dimensions, could potentially alter the outcomes.

Supplementary Materials: The following supporting information can be downloaded at: https://www.mdpi.com/article/10.3390/fluids8110291/s1, Video S1: Supplementary video.mov.

Author Contributions: Conceptualization, S.J.; methodology, J.H., E.H. and S.J.; formal analysis, J.H. and E.H.; investigation, J.H. and S.J.; writing—original draft preparation, J.H., E.H. and S.J.; writing—review and editing, J.H., E.H. and S.J.; supervision, S.J.; project administration, S.J.; funding acquisition, S.J. All authors have read and agreed to the published version of the manuscript.

Funding: This research was partially funded by Nilfisk Co. and the NSF CBET-2002714.

Data Availability Statement: The data presented in this study are openly available at https://osf.io/yc8zg/ accessed on 26 October 2023 at DOI 10.17605/OSF.IO/YC8ZG.

Acknowledgments: The authors thank Dave Wood and Gita Rao at Nilfisk Co. for financial support.

Conflicts of Interest: The authors declare no conflict of interest. The funder had no role in the design of the study; in the collection, analyses, or interpretation of data; in the writing of the manuscript; or in the decision to publish the results.

References

1. Lamminen, M.O.; Walker, H.W.; Weavers, L.K. Mechanisms and factors influencing the ultrasonic cleaning of particle-fouled ceramic membranes. *J. Membr. Sci.* **2004**, *237*, 213–223. [CrossRef]
2. Brennen, C.E. *Cavitation and Bubble Dynamics*; Oxford University Press: Oxford, UK, 1995; Volume 44.
3. Wang, J.; Li, S.; Gu, J.; Zhang, A.-M. Particle propulsion from attached acoustic cavitation bubble under strong ultrasonic wave excitation. *Phys. Fluids* **2023**, *35*, 042009.
4. Li, S.; Zhang, A.-M.; Wang, S.; Han, R. Transient interaction between a particle and an attached bubble with an application to cavitation in silt-laden flow. *Phys. Fluids* **2018**, *30*, 082111. [CrossRef]
5. Ge, M.; Petkovšek, M.; Zhang, G.; Jacobs, D.; Coutier-Delgosha, O. Cavitation dynamics and thermodynamic effects at elevated temperatures in a small Venturi channel. *Int. J. Heat Mass Transf.* **2021**, *170*, 120970. [CrossRef]
6. Ge, M.; Sun, C.; Zhang, G.; Coutier-Delgosha, O.; Fan, D. Combined suppression effects on hydrodynamic cavitation performance in Venturi-type reactor for process intensification. *Ultrason. Sonochem.* **2022**, *86*, 106035. [CrossRef] [PubMed]

7. Ge, M.; Manikkam, P.; Ghossein, J.; Subramanian, R.K.; Coutier-Delgosha, O.; Zhang, G. Dynamic mode decomposition to classify cavitating flow regimes induced by thermodynamic effects. *Energy* **2022**, *254*, 124426. [CrossRef]
8. Ge, M.; Zhang, G.; Petkovšek, M.; Long, K.; Coutier-Delgosha, O. Intensity and regimes changing of hydrodynamic cavitation considering temperature effects. *J. Clean. Prod.* **2022**, *338*, 130470. [CrossRef]
9. Bremond, N.; Arora, M.; Ohl, C.-D.; Lohse, D. Cavitation on surfaces. *J. Phys. Condens. Matter* **2005**, *17*, S3603–S3608. [CrossRef]
10. Tomita, Y.; Robinson, P.B.; Tong, R.P.; Blake, J.R. Growth and collapse of cavitation bubbles near a curved rigid boundary. *J. Fluid Mech.* **2002**, *466*, 259–283. [CrossRef]
11. Gaze, M.J.; Emmony, D.C. Laser generated cavitation near a curved surface. In Proceedings of the 2000 Conference on Lasers and Electro-Optics Europe (Cat. No.00TH8505), Nice, France, 10–15 September 2000. [CrossRef]
12. Vanden-Broeck, J.-M. Numerical solutions for cavitating flow of a fluid with surface tension past a curved obstacle. *Phys. Fluids* **1984**, *27*, 2601–2603. [CrossRef]
13. Vanden-Broeck, J.-M. The influence of surface tension on cavitating flow past a curved obstacle. *J. Fluid Mech.* **1983**, *133*, 255–264. [CrossRef]
14. Poulain, S.; Guenoun, G.; Gart, S.; Crowe, W.; Jung, S. Particle motion induced by bubble cavitation. *Phys. Rev. Lett.* **2015**, *114*, 214501. [CrossRef] [PubMed]
15. Leonov, K.; Akhatov, I. Dynamics of an externally driven cavitation bubble in an elastic microconfinement. *Phys. Rev. E* **2021**, *104*, 015105. [CrossRef] [PubMed]
16. Wentzell, R.A. Dynamics of cavitation bubbles. *Can. J. Phys.* **1976**, *54*, 454–459. [CrossRef]
17. Brujan, E.-A.; Nahen, K.; Schmidt, P.; Vogel, A. Dynamics of laser-induced cavitation bubbles near an elastic boundary. *J. Fluid Mech.* **2001**, *433*, 251–281. [CrossRef]
18. Brujan, E.-A.; Nahen, K.; Schmidt, P.; Vogel, A. Dynamics of laser-induced cavitation bubbles near an elastic boundary used as a tissue phantom. In *AIP Conference Proceedings*; AIP: College Park, MD, USA, 2000. [CrossRef]
19. Xu, W.; Zhai, Y.; Luo, J.; Zhang, Q.; Li, J. Experimental study of the influence of flexible boundaries with different elastic moduli on cavitation bubbles. *Exp. Therm. Fluid Sci.* **2019**, *109*, 109897. [CrossRef]
20. Knapp, R.T. Recent investigations of the mechanics of cavitation and cavitation damage. *Wear* **1958**, *1*, 455. [CrossRef]
21. Vincent, O.; Marmottant, P.; Gonzalez-Avila, S.R.; Ando, K.; Ohl, C.-D. The fast dynamics of cavitation bubbles within water confined in elastic solids. *Soft Matter* **2014**, *10*, 1455–1461. [CrossRef]
22. Jyoti, K.; Pandit, A. Water disinfection by acoustic and hydrodynamic cavitation. *Biochem. Eng. J.* **2001**, *7*, 201–212. [CrossRef]
23. Carolyn, M. *Treatise on Materials Science and Technology*; Academic Press: Cambridge, MA, USA, 1979; Volume 16.
24. Azar, L. Cavitation in ultrasonic cleaning and cell disruption. *Control. Environ.* **2009**, *2*, 14–17.
25. Teran, L.A.; Laín, S.; Jung, S.; Rodríguez, S.A. Surface damage caused by the interaction of particles and a spark-generated bubble near a solid wall. *Wear* **2019**, *438–439*, 203076. [CrossRef]
26. Teran, L.A.; Rodríguez, S.A.; Laín, S.; Jung, S. Interaction of particles with a cavitation bubble near a solid wall. *Phys. Fluids* **2018**, *30*, 123304. [CrossRef]
27. Preece, C.M.; Macmillan, N.H. Erosion. *Annu. Rev. Mater. Sci.* **1977**, *7*, 95–121. [CrossRef]
28. Sinha, N. Megasonic Cleaning with Controlled Boundary Layer Thickness and Associated Systems and Methods. U.S. Patent 8156950, 9 January 2008.
29. Jung, S.; Eifert, J. Cleaning Fruits Using Micro-Bubbles. *Food Online*. April 2016. Available online: http://www.foodonline.com/doc/cleaning-fruits-using-micro-bubbles-0001 (accessed on 25 April 2016).
30. Esmaili, E.; Shukla, P.; Eifert, J.D.; Jung, S. Bubble impact on a tilted wall: Removing bacteria using bubbles. *Phys. Rev. Fluids* **2019**, *4*, 043603. [CrossRef]
31. Lee, J.J.; Eifert, J.D.; Jung, S.; Strawn, L.K. Cavitation Bubbles Remove and Inactivate *Listeria* and *Salmonella* on the Surface of Fresh Roma Tomatoes and Cantaloupes. *Front. Sustain. Food Syst.* **2018**, *2*, 61. [CrossRef]
32. Hooshanginejad, A.; Sheppard, T.J.; Manyalla, J.; Jaicks, J.; Jung, S. Cleaning effect of bubbles impacting tilted walls under acoustic waves. In Proceedings of the ASME FEDSM Fluids Engineering Division Summer Meeting, Toronto, ON, Canada, 3–5 August 2022; p. 86897.
33. Hooshanginejad, A.; Baskota, A.; Jung, S. Backflipping motion of air bubbles colliding with a tilted wall. *arXiv* **2022**, arXiv:2208.14486.
34. Hooshanginejad, A.; Sheppard, T.; Xu, P.; Manyalla, J.; Jaicks, J.; Esmaili, E.; Jung, S. Removing proteins or bacteria on a tilted surface using air bubbles. *arXiv* **2022**, arXiv:2206.03309.
35. Busnaina, A.A.; Kashkoush, I.I.; Gale, G.W. An Experimental Study of Megasonic Cleaning of Silicon Wafers. *J. Electrochem. Soc.* **1995**, *142*, 2812–2817. [CrossRef]
36. Kim, W.; Kim, T.-H.; Choi, J.; Kim, H.-Y. Mechanism of particle removal by megasonic waves. *Appl. Phys. Lett.* **2009**, *94*, 081908. [CrossRef]
37. Reuter, F.; Mettin, R. Electrochemical wall shear rate microscopy of collapsing bubbles. *Phys. Rev. Fluids* **2018**, *3*, 063601. [CrossRef]
38. Reuter, F.; Gonzalez-Avila, S.R.; Mettin, R.; Ohl, C.-D. Flow fields and vortex dynamics of bubbles collapsing near a solid boundary. *Phys. Rev. Fluids* **2017**, *2*, 064202. [CrossRef]
39. Reuter, F.; Mettin, R. Mechanisms of single bubble cleaning. *Ultrason. Sonochem.* **2016**, *29*, 550–562. [CrossRef]

40. Zeng, Q.; An, H.; Ohl, C.-D. Wall shear stress from jetting cavitation bubbles: Influence of the stand-off distance and liquid viscosity. *J. Fluid Mech.* **2022**, *932*, A14. [CrossRef]
41. Reuter, F.; Lauterborn, S.; Mettin, R.; Lauterborn, W. Membrane cleaning with ultrasonically driven bubbles. *Ultrason. Sonochem.* **2017**, *37*, 542–560. [CrossRef]

fluids

MDPI

Article

Cavitation Strength, Acoustic Nonlinearity, and Gas Bubble Distribution in Water

Alexey V. Bulanov *, Ekaterina V. Sosedko, Vladimir A. Bulanov and Igor V. Korskov

V.I. Il'ichev Pacific Oceanological Institute Far Eastern Branch of the Russian Academy of Sciences, 43 Baltiyskaya Str., 690041 Vladivostok, Russia; s_kat@mail.ru (E.V.S.); bulanov@poi.dvo.ru (V.A.B.); i_korskov@poi.dvo.ru (I.V.K.)

* Correspondence: a_bulanov@me.com; Tel.: +7-(423)-237-4913; Fax: +7-(423)-231-2573

Abstract: The acoustic properties of real liquids are largely related to the phase inclusions contained in them, of which gas bubbles are the most common. The aim of the work was to find the relationship between the nonlinear acoustic parameter and the cavitation strength of the liquid with the distribution of bubbles in the liquid, which has so far been poorly studied. The theoretical studies of the parameter of acoustic nonlinearity and the cavitation strength of a liquid with bubbles were carried out within the framework of the homogeneous approximation of a micro-homogeneous liquid; the relationship of these parameters with the bubble distribution function was established, and the typical values of these parameters for different concentrations of bubbles were calculated. Experimental measurements of the parameter of acoustic nonlinearity and the cavitation strength in the upper layer of seawater were carried out; these measurements were consistent with the theoretical estimates. A connection was established between the thresholds of acoustic and optical cavitation—the optical breakdown of a liquid by laser radiation. The results obtained can find practical application in the measurement of the cavitation strength of seawater at great depths in the sea, and the use of an optoacoustic method associated with the use of optical cavitation is proposed.

Keywords: nonlinear acoustic parameter; cavitation strength; optical breakdown; seawater; bubbles; laser radiation

1. Introduction

The important properties of liquids include their nonlinear properties [1–4], which are characterized by a nonlinear acoustic parameter, as well as cavitation strength—a discontinuity of the continuity of a liquid at high intensities in an acoustic wave. The manifestation of nonlinear effects is greatly facilitated by the presence of various nuclei in the liquid–gas bubbles, foreign particles, and other inclusions of various origins [3–6]. For practical applications, it is important to study the properties of real liquids with inclusions in them. Bubbles occupy a special position among the inclusions of various natures contained in liquids, which, as a rule, most significantly affect the acoustic characteristics of liquids [3–5,7–13]. Such characteristics usually include linear acoustic characteristics—the speed of sound, the coefficient of absorption, and the scattering of sound [1,6].

The study of linear acoustic characteristics in liquids began to be studied at the very beginning of the history of acoustics. In the second half of the 20th century, the intensive study of the nonlinear acoustic properties of liquids was originated [1–3]. A special interest in nonlinear acoustics appeared in connection with the creation and implementation of fundamentally new nonlinear acoustic parametric emitters with the unique properties of broadband and high radiation directivity [1–4,14]. However, it turns out that the nonlinear characteristics of liquids are among the parameters that strongly affect the efficiency of such radiators. These primarily include the cavitation strength of liquids, which fundamentally limits the possible achievable power of emitters due to the appearance of a discontinuity in the continuity of the liquid near the emitting surfaces at high sound intensities [4–6,14].

Another important nonlinear characteristic of a liquid that affects the efficiency of any acoustic systems operating at high energy densities is the parameter of the acoustic nonlinearity of the liquid [1–4,14]. In addition, it turns out that this parameter can vary greatly in the presence of various micro-inhomogeneities in the liquid and especially in the presence of bubbles [3,4,7,14,15]. Therefore, along with the tasks of directly studying the nonlinear properties of liquids, the studies of bubble distributions in real liquids are becoming important [7–13].

The studies of the functions of bubble size distribution in liquid media have a long history; however, to date, there remain questions related to both the diagnosis of bubbles and the concentration of bubbles; these issues are mainly related to natural liquids—water in various reservoirs, including seawater [8–13]. Thus, the characteristics of the liquid indicated in the title of the article turn out to be mutually related by practical application, and it is of interest to study them comprehensively in relation to real liquids. At the same time, the relationship of these different nonlinear acoustic characteristics of micronuclear liquids has so far been poorly studied [1,4,5,14,15]. Therefore, the issues related to the study of the nonlinear characteristics of micro-homogeneous liquids are among the fundamental issues of modern acoustics, which are also in demand in practice.

Unfortunately, the nonlinear characteristics of liquids are difficult to study in simple ways [1–5]. These characteristics include the cavitation strength or tensile strength of a liquid [4,5,16,17]. Cavitation strength is associated with the presence of the nuclei of a new phase in the liquid. For water in marine conditions, the issue of cavitation strength has turned out to be particularly difficult. It turns out that the presence of the phase inclusions of the various natures specific to marine conditions (gas bubbles, suspensions, plankton, etc.) leads to an unusual behavior of the cavitation strength of seawater. In various regions of the world ocean, on the one hand, it turns out that it depends on geographical latitude; on the other hand, it turns out to be unusually low, even having a value less than hydrostatic pressure [5,15,16]. Therefore, one of the issues to be studied in this work is the question of the relationship of cavitation strength with the bubbles distributed in a liquid.

The nonlinear acoustic parameter ε is also very sensitive to the presence of micro-inhomogeneities in water [1–4]. Therefore, in particular, it can be used to diagnose a real liquid, including seawater. Usually, measurements of the parameters of the thermodynamic state of a liquid and the speed of sound—the first derivative $c = (\partial\rho/\partial P)_S{}^{-1/2}$ (where ρ is density, P is pressure, and S is entropy)—are used to study the ocean by acoustic methods. However, the entire history of the development of physics indicates that not only the first derivatives but also the following derivatives of the parameters of the medium can also turn out to be important characteristics that reveal the state of the medium. We believe that for these purposes, we can try to use a parameter ε associated with the second derivative of the equation of state $\Gamma = \rho\left(\partial c^2/\partial P\right)_S$, which can become an informative feature for diagnosing a real liquid, including seawater. The nonlinear acoustic parameter is related to the ratio [1–4]:

$$\varepsilon = 1 + (\rho/2)\left(\partial c^2/\partial P\right)_S \equiv 1 + \Gamma/2 \quad (1)$$

In liquids containing various phase inclusions, the nonlinearity parameter may increase significantly. For liquids containing gas bubbles, the nonlinearity parameter ε will depend on the size distribution of the bubbles, as well as on the dynamic characteristics of the bubbles at different frequencies of acoustic disturbances [3,4,7]. Despite the importance of the nonlinearity parameter for the marine environment, information on its measurements at sea is very scarce [1–4,7,15,18]. An important task to be solved in this work is the establishment of the relationship of the nonlinear parameter ε with the distribution of bubbles and the cavitation strength.

The issue of practical methods for measuring cavitation strength in marine conditions, especially at great depths, remains relevant. As a rule, acoustic methods are used, and mainly low-frequency methods (no more than a dozen kHz) are used [4,5,16]. However, due to the successful application of nonlinear parametric emitters in oceanographic research, the question of the cavitation strength of seawater under the influence of relatively high-frequency pumping, which can be hundreds of kHz, up to units of MHz, remains open [3,4,14]. In this connection, the question of the behavior of the acoustic nonlinearity parameter at the corresponding frequencies also arises; this behavior is directly related to the efficiency and stability of the characteristics of the parametric radiators.

Due to the difficulties involved in measuring cavitation strength by acoustic methods, other methods are being sought. These include optoacoustic methods [17,19,20]. Under the action of laser radiation in a liquid, an optical breakdown occurs, along with the appearance of cavities filled with plasma in the liquid, a kind of optical cavitation, accompanied by a strong sound generation effect [19,20]. The spectrum of such radiation is directly related to the breakdown dynamics—the formation and dynamics of a breakdown cavity in a liquid. The interaction process is short-term, but at the same time, most of the laser radiation (up to 40–50%) is converted into acoustic radiation [20–24].

To date, the relationship between the optical breakdown thresholds and the cavitation strength of a real liquid has been poorly studied; therefore, it turns out to be relevant to study the mechanisms of optical breakdown generation in a liquid (especially in seawater) and to establish a connection with cavitation strength [22–24]. In such conditions, the question arises about the possibility of the practical application of laser radiation to study the nonlinear properties of a liquid—cavitation strength, distribution of the nuclei of a new phase, etc.

In this paper, the features of the nonlinear acoustic characteristics of seawater containing bubbles, as well as the characteristics of cavitation strength and the possibility of evaluating the cavitation strength of water by the optoacoustic method, are discussed.

2. Theoretical Foundations

2.1. Methods of the Acoustic Nonlinearity Measuring

As the main approximation, we consider a micro-inhomogeneous liquid in a homogeneous approximation, i.e., we consider the wavelength of sound λ to be much larger than the average distance l between heterogeneous inhomogeneities (bubbles, suspensions, plankton, etc.), which, in turn, is much larger than the radius of inhomogeneity R, i.e., $R << l << \lambda$. We also require that the sound scattering sections on the inhomogeneities σ (including the resonance of the bubbles) do not overlap, i.e., $l > \sqrt{\sigma}$. In addition, we consider the amplitudes of the pumping waves P_1 and P_2 with frequencies ω_1 and ω_2 that are not too large; thus, we can ignore the effects of nonlinear attenuation and use the constant pumping approximation [3,14]. Under the assumptions made, the equation describing the generation of a quasi-plane wave of difference frequency $\Omega = \omega_1 - \omega_2$ can be written as [3,7,18]:

$$\nabla^2 P_\Omega - \frac{1}{c^2}\frac{\partial^2 P_\Omega}{\partial t^2} = -\frac{\varepsilon_e}{\rho c^4}\frac{\partial^2}{\partial t^2}(P_1{}^* P_2) \quad (2)$$

Here, ε_e is an effective nonlinear parameter that takes into account the nonlinearity of a pure liquid and the nonlinearity introduced by micro-inhomogeneities.

An important parameter for determining ε is the distance at which nonlinear effects develop—the rupture distance in the wave. The nonlinear acoustic parameter ε is directly related to the Riemann solution [3] in the evolution of simple waves, according to which the propagation velocity of a simple wave is $c = c_0 + \varepsilon v$, where c_0 is the adiabatic speed of sound and v is the velocity of particles in the wave. The appearance of the dependence of the wave propagation velocity on its amplitude leads to distortions of the wave profile up to the formation of shock waves. The distance at which a plane harmonic wave degenerates into a shock wave is commonly called the rupture distance r^*, which is determined by the ratio [3] $r^* = 1/\varepsilon k M$, where $k = \omega/c$ is the wave number and $M = v/c = P/\rho c^2$ is the Mach number. By measuring the distance r^* at which nonlinear harmonics appear in the wave, it is possible to determine the nonlinear acoustic parameter ε by the formula [25]:

$$\varepsilon = \rho c^3/(\omega P_\omega r^*). \quad (3)$$

In practice, a relative method of measuring a nonlinear acoustic parameter is often used, which consists of the preliminary calibration of the meter in a known medium and then the calculations of ε according to the formula $\varepsilon = \varepsilon_0 \cdot (U_\Omega / U_{\Omega 0})$, where ε_0 and $U_{\Omega 0}$ are the values corresponding to the reference sample [2,3] and U_Ω is the amplitude of the signal in real measurements.

A more universal method, which allows the measurement of the frequency features of the acoustic nonlinearity parameter, is to measure the amplitude of the waves of the difference frequency P_Ω and pumping P_ω at a different distance r. The basis of the method is based on the solution in [3], which can be written as:

$$\frac{v_{2\omega}}{v_\omega} = \frac{1}{2\alpha_\omega r^*}\left(e^{-2\alpha_\omega r} - e^{-4\alpha_\omega r}\right) \tag{4}$$

where $r^* = 1/\varepsilon k M$ is the rupture distance and α_ω is the sound absorption coefficient at the frequency ω. It can be seen from Formula (4) that the amplitude of the second harmonic increases to a distance $r_m = \ln 2/(2\alpha_\omega)$, where it has a maximum equal to $v_{2\omega}/v_\omega = 1/(8\alpha_\omega r^*)$, and then abruptly decays, obeying the exponential law. The latter solution is valid when $\alpha_\omega r_* > 1$, i.e., when the attenuation length is less than the rupture length. Very often, the opposite case occurs when $\alpha_\omega r_* < 1$. Then, the solution is valid only at small distances $r < r_m$, when nonlinear effects do not have time to develop. In this case, using $\alpha_\omega r << 1$, the following simple expression $v_{2\omega}/v_\omega = r/r^*$ is obtained. We consider the behavior of the wave only on a linear section $r < r^*$, while it is more convenient to switch to pressure $P = \rho c v$; then, we obtain:

$$P_{2\omega} = \left(\varepsilon_{2\omega}\omega/\rho c^3\right) P_\omega r. \tag{5}$$

In the more complex and the most practically important case of using a biharmonic signal with the frequencies ω_1 and ω_2, it can be shown that in a linear section $r < r^*$, the generation of a signal with a difference frequency $\Omega = \omega_1 - \omega_2$ is described by the formula:

$$P_\Omega = \left(\varepsilon_\Omega/\rho c^3\right)_e \Omega {P_\omega}^2 r \tag{6}$$

Parametric emitters (PE), combining broadband with the maintenance of high directivity in a large frequency range, have recently acquired an important role in hydroacoustics [3,14]. The calculation of the amplitude of the difference frequency wave can be obtained using Equation (2), which includes the acoustic nonlinearity parameter in the right part. Thus, the efficiency of PE is related to the magnitude; therefore, PE can be used to determine the magnitude of a nonlinear acoustic parameter [3,18,25].

As a result of solving Equation (2) under the assumptions made above, it is possible to obtain several limiting expressions for the difference frequency wave field, from which it is possible to obtain the corresponding expressions and to allow, in practice, the calculation of the nonlinear parameter by the following formulas:

(a) Valid for the far field in the Berktay mode of a parametric emitter [3,14]:

$$\varepsilon = 4P_\Omega(r) r e^{\alpha_\Omega r} / [D_\Omega R_F P_0 \ln[2\alpha_\omega R_F \gamma_E/(\Omega/\omega)]], \tag{7}$$

where $R_F = k_\omega d^2/8$ is the Fraunhofer parameter; $D_\Omega = k_\Omega d/4$; $\gamma_E = 1.78$ is the Euler constant; $P_0 = P_{\omega 1} P_{\omega 2}/\rho c^2$; and α_ω and α_Ω are the damping coefficients of the pump wave and the difference frequency wave, respectively.

(b) In conditions when the nonlinear interaction zone is determined not by the spherical divergence of the beams, but by attenuation at the pumping frequency, which corresponds to the Westervelt regime, the nonlinear parameter ε is determined by the formula:

$$\varepsilon = 2\alpha_\omega r P_\Omega(r) e^{\alpha_\Omega r} / \left[{D_\Omega}^2 P_0 [1 - \exp(2\alpha_\omega r)]\right], \tag{8}$$

which, along with the above values, also includes the sound absorption coefficient at the pump frequency α_ω, which was previously measured in each experiment.

The most practical method is considered to be the Berktay mode. In this mode, most of the PE used in practice operate in the pumping frequency range of 100–300 kHz. Here, it is necessary to take into account the divergence of the biharmonic pumping beam in the far field with relatively weak absorption at a high frequency. In this case, the nonlinear parameter can be determined by the following formula [25]:

$$\varepsilon = A(\omega,\Omega)\frac{P_\Omega r}{P_{\omega 1}P_{\omega 2}},\ A(\omega,\Omega) = \frac{4\rho c^3}{R_{F\omega}^2\Omega\ln(2\gamma_E N_A^{-2})}. \quad (9)$$

where $P_{\omega 1}$, $P_{\omega 2}$, P_Ω are the pressure amplitudes of the pumping waves with frequencies ω_1 and ω_2, and the difference frequency Ω ($\omega_{1,2} = 2\pi f_{1,2}, \Omega = 2\pi F,\ F = f_1 - f_2$); $N_A = (\alpha_\omega R_{F\omega})^{1/2}$; γ_E=1.78 is the Euler constant; $R_{F\omega} = k_\omega d^2/8$ is the length of the near zone at the frequency ω; $k_\omega = \omega/c$; and d is the aperture of the emitter.

2.2. Effective Parameters of a Micro-Inhomogeneous Liquid

The basic assumptions of the homogeneous continuum model, formulated in Section 2.1, make it possible to write the effective parameters of a micro-homogeneous fluid without detailed knowledge of its structure. Thus, the effective density ρ_e is obtained directly from the condition of preserving the mass of a unit volume of a micro-homogeneous liquid in the form of [3]:

$$\rho_e = \rho(1-x) + \rho' x,\ \ x = \frac{4}{3}\pi\int_{R_{\min}}^{R_{\max}} R^3 g(R)dR, \quad (10)$$

where ρ and ρ' are, respectively, the density of liquid and gas in phase inclusions (bubbles, suspensions, etc.); the strokes hereafter refer to the phase inclusions; x is the volume concentration of the inclusions. Here, $g(R)$ is the size distribution function of the phase inclusions; it is defined in terms of the amount of phase inclusions per radius interval dR and is related to the concentration N (the total amount of inclusions per unit volume of liquid) by the formula $N = \int_0^\infty g(R)dR$ [3]. The dimension $g(R)$ is then defined as cm^{-4} or m^{-4}; however, usually in practice, the number of bubbles in the volume of m^3 falling within the radius range of 1 μm is taken to determine $g(R)$, so that $[g(R)] \to m^{-3}mkm^{-1}$ [3,4,18]. The effective compressibility of a micro-inhomogeneous liquid β_e, taking into account the resonant and relaxation characteristics of the inclusions, is [18]:

$$\beta_e = \frac{\beta}{\gamma}\left\{1 - x + \hat{X}\frac{\beta'/\gamma'}{\beta/\gamma}\left[1 - \left(1 - \frac{\rho'}{\rho}\right)\left(1 - \frac{\mathcal{K}(R,\omega)}{\beta'/\gamma'}\right)\right]\right\}. \quad (11)$$

Here, the value $\hat{X}f$ means an integral expression $\hat{X}f = \frac{4}{3}\pi\int_{R_{\min}}^{R_{\max}} dRg(R)R^3 f(R)$, from which, in the case of constancy f, we have $\hat{X}f = xf$. The value of $\gamma = C_P/C_V$ is the adiabatic constant; C_P and C_V are the heat capacity at constant pressure and volume, respectively; $\beta = (1/\rho)(\partial\rho/\partial P)_T$ is the coefficient of isothermal compressibility, which differs from the coefficient of adiabatic compressibility $\beta/\gamma = (1/\rho)(\partial\rho/\partial P)_S = \beta - \alpha^2 T/\rho C_p$ by an amount γ, where T, P, and S are the temperature, pressure, and entropy; and $\alpha = (1/\rho)(\partial\rho/\partial T)_P$ is the coefficient of thermal expansion at constant pressure. The compressibility of the inclusion $(1/V_R)\partial V_R/\partial R = (3/R)\partial R/\partial P \equiv \mathcal{K}$ differs from the adiabatic and isothermal compressibility and takes into account the resonant properties of the inclusion (for example, the resonant frequency of bubbles ω_0) and thermal relaxation, depending on the frequency of the sound and the size of the inclusion in the form of [18]:

$$\mathcal{K} = \frac{K(R,\omega)}{q(R,\omega)}, \quad (12)$$

$$K(R,\omega) = \frac{\beta'}{\gamma'} + \alpha' d'_S \varphi(k'_T R),\ q(R,\omega) = 1 - \frac{\omega^2}{{\omega_0}^2} - i\delta(R,\omega), \quad (13)$$

$${\omega_0}^2 = \frac{3}{\mathrm{Re}(K)\rho R^2}\left[1 - \frac{\mathrm{Re}(K)}{3}\frac{2\sigma}{R}\right], \quad (14)$$

$$\delta(R,\omega) = \delta_T(R,\omega) + \delta_r(R,\omega) + \delta_\eta(R,\omega) = \frac{\mathrm{Im}(K)}{\mathrm{Re}(K)} + \frac{\omega^2}{{\omega_0}^2}\left[\frac{\omega R}{c} + \frac{4\nu}{\omega R^2}\right], \quad (15)$$

$$\varphi(k'_T R) = 3\frac{(k'_T R)cth(k'_T R) - 1}{(k'_T R)^2}, \quad (16)$$

where $k'_T = \sqrt{i\omega/\chi'}$ is the wave number of the heat wave; χ' is the coefficient of thermal conductivity of the gas; $d'_S = (dT'/dP')_{S'} = \alpha' T'/\rho' C'_{P'}$ is the slope of the adiabatic curve; ν is the coefficient of kinematic viscosity of the liquid; and σ is the coefficient of surface tension for the inclusion. The compressibility of the inclusion $\mathcal{K} = \mathcal{K}(R,\omega)$ in the limit of the small sizes tends to isothermal compressibility β'; at large sizes, it tends to adiabatic compressibility $\beta'_S = \beta'/\gamma'$, while at resonance $\omega = \omega_0$ the absolute value of compressibility increases sharply, by a factor of approximately $1/|\delta(R,\omega_0)|$.

A generalization of Wood's formula for the effective speed of sound $\tilde{c}_e$ in a micro-inhomogeneous liquid is written in the form [3,14,18] $\tilde{c}_e = [\rho_e \beta_e]^{-1/2}$, where the real part $c_e = \mathrm{Re}(\tilde{c}_e)$ and the imaginary part $\alpha = \omega \mathrm{Im}(1/\tilde{c}_e)$ determine the phase velocity c_e and the absorption coefficient α of the pressure wave in the form:

$$c_e = c\mathrm{Re}\left[\left\{1 - x + \frac{\beta'_S}{\beta_S}\hat{x}\left[1 - \left(1 - \frac{\rho'}{\rho}\right)\left(1 - \frac{\mathcal{K}(R,\omega)}{\beta'_S}\right)\right]\right\}\left(1 + \frac{\rho' - \rho}{\rho}x\right)\right]^{-1/2}, \quad (17)$$

$$\alpha = \frac{\omega}{c}\mathrm{Im}\left[\left\{1 - x + \frac{\beta'_S}{\beta_S}\hat{x}\left[1 - \left(1 - \frac{\rho'}{\rho}\right)\left(1 - \frac{\mathcal{K}(R,\omega)}{\beta'_S}\right)\right]\right\}\left(1 + \frac{\rho' - \rho}{\rho}x\right)\right]^{1/2} + \alpha_0, \quad (18)$$

where α_0 is the absorption coefficient of sound in pure liquid without any inclusions. In particular, for a liquid with bubbles, for which $\rho' << \rho$ and $\beta << \beta'$, Formulas (17) and (18) are simplified:

$$c_e = c\mathrm{Re}\left[\left\{1 + \hat{x}\frac{\mathcal{K}(R,\omega)}{\beta_S}\right\}(1 - x)\right]^{-1/2} \quad (19)$$

$$\alpha = \frac{\omega}{c}\mathrm{Im}\left[\left\{1 + \hat{x}\frac{\mathcal{K}(R,\omega)}{\beta_S}\right\}\right]^{1/2} + \alpha_0 \quad (20)$$

Depending on the volume concentration of bubbles x, simpler formulas can be obtained:

$$c_e \approx \begin{cases} c\left(1 - xc^2/2c'^2\right), & x << x_* \sim 0.05 \\ c'/\sqrt{x}, & 1 > x > x_* = c'^2/c^2_T \end{cases}, \quad (21)$$

$$\alpha \approx \begin{cases} \omega(\rho c/2)\mathrm{Im}\left[\hat{x}\mathcal{K}(R,\omega)\right], & x << x_* \\ \omega\sqrt{\rho}\mathrm{Im}\sqrt{\hat{x}\mathcal{K}(R,\omega)}, & 1 > x > x_* \end{cases} + \alpha_0. \quad (22)$$

Taking into account Formula (1) for the nonlinear parameter ε and using Formulas (12)–(16), it is possible to calculate the effective nonlinear parameter ε_e of a micro-inhomogeneous fluid with inclusions. Previously, using Formulas (12)–(16), it was possible to calculate the derivative $\partial\mathcal{K}/\partial P'$, which, in the case of non-resonant bubbles ($\mathcal{K} \approx \beta'_S$), is equal to

$$\partial\mathcal{K}/\partial P' = \partial\beta'/\partial P' \approx -\gamma/P'^2 = -\beta'_S/P', \quad (23)$$

where β'_S is the adiabatic compressibility of gas in bubbles $\beta'_S = \gamma/P$. In the general case of a liquid with bubbles, the expression for the nonlinear parameter takes the form:

$$\varepsilon_e = \varepsilon\left\{1+\frac{\varepsilon'\beta'_S{}^2}{\varepsilon\beta_S{}^2}\hat{X}\left[1-\frac{1}{\varepsilon'\beta'_S{}^2}\frac{\partial\beta'}{\partial P'}\left(1-\frac{\partial\mathcal{K}}{\partial P'}/\frac{\partial\beta'}{\partial P'}\right)+\frac{2}{\varepsilon'}\frac{\mathcal{K}}{\beta'_S}\left(1-\frac{\mathcal{K}}{\beta'_S}\right)\right]\right\}\left[1+\hat{X}\frac{\mathcal{K}}{\beta_S}\right]^{-2}, \tag{24}$$

where $\varepsilon' = (\gamma'+1)/2$. In the case of non-resonant bubbles, the following expression can be obtained:

$$\varepsilon_e \approx \varepsilon\frac{1+x(\varepsilon'\beta'_S{}^2/\varepsilon\beta_S{}^2)}{[1+x(\beta'_S/\beta_S)]^2}, \tag{25}$$

which has a maximum at $x_m = 2(\varepsilon\beta_S{}^2/\varepsilon'\beta'_S{}^2)$ and is approximately equal to $(\varepsilon_e/\varepsilon)_m \approx (3/4)(\varepsilon'\beta'_S/\varepsilon\beta_S)^2$.

By making further simplifications and leaving only the resonant characteristics and the main contribution to the scattering amplitude associated with the monopole component of the inclusions vibrations, it is possible to calculate a parameter ε_e that will depend on the structure of the medium, as well as on the dynamic properties of the inclusions. In the case of gas bubbles, the value ε_e is defined as:

$$\varepsilon_e \approx \varepsilon\left(1+\hat{X}\frac{\mathcal{K}^2\varepsilon'}{\beta_S{}^2\varepsilon}\right)\left[1+\frac{\mathcal{K}}{\beta_S}\right]^{-2}. \tag{26}$$

The contribution to the sound absorption coefficient α_b caused by bubbles can be presented in the following form, taking into account Formula (20) [18,26]:

$$\alpha_b \approx \frac{\omega}{c}\mathrm{Im}\sqrt{1+\hat{X}\frac{\mathcal{K}}{\beta_S}} \approx \frac{\omega}{c}\mathrm{Im}\left[1+\frac{4\pi}{3}\frac{\rho c^2}{\gamma P_0}\int_0^\infty\frac{g(R)dR}{q(R,\omega)}\right]^{1/2}. \tag{27}$$

As can be seen from Formulas (26) and (27), to determine the acoustic characteristics of a liquid with bubbles, the form of the function *g*(*R*) in the widest possible range of variation of *R* is important. The data on the sound scattering at various frequencies, including full-scale measurements in the near-surface layer of the sea saturated with bubbles, revealed the structure of the bubble size distribution function *g*(*R*), which, according to [18,26], can be represented by the formula:

$$g(R,z) = A_g R^{-n(z)}\exp\left[-m\left(\frac{R_p}{R}+\frac{R}{R_m}\right)\right]e^{-z/L}, \tag{28}$$

where for the sea conditions $L \approx (2-4)\cdot 10^{-3}U_{10}{}^{2.5}$ (here, L is given in meters, and wind speed U_{10} at an altitude of 10 m is given in m/s), the indicator m depends on the state of the sea; m changes from 1 up to 3, but for moderate and calm waves $m \approx 1$. According to Formula (28), the function $g(R,z)$ has a maximum, which is located at $R_p > 10$ microns, while the value R_p depends on the depth. At the same time, for $R > R_p$ there is a power dependence of the bubble size distribution function with an exponential decline with depth. Thus, Formula (28) takes into account the decline of the function *g*(*R*) at small *R*, the presence of a maximum at $R = R_p$, and the limitation of the spectrum from above by the maximum bubble size R_m. The advantage of such a record *g*(*R*) is the practicality and speed of the calculations of various parameters of the medium [18]. It is also important that the exponent $n \sim 3.3$ and the critical dimensions R_m and R_P are natural parameters that follow from the Garrett–Lee–Farmer theory (GLF) [27]. The measurements of *g*(*R*) on a large factual material under similar conditions of moderate sea conditions give values n in the interval $n \approx 3.3 \div 3.8$ [9–13,18,26–29], which is close enough to the estimate $n \sim 3.3$ obtained for the inertial interval between the sizes R_m and R_P, which follow from the theory of GLF [27].

2.3. *Cavitation Strength and Nonlinear Acoustic Parameter of the Liquid, General Relations*

The relationship between the cavitation strength ΔP_k and the nonlinear acoustic parameter ε of the liquid was discussed in [30], in which a dependence of the following form was obtained:

$$\Delta P_k = \frac{1}{2\sqrt{3}} \frac{1}{(\varepsilon - 1)\beta} \quad (29)$$

where $\Delta P_k = P_0 - P_k$; P_0 is the hydrostatic pressure in the liquid; P_k is the threshold pressure of cavitation in the liquid; and β is the compressibility of the liquid. For a pure liquid, the expression ΔP_k essentially represents the intramolecular pressure from the van der Waals equation of state and, by taking into account the mechanism of thermal heterophase fluctuations, can be written as [17–20]:

$$\Delta P_{k0} = \left(16\pi\sigma^3/3kT\ln(C/J)\right)^{1/2}, \quad (30)$$

where σ is the surface tension coefficient, k is the Boltzmann constant, J is the intensity of nucleation, i.e., the number of growing embryos of a new phase per unit volume per unit time, which usually takes the value $J \sim 1\ \text{cm}^{-3}\text{s}^{-1}$, then $\ln(C/J) \approx 70$–78. For water, $\Delta P_{k0} \approx 140$ MPa, and from (30), it follows that $\varepsilon \approx 5$, which is consistent with the values for pure water.

Equation (29) is also generalized to the case of a micro-inhomogeneous liquid containing phase inclusions. Then, in Equation (29), the parameters ε and β should be replaced everywhere by ε_e and β_e—effectively the nonlinear parameter and compressibility of the liquid with inclusions, as determined by Equations (11) and (24)–(26). Taking into account these dependencies, we can write the following formula for cavitation strength:

$$\Delta P_k = \Delta P_{k0}\left[1 + x\frac{\beta'}{\beta\delta}\right] / \left[1 + x\frac{\gamma'\beta'^2}{\varepsilon\beta^2\delta^2}\right], \quad (31)$$

where x is the volume concentration of bubbles and δ is the attenuation constant of resonant bubbles at the frequency ω. It follows from (31) that when $x > x_* = \varepsilon\beta^2\delta^2/\gamma'\beta'^2 \approx 10^{-10}$, we have:

$$\Delta P_k = \Delta P_{k0}(x_*/x). \quad (32)$$

At high concentrations of bubbles, $x > x_{**} = \beta\delta/\beta' \approx 10^{-5}$, the cavitation strength tends to a minimum value:

$$\Delta P_{k,min} = \Delta P_{k0}\frac{\varepsilon\beta\delta}{(2\varepsilon' - 1)\beta'} = \Delta P_{k0}\frac{\varepsilon}{(2\varepsilon' - 1)}x_{**} \approx 10^{-5}\Delta P_{k0}, \quad (33)$$

from which we obtain $\Delta P_{k,min} \approx 10^4$ Pa.

Figure 1 shows a typical dependence $P_k(x)$ in a wide range of values for water at 20 °C at different frequencies of the acoustic field causing cavitation. The hydrostatic pressure is 0.1 MPa. It can be seen that with an increase in frequency, the cavitation strength increases, and with an increase in the concentration of bubbles, it first decreases sharply and then, when the concentration exceeds the value of 10^{-4}, stabilization occurs—it strives for a constant value of cavitation strength, regardless of the concentration of bubbles.

Let us consider the concentration dependence $P_k(x)$ at different pressures as a model for studying the behavior of the cavitation threshold at different depths in the sea. Figure 2 shows a typical dependence of low-frequency cavitation strength $P_k(x)$ in a wide range of values x at various hydrostatic pressures. It can be seen that with an increase in hydrostatic pressure, the cavitation strength increases.

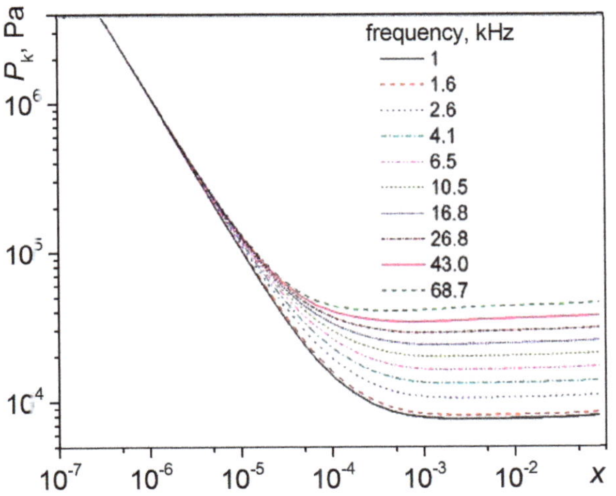

Figure 1. Dependence $P_k(x)$ for water at 20 °C at different frequencies of the acoustic field.

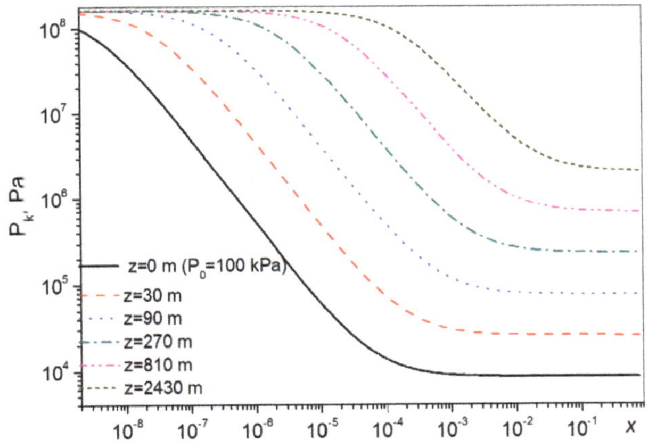

Figure 2. Dependence $P_k(x)$ at different hydrostatic pressures.

2.4. Rectified Gas Diffusion and Bubble Growth Thresholds—Gas Cavitation Threshold

In the previous section, we talked about the threshold of the so-called vapor cavitation in a liquid, in which the threshold is determined by the creation of a critical nuclei and its further growth in the liquid without taking into account the diffusion of gas into the bubble under the influence of sound. Often, in practice, the liquid contains gas bubbles that have formed in the liquid due to the presence of dissolved gas in it. The threshold for the formation of a critical nuclei and its further growth under these conditions depends on the concentration of the dissolved gas relative to the equilibrium concentration of dissolution in the liquid of this gas [16,18,19].

To calculate the threshold, it is necessary to calculate the compressibility of the bubble $\mathcal{K}$, taking into account the effects of gas diffusion. In the expression for the compressibility of inclusion $\mathcal{K}$, the terms related to relaxation due to the processes of gas diffusion exchange in the dynamics of bubbles should be added, when periodically alternating processes of gas exchange occur through the surface of the bubble with gas dissolved in the liquid. In this case, the refined expressions for compressibility $\mathcal{K}$ should be added to Equations (12)–(16). To do this, we use small values of the typical equilibrium concentration of dissolved gas in a liquid $c_\sigma \approx 10^{-5}$–10^{-6}. In addition, attention should be paid to the smallness of the diffusion coefficient $D_c \approx 10^{-9}$ m^2/s; finally, we obtain:

$$\mathcal{K}_c = \frac{K_c(R,\omega)}{q(R,\omega)},\ K_c(R,\omega) = K(R,\omega)\left[1 - 3c_\sigma \frac{1 - ik_\sigma R}{(k_\sigma R)^2}\right], \tag{34}$$

where $k_\sigma = (i\omega/D_c)^{1/2}$. It can be seen from (34) that the additional term begins to play a significant role only at radii smaller by about an order of magnitude of the diffusion wavelength in a liquid, i.e., at $R < 0.1\lambda_\sigma = (2D_c/\omega)^{1/2}$. Thus, it turns out that at high frequencies, the contribution of diffusion to the intrinsic compressibility of the gas bubble can be neglected. However, at low frequencies with a slow gas diffusion process for small bubbles, it can play a significant role, as can be seen from Formula (34).

Solving the equations of the dynamics of a vapor–gas cavity in a liquid with dissolved gas averaged over the period of the sound field, it is possible to obtain expressions for the average values of the physical quantities in a liquid and in a bubble [18]. The crosslinking of the obtained values at the boundary makes it possible to determine the change in the time averages of the values set on the surface of the vapor–gas bubble. Note that these changes are quadratic in the amplitude of the sound field.

$$P_k = \left\{\left[\Delta c_0 + 3c_\sigma\left(\frac{\partial c_\sigma}{\partial T}\right)_\sigma \Delta T_0\right] / \left[B_q(R,\omega) + B_v(R,\omega)\right]\right\}^{1/2}, \tag{35}$$

where in the form $\Delta c_0 = c_0 - c_\sigma$ and $\Delta T_0 = T_\sigma - T_0$ are indicated by supersaturation with gas and overheating of the liquid on the surface of the bubble. The mechanisms of rectified mass transfer, which are quadratic in the amplitude of the sound field, are contained in the sum $B_q(R,\omega) + B_v(R,\omega)$ and discussed in a number of papers [4,17–19].

Figures 3 and 4 show the growth thresholds $P_k(R,\omega)$ depending on the radii of the bubbles and the frequency of the sound field in water at different water temperatures (at different gas concentrations in the bubbles), as calculated by Formula (35) based on the theory presented in [18]. It follows from Figure 3 that near the Minnaert resonance and the second maximum of the function $\mathcal{K}_c(R,\omega)$, the dependence $P_k(R,\omega)$ takes minimal values. It should be noted that the threshold $P_k(R,\omega)$ in this area may be significantly lower than the known Blake threshold [4,16], as determined by the approximate formula $P_k = 2\sigma/R$ and represented in Figure 3b by a dashed purple line. The growth threshold $P_k(R,\omega)$ depends in a complex way on the parameters of the medium and the frequency of the external field.

Figure 4 shows the dependences of the bubble growth thresholds at different concentrations of gas c_0 dissolved in the liquid, respectively; these concentrations are below the equilibrium concentration for a given temperature and external pressure in the liquid and equal to the equilibrium concentration c_σ. Figure 4 shows that the threshold for the growth of the vapor–gas bubbles decreases sharply with an increase in the concentration of gas dissolved in the water. Thus, the mechanism of the rectified diffusion of gas simultaneously with the mechanism of rectified heat transfer turns out to be essential for vapor–gas bubbles, and therefore, the concentration of gas dissolved in a liquid should always be taken into account when comparing theoretical and experimental results.

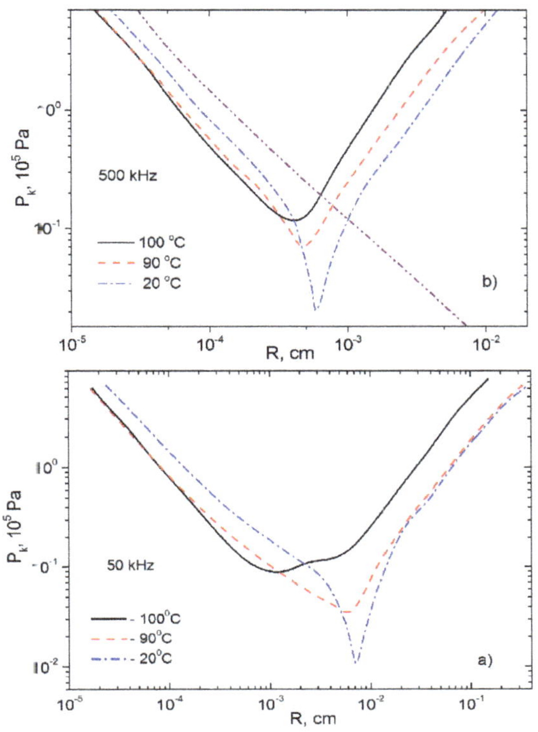

Figure 3. Thresholds for the growth of vapor–gas bubbles in water at frequencies of 50 kHz—(**a**) and 500 kHz—(**b**); the concentration of dissolved gas in water is below equilibrium concentration at different temperatures $c_0/c_\sigma = 0.8$.

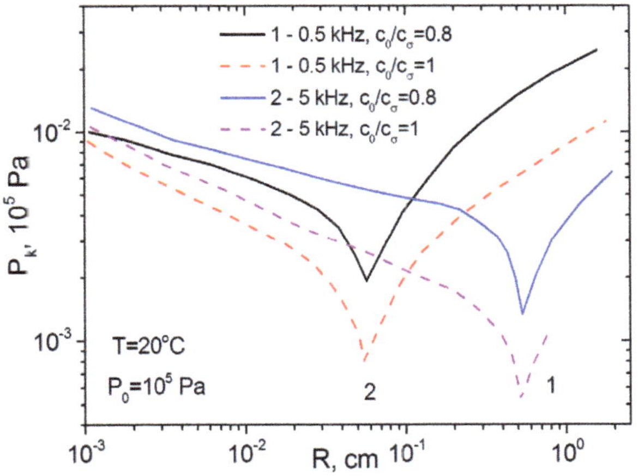

Figure 4. Thresholds for the growth of vapor–gas bubbles in water at frequencies of 0.5 kHz and 5 kHz at different concentrations of dissolved gas in water (solid lines—$c_0/c_\sigma = 0.8$, dashed lines—$c_0/c_\sigma = 1$).

2.5. Experimental Methods and Equipment

The measurements of the nonlinear parameter, according to the method described above and according to Equation (9), were carried out for the first time in the V.I. Il'ichev Pacific Oceanological Institute Far Eastern Branch of the Russian Academy of Sciences (POI FEB RAS) on 12 and 16 expedition cruises of the RV "Academician Alexander Vinogradov" (1988, 1990) in the frequency range from 4 to 40 kHz at various depths [18]. The experimental setup for measuring the nonlinear parameter in various waters of the voyage included an outboard part with receiving and transmitting acoustic antennas and a measuring electronic part, which were connected to each other by connecting cables. The outboard part was a square platform with a side length of 1 m made of 30 cm thick foam, to which an acoustic parametric antenna was attached using a thin halyard. The length of the antenna suspension could vary. The radiation in the working position occurred upwards, towards the surface. The receiving antenna recorded the signal reflected from the water surface. A raft with an antenna could be released from the side of the vessel on an exhaust halyard at a distance of up to 150 m. The sending signal and the echo signals from the receiving antenna were transmitted via separate cables to the electronic measuring part on board the vessel.

Subsequently, to study the distribution of nonlinearity at great depths, a nonlinear acoustic probe was created; the main elements of the probe were a parametric emitter and a cell of a certain length, inside which a biharmonic sound pulse propagated [25], emitted by a parametric emitter. To study the fine structure of the near-surface layer of seawater, the probe made measurements in a small layer of water limited by the cell length, while in the process of lowering (or lifting), a layer-by-layer measurement of a nonlinear parameter, the volume scattering coefficient, and sound attenuation occurred. Thus, the researcher obtained a fine structure of water with a high spatial resolution.

The essence of the measurements in the probe is that the emitted biharmonic pulse propagates in a space bounded on the one hand by a reflecting plate and on the other hand by the emitter itself. The pulse is repeatedly reflected from the plate and the emitter, which makes a large run in a small volume of space, while the accumulation of nonlinear effects occurs, particularly in the growth of a non-linearly generated wave of difference frequency. The selective amplifier of the receiving path allocates the difference frequency, digitizes it, and transmits it to the processing processor. The attenuation coefficient is measured by the decay of the amplitude of the repeatedly reflected pulses of the pumping frequency.

Figure 5 shows the functional diagram of the probe and the frequency spectrum on the receiving hydrophone when the parametric emitter emits a biharmonic signal of 698 kHz and 716 kHz. The submerged probe itself is a 70 cm long rod, at one end of which a parametric emitter is fixed; the radiation axis of the emitter is directed along the axis of the rod towards the reflecting plate fixed at the opposite end. A parametric piezoceramic emitter with a resonant frequency of 650 kHz has a diameter of 66 mm and a directivity characteristic width of 2 degrees. A digital programmable arbitrary waveform generator GSPF-053 was used in the radiation path, the signals of which were amplified by a U7-5 power amplifier and additionally by a radiation unit with a built-in signal switch that allows the receiving of a pulse reflected from the plate in pauses between parcels. The operating frequency range of the pump is in the range of 650–750 kHz. The reception path of the nonlinearity meter is based on the selective amplifier SN-233, which allows the qualitative separation of the acoustic signals of the difference frequency of the nonlinearity meter from the pumping signals and has a gain of up to 10^6. Data entry into the computer was carried out using a 12-bit multichannel ADC board L783, manufactured by L-Card with a maximum quantization frequency of 3 MHz. Other ADCs were also used in the experiments. Their parameters are shown in Table 1. Table 1 presents the main parameters of the typical experimental equipment used for conducting the experiments.

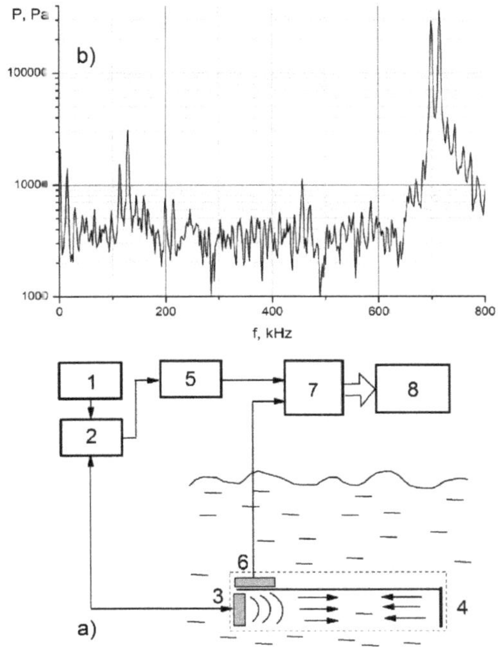

Figure 5. Functional diagram of the probe for measuring the nonlinear acoustic parameter of water (**a**) and the frequency spectrum on the receiving hydrophone (**b**) when emitting a biharmonic signal of 698 kHz and 716 kHz: 1—path of generation and emission of parametric pulses; 2—switch of radiation-reception signals; 3—parametric emitter; 4—reflecting plate; 5—selective difference frequency signal amplifier; 6—depth sensor; 7—two-channel ADC; 8—computer.

Table 1. Main parameters of experimental equipment.

Type	**Main Characteristics**
Acoustic emitters	650 kHz emitter: directivity 2°, power 1 kW.
Equipment for the emission of acoustic signals	1. Digital signal generator GSPF-053 (Rudnev and Shilyaev, Moscow, Russia): 12 digits, frequency up to 10 MHz, voltage 5 V. 2. Broadband power amplifier U7-5(Etalonpribor, Moscow, Russia): frequency up to 3 MHz, power 20 W, voltage 20 V. https://www.etalonpribor.ru/catalog (accessed on 10 April 2022). 3. BIP power amplifier (V.I. Il'ichev Pacific Oceanological Institute, Vladivostok, Russia): frequency up to 1 MHz, power 600 W, voltage up to 400 V.
Equipment for receiving acoustic signals	1. Selective nanovoltmeter SN-233(Pribor-59, Perm, Russia): standard frequency up to 150 kHz, extended frequency up to 200 kHz; selectivity: (a) wide band, (b) 20 dB/octave, (c) 36 dB/octave, (d) 54 dB/octave; maximum gain 10^6. (https://pribor-59.ru/selectivnye-voltmetry accessed on 18 May 2021). 2. Third-octave filters RFT01018 (Robotron, German Democratic Republic (East Germany)): frequency up to 200 kHz, switchable selectivity: (a) octave, (b) 1/3 octave; maximum gain 10. 3. Brüel & Kjær, Copenhagen, Denmark type 2650 amplifier: frequency up to 200 kHz, LF and HF filters: 30 kHz and 2 kHz, maximum gain 10^2 4. Microphone amplifier RFT00011 (Robotron, German Democratic Republic (East Germany)): frequency up to 1 MHz, maximum gain 10^6
Equipment for recording acoustic signals	1. La2 USB ADC (Rudnev and Shilyaev, Moscow, Russia): quantization frequency up to 400 kHz, 14 digits, 16 channels, maximum gain 10^3. https://rudshel.ru/show.php?dev=35 (accessed on 20 October 2023) 2. ADC E20-10 (L-Card, Moscow, Russia): quantization frequency up to 10 MHz, 14 digits, 4 channels, maximum gain 10^2. https://www.lcard.ru/products/external/e20-10 (accessed on 20 October 2023) 3. ADC board L783. (L-Card, Moscow, Russia): quantization frequency up to 2.8 MHz, 12 digits, 32 channels, maximum gain 10^2. https://www.lcard.ru/products/boards/l-783 (accessed on 20 October 2023)

2.6. Measurement Results of a Nonlinear Parameter

2.6.1. Measurements of a Nonlinear Parameter in the Near-Surface Layer of the Ocean

Figure 6 shows the results of the measurements of the nonlinear acoustic parameter of the near-surface layer of the waters of the North Pacific at various frequencies (averaged data) obtained on a number of expeditions of the RV "Academician A. Vinogradov" in 1988 and 1990 [18]. The frequency dependences of the nonlinear parameter of seawater, approximating the experimental values by the least squares method, can be written as the power dependence $\varepsilon_e = A_\varepsilon F^n$. The frequency dependence $\varepsilon_e(F)$ is due to the presence of bubbles in the near-surface layer of the sea, the nonlinear oscillations of which cause the dependence on the frequency of the nonlinear parameter. As a rule, with a decrease in the difference frequency F, the value of the nonlinear parameter $\varepsilon_e(F)$ increases. At the same time, the values ε_e can reach the values $\varepsilon_e > 100$, which are 20–30 times higher than the values ε for pure (without inclusions) seawater.

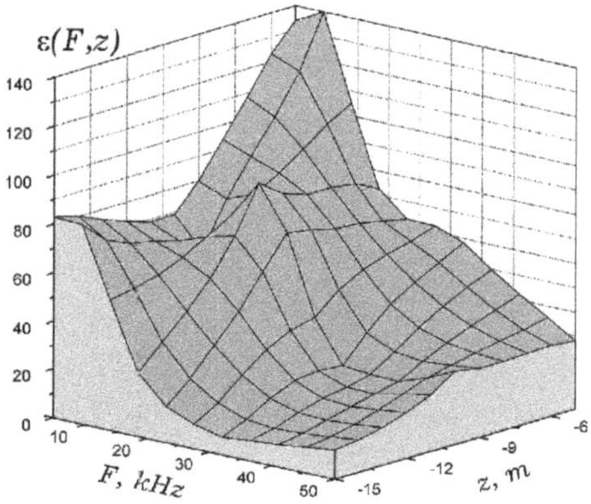

Figure 6. Nonlinear parameter of the subsurface layer of the subarctic waters of the North Pacific (averaged data).

The function $\varepsilon_e(F)$ depends on the concentration of bubbles in the near-surface layer, the value of which in turn varies depending on the depth. At the same time, the value of the nonlinear parameter also changes. Figure 6 shows a near-surface layer with a characteristic thickness of h~5–10 m, depending on the frequency. Summarizing the results discussed above, we can propose the following empirical dependences of the nonlinear parameter on frequency and on depth:

$$\varepsilon_e(F,z) = A_\varepsilon F^n \exp(-z/h). \tag{36}$$

The obtained values reflect our understanding of the influence of the near-surface layers of bubbles on the magnitude of the nonlinear parameter of seawater.

2.6.2. Measurements of a Nonlinear Parameter in the Upper Ocean Layer

A nonlinear acoustic probe was used in the experiments to study the nonlinearity of seawater in the upper layer of the northern Indian Ocean to the depth of the thermocline location of about 100 m during a circumnavigation expedition on the sailboat "Nadezhda" in 2003 [18].

Figure 7 shows a typical sounding schedule conducted in the center of the Arabian Sea of the Indian Ocean at one of the stations. Curve 3 shows the relative change in the nonlinear acoustic parameter of seawater $\varepsilon_{\exp}(z)$ from depth; curve 1 shows the relative change in temperature with depth. From the value of the nonlinear parameter on the surface of the sea, $\varepsilon_0 = 6.1$, it can be seen that the nonlinear parameter changes significantly with depth. As an estimate of curve 2, the distribution of the calculated parameter $(\varepsilon/\varepsilon_0)_{th}$

is presented; it was obtained on the basis of data from UNESCO [31] for the speed of sound $c(T, P, s)$ as a function of temperature T, pressure P, and salinity s.

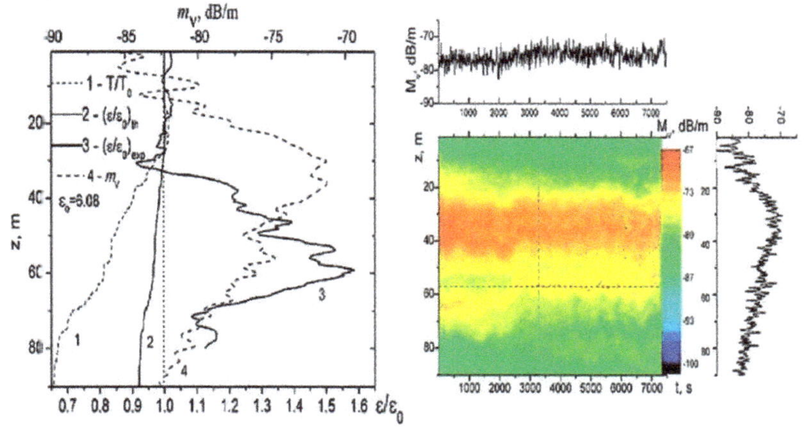

Figure 7. The dependences of the nonlinear parameter $\varepsilon_{exp}(z)$ on the depth z, significantly exceeding the calculated dependences $\varepsilon_{th}(z)$ for pure water and indicating the presence of an additional contribution associated with the presence of micro-inhomogeneities in the sea and recorded using the sound scattering coefficient M_V: 1—temperature $T(z)/T_0(z=0)$; 2—$\varepsilon_{th}(z)$; 3—$\varepsilon_{exp}(z)$ at a frequency of 15 kHz (pumping 700 kHz); 4—$M_V(z)$ at a frequency of 100 kHz.

From the comparison of the calculated and experimental results for $(\varepsilon/\varepsilon_0)_{th}$ and $(\varepsilon/\varepsilon_0)_{exp}$ (curves 2 and 3), it can be seen that these results differ significantly, which indicates that the nonlinearity in seawater is mainly due to the presence of the micro-heterogeneities of various origins in it. Curve 4 in Figure 7 on the right shows the measurements of the sound scattering coefficient m_V at a frequency of 100 kHz. A comparison of the obtained results shows that the change in the nonlinearity parameter diverges from the maximum absolute values of the sound scattering coefficients m_V, and a significant change is observed slightly below the maximum horizon m_V; it coincides with the position of the internal wave, which is usually present at the boundaries of the sound scattering layers, in places of large gradients of the sound scattering coefficient.

2.6.3. Measurements of a Nonlinear Parameter on the Shelf of the Sea of Japan

On the sea shelf, the behavior of the characteristics in the near-surface layer is characterized by great variability. Measurements of nonlinearity were carried out using a nonlinear acoustic probe in Vityaz Bay from aboard a boat that was adrift, according to the method described in [25]. The meter was lowered manually from the side; at the same time, the signal of the difference frequency and the depth of immersion were recorded. The signal with frequencies of 698 and 718 kHz was emitted periodically with an interval of 40 ms. The repeatedly reflected signal was recorded by the computer via the ADC board. The nonlinear acoustic parameter was determined using Equation (6).

Figure 8 shows the dependence of the nonlinearity parameter on the depth in the water area of Vityaz Bay (the Sea of Japan) when lowering the meter from the surface to a depth of 21 m, in the absence of sea waves. The results indicate the presence of a pronounced near-surface layer up to 5–10 m thick with increased nonlinearity, which is observed even in relatively calm water without wave collapse and the formation of bubble clouds.

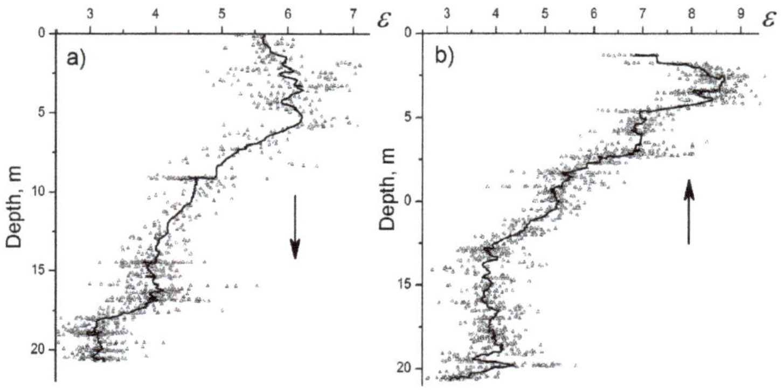

Figure 8. Dependences of the nonlinearity parameter on the depth in the water area of Vityaz Bay (the Sea of Japan) with different soundings: (**a**)—from the surface to a depth of 21 m, (**b**)—from a depth of 21 m upwards.

The most impressive results on the acoustic properties of the near-surface layer of the sea are the results obtained during periods of strong wind over the sea. The effects of the collapse of wind waves produce numerous bubble clouds, which, along with nonlinear properties, are effective absorbers of the energy of the sound waves propagating in the sea. The experimental data on the concentration of bubbles in the near-surface layers of the seawater and Equation (28) allow us to determine additional acoustic nonlinearity, as well as additional sound attenuation introduced by bubbles distributed in the water. Figure 9 shows changes in the bubble concentration over time, the sound absorption coefficient at a frequency of 138 kHz, and the parameter of acoustic nonlinearity of seawater in the near-surface layer of bubbles. It can be seen that the acoustic parameters vary widely with collapses of surface waves and strong wind, leading to modulations of the acoustic properties in the near-surface layer of the sea.

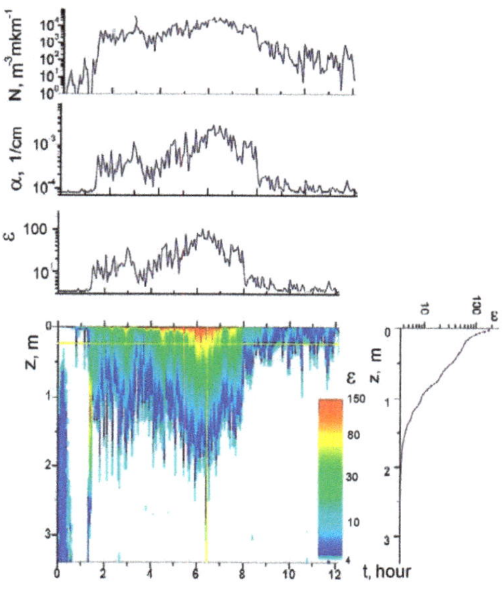

Figure 9. The time change in the near-surface layer of the sea with a sharp change in wind strength and the manifestation of the relationship of acoustic nonlinearity ε, sound absorption α, and bubble distribution in the near-surface layer of the sea (in $m^{-3}mkm^{-1}$) at a depth of z, marked by a horizontal line in the lower figure.

2.7. Acoustic Cavitation Criteria and Cavitation Strength of Seawater

The experimental studies of the cavitation strength of seawater were carried out using an acoustic radiator in the form of a hollow cylinder with a resonant frequency of 10 kHz. Cavitation was recorded by acoustic noises inherent in the cavitation regime [5,18]. The noises were recorded using measuring hydrophones of the company "Akhtuba" (operating frequency band 0.01–300,000 Hz) and the company Bruel & Kjaer, type 8103 (operating frequency band 0.01–200,000 Hz). The signals were recorded digitally using a multichannel 14-bit E20-10 card from L-Card with a maximum digitization frequency of 5 MHz. High voltage was applied to the emitter at a resonance frequency of 10.7 kHz using a Phonic XP 5000 type power amplifier with a maximum power of 2 kW and adjustable inductance compensating for the capacitive load at the resonance frequency. When probing in marine conditions, the hydrophone was attached from the outside of the concentrator near the free end. The relationship between the acoustic characteristics measured by the hydrophone outside and inside the radiator was previously established. Appropriate amendments were subsequently made to the readings of the external hydrophone during the experiments in marine conditions.

When conducting the cavitation studies, special attention was focused on studying the dependence of the cavitation threshold on various criteria for detecting a discontinuity in seawater: by the nonlinearity of the radiated power curve at the frequency of the radiated signal ω_s; by the second harmonic $P_{2\omega}$; and by the total higher harmonics $\sum_{\omega>\omega_s} P_\omega$, as well as by the subharmonics $P_{\omega/2}$ and $P_{3\omega/2}$ [5,18].

Figure 10 shows the dependences of the various spectral components of acoustic noise on the emitter voltage: the total noise for $\omega > \omega_s$ $\sum_{\omega>\omega_s} P_\omega(U_{source}(t))$ and the subharmonic signal $P_{3\omega/2}(U_{source}(t))$ at a frequency of $(3/2)\omega_s$, as well as higher harmonics noise for $\omega > 5\omega_s$ $\sum_{\omega>5\omega_s} P_\omega(U_{source}(t))$. The depth at which the model of the cavitation strength meter was located was 3 m.

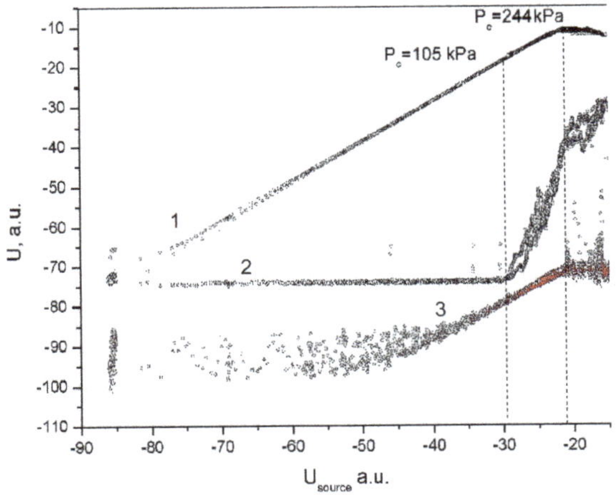

Figure 10. The dependence of the spectral components of acoustic noise on the voltage at the emitter $U_{source}(t)$: 1 is the total noise for $\omega > \omega_s$ – $\sum_{\omega>\omega_s} P_\omega(U_{source}(t))$, 2 is the subharmonic signal $P_{3\omega_s/2}(U_{source}(t))$, 3 is the higher harmonics noise for $\omega > 5\omega_s$ – $\sum_{\omega>5\omega_s} P_\omega(U_{source}(t))$.

It can be seen from Figure 10 that it is possible to clearly distinguish two cavitation thresholds that differ by more than two times: by the curve 3 bend $\sum_{\omega>5\omega_s} P_\omega(U_{source}(t))$ and by the beginning of the asymptotics of all the listed curves and, especially, the curve $\sum_{\omega>\omega_s} P_\omega(U_{source}(t))$. The first threshold of 105 kPa corresponds to the beginning of cavitation, and the second threshold of 244 kPa corresponds to the beginning of violent cavitation, accompanied by a sharp decrease in acoustic impedance. Thus, the criterion of the cavitation threshold is to a certain extent quite conditional. Nevertheless, from the standpoint of the precise detection of the beginning of cavitation, as the beginning of the discontinuity of the continuity of the liquid and the beginning of the formation of bubbles in the liquid, the cavitation strength of the liquid in this example can be considered as the first threshold, which is $P_{c1} = 105$ kPa.

Figure 11 shows the time dependences of various spectral components of acoustic noise: a subharmonic signal $P_{3\omega/2}(t)$ at a frequency of $(3/2)\omega_s$, total harmonics $P_{\Sigma\omega}(t) = \sum_{\omega} P_\omega(t)$ in the frequency range from 1 kHz to 1 MHz, and higher harmonics $P_{\Sigma 6\omega}(t) = \sum_{\omega > 5\omega_s} P_\omega(t)$, starting from the 6th harmonic 64.2 kHz. The depth at which the model of the cavitation strength meter was located was 3 m. It can be seen that the first threshold of $P_{c1} = 105$ kPa corresponds to the beginning of cavitation, and the second threshold of $P_{c2} = 244$ kPa, located in the asymptotic section $P_{\Sigma\omega}(t)$, corresponds to the beginning of violent cavitation.

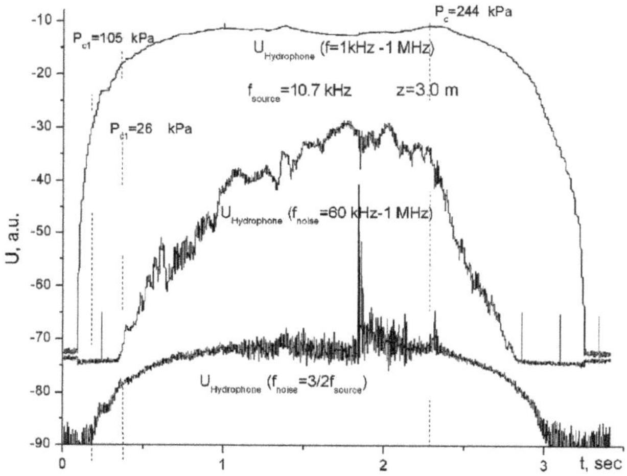

Figure 11. Dependences on the time of the subharmonic signal $P_{3\omega/2}(t)$, total harmonics $P_{\Sigma\omega}(t)$ in the frequency range from 1 kHz to 1 MHz, and higher harmonics $P_{\Sigma 6\omega}(t)$, starting from the 6th harmonic.

The experimental studies of the cavitation strength of seawater were carried out in the autumn period in the Vityaz Bay of the Peter the Great Gulf of the Sea of Japan. Figure 12 shows the temperature and salinity distributions of seawater depending on the depth. It can be seen that there is a clearly defined upper mixed layer with quasi-homogeneous temperature and salinity, extending to a depth of about 6–8 m. Below is a pronounced jump layer characterized by the high vertical gradients of the hydrophysical parameters of seawater.

Figure 13 shows the cavitation strength of seawater as a function of depth, measured in a series of experiments in the same place in Vityaz Bay, the hydrology of which corresponds to Figure 12. The voltage continuously changed during probing. So, the measurements of each point of cavitation strength were carried out in a certain depth range of about 0.5 m. The individual points in Figure 13 correspond to the specified depth intervals. The data on the first cavitation threshold P_{c1} were taken as a cavitation criterion. The measurement errors of the cavitation strength are indicated on the graph and partly reflect the statistical nature of acoustic cavitation.

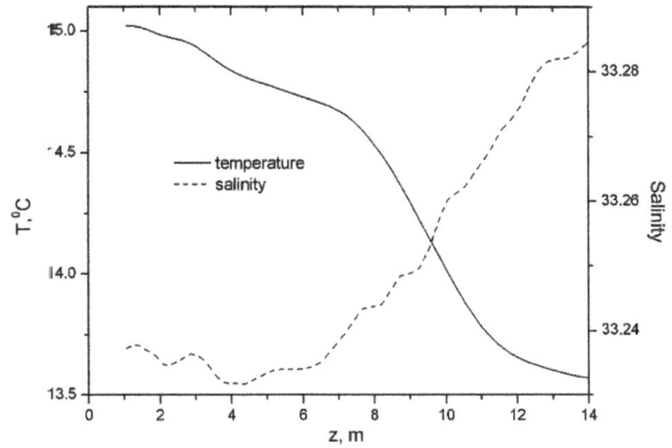

Figure 12. Distribution of temperature and salinity of seawater depending on depth.

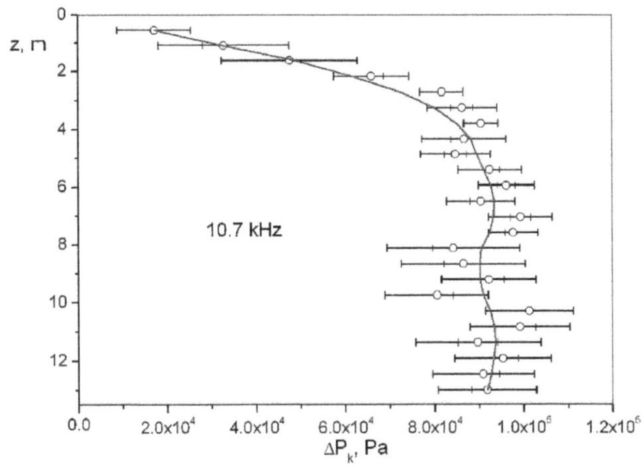

Figure 13. The cavitation strength of seawater depending on the depth: the points at different depths show the measurement results with measurement errors, the standard deviation of which is marked with the error bars; the solid line is the average cavitation strength in the depth.

Figure 13 shows that the cavitation strength of seawater significantly depends on the depth in the subsurface layer up to 6 m thick; then, the dependence on depth is weakly expressed. We associate the obtained results on the decrease in the cavitation strength of seawater in the near-surface layer with the presence of gas bubbles that are always present in this layer. Referring to the theoretical results for the cavitation strength of water with bubbles shown in Figure 1, it can be seen that the experimentally detected decrease to 20 kPa of the cavitation strength of water in the immediate vicinity of the sea surface, which is shown in Figure 13, can be explained by the presence of air bubbles with a total volume concentration of 1.2×10^{-4}.

2.3. *Cavitation in Liquid Caused by Optical Breakdown*

Experimental complex

The experimental studies of aqueous solutions were carried out on the basis of complexes, including Nd:YAG "Brilliant", "Brio", and "Ultra" lasers, with the following radiation parameters: wavelength 532 nm, pulse duration 10 ns, pulse energy up to 180 mJ, varying in the modulated Q-factor mode, and pulse repetition rate of 1–15 Hz [21–24,32–34].

A typical scheme of the experimental setup is shown in Figure 14. The laser provided a pulsed mode of plasma generation on the surface of the aqueous solutions. The power density of the laser radiation was further increased due to the sharp focusing of the radiation in the desired location (in the liquid, on the surface, or near the surface of the liquid), using lenses with different focal lengths: F = 40 mm, 75 mm, and 125 mm. The optical breakdown was recorded using an optical multichannel spectrum analyzer, the Flame Vision PRO System, with a time resolution of 3 ns, i.e., an optical breakdown occurred in the focusing area, the radiation of which was directed by means of a quartz lens or a light guide to the entrance slit of the Spectra Pro spectrograph coupled with a strobed CCD camera. This scheme provided a delay in the recording of the pulse relative to the beginning of the optical breakdown and varied the exposure time of the signal from 10 ns to 50 microseconds from the beginning of the laser breakdown. Taking into account the variation in delays and exposures, the necessary optimal conditions for recording optical breakdown inside the liquid were found.

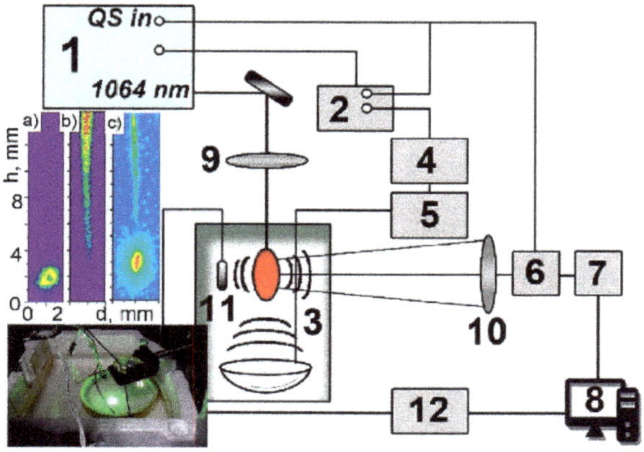

Figure 14. The scheme of the experiment: 1—laser, 2—control pulse switch, 3—piezoceramic ultrasonic emitter (the photo of the emitter is shown on the left), 4—pulse generator of arbitrary shape GSPF 053, 5—power amplifier type U7-5 or Pioneer GM-A3702, 6—delay generator, 7—CCD camera, 8—E20-10 ADC board, 9, 10—rotary mirror and lens, 11—Brüel & Kjær hydrophone type 8103, 12—computer, red oval near hydrophone—breakdown area. Images of laser breakdown of liquid for various types (**a–c**) are shown on the left.

Methods of conducting experiments

The optical part of the experiments is traditional, and they were carried out according to the following scheme [32,33]. Laser radiation (1) was focused into a liquid using a rotary mirror and a lens (9). The plasma radiation of the optical breakdown was projected by a lens (10) onto the input slit of a monochromator (7) coupled to a CCD camera. The control was carried out by a computer (8). Various types of breakdown in water were achieved by focusing laser radiation using various lenses. The breakdown occurred either in the depth of the water or in the near-surface layers or in a combination of these two types. Depending on the types of breakdown, different resolutions of the spectral lines were realized.

To analyze the breakdown dynamics and study the parameters of the acoustic wave initiated by optical breakdown, a Brüel & Kjær type 8103 hydrophone was used as a broadband acoustic receiver, the calibration of which at high frequencies was expanded to 800 kHz. Acoustic information was digitized and recorded using a multichannel I/O board from L-Card with a maximum digitization frequency of ~5 MHz. To control ultrasound, an arbitrary pulse generator GSPF 053, a power amplifier, and a resonant cylindrical radiator [34–36] were used, inside which a liquid breakdown occurred in a converging ultrasound field.

2.9. Acoustic Emission Caused by Exposure to Laser Radiation

Experimental data on the spectral density of acoustic emission were obtained under various modes of breakdown in water: for surface breakdown, for breakdown in the water column, and for mixed breakdown. Different modes of breakdown in water were implemented with the different focusing of laser radiation by lenses. As a result, the breakdown occurred either in the water column or in the near-surface layers of the water, or a mixed breakdown was observed, which is a combination of the above types of breakdown. As a result, it turns out that the acoustic emission and the values of the spectral densities of sound differ significantly depending on the nature of the optical breakdown, which can be seen from Figure 15.

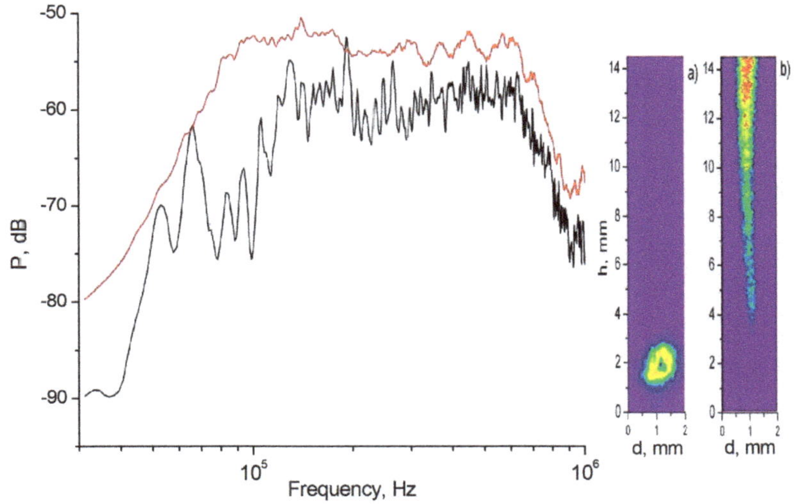

Figure 15. The spectrum of acoustic emission in various separate forms of laser breakdown of water: on the surface (**b**)—black line and in the water column (**a**)—red line.

Simultaneously with the optical breakdown of a liquid in a liquid, a disturbed density region is formed in the vicinity of the breakdown zone, which leads to the formation of an acoustic pulse. Figure 16 shows a recording of a typical acoustic pulse accompanying a liquid breakdown. The distance from the breakdown area to the hydrophone was about 10 cm. Figure 16 shows that a relatively low-frequency component reaches the hydrophone.

The measurement of acoustic emission was used to study the dependence of the efficiency of sound generation on the energy of a laser pulse and its focusing in a liquid.

Figure 17 shows the spectral characteristics of an acoustic wave generated in a liquid by an optical breakdown depending on the energy of the laser pulse. It can be seen from Figure 17 that shifts of the low-frequency maximum to the region of lower frequencies are observed with an increase in the pulse energy. The high-frequency spectral maximum, which is not displaced at all laser pulse energies, is probably related to the natural resonance of the hydrophone at a frequency of ~300 kHz.

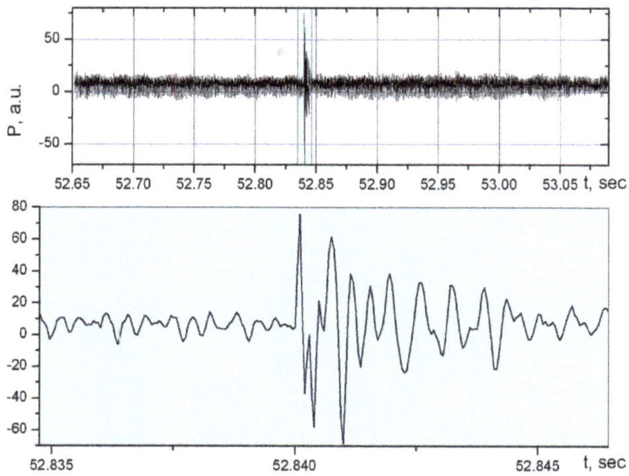

Figure 16. The shape of the acoustic pulse from the breakdown zone received by the hydrophone in the liquid.

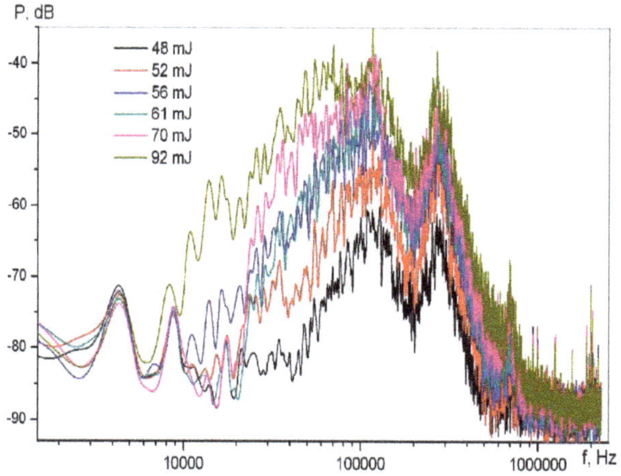

Figure 17. Features of acoustic emission with an increase in the energy of the laser pulse E.

Figure 18 shows the dependence of the sound pressure P on the leading edge of the acoustic pulse on the energy E. Let us analyze the dependence of acoustic emission on the dynamics of bubbles. The method of analysis is as follows. The total energy of the acoustic pulse E_{ac} is calculated as:

$$E_{ac} = \int\limits_{V_{imp}} \langle P(r,t)v(r,t)\rangle dV, \tag{37}$$

where $v(r,t) = -\int_{-\infty}^{t} \nabla_r P(r,\xi)d\xi$, the angle brackets for the function $f(t)$ mean averaging over the duration of the pulse τ of the type $\langle f(t)\rangle = (1/\tau)\int_t^{t+\tau} f(x)dx$. At distances r greater than the radius of the bubble R, $r >> R$, where acoustic emission is registered, the wave front in the registration area can be considered flat, from which we have a relationship between the pressure and the velocity of particles in the wave in the form [3]

$v(r,t) \approx P(r,t)/\rho c$, where ρ is the density and c is the speed of sound in the medium. As a result, we have the total energy of the acoustic pulse E_{ac} in the form

$$E_{ac} = \frac{2\pi r^2}{\rho c}\int_0^{\tau} P^2(r,t)dt. \tag{38}$$

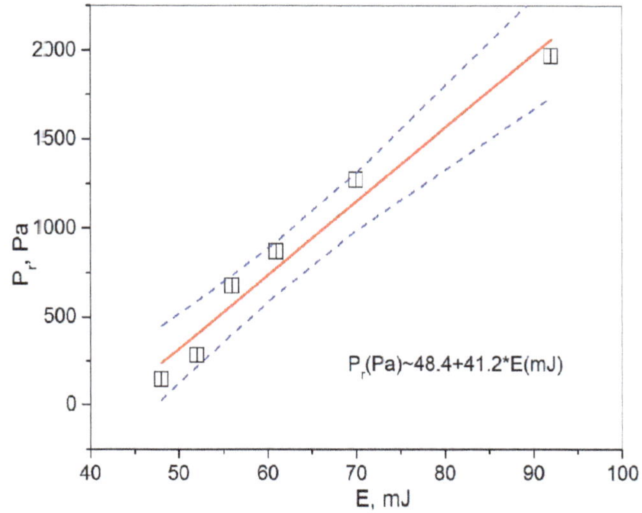

Figure 18. The dependence of the sound pressure P on the leading edge of the acoustic pulse on the energy E: experimental points, linear approximation by the formula indicated on the graph (solid line) and 96% confidence interval of approximation (dashed lines).

Along with the obtained estimates of the total energy of the emitted acoustic pulse, it is of interest to try to solve the inverse problem—to restore the dynamics of the bubble according to the acoustic emission data. The theoretical basis is a formula for the distribution of pressure in the radiated wave from a spherical bubble as a source of monopole radiation, which can be written in the form [3]

$$P_r(t) = \rho\left(R\ddot{R} + 2\dot{R}^2\right)(R/r), \tag{39}$$

where $\dot{R} = dR/dt = U(t)$. By solving a nonlinear differential equation with respect to the function *R(t)*, while considering the function *P(t)* known on the basis of the experimental data in the received acoustic pulse, it is possible to calculate the function *R(t)*, the velocity of the bubble wall *U(t)*, and the intensity in the acoustic wave:

$$I = \left\langle P_R(t)\dot{R}(t)\right\rangle,\ P_R(t) = (r/R)P_r(t) \tag{40}$$

Figure 19 shows these dependencies, which show that the acoustic data can reproduce the function *R(t)*, which is consistent with the characteristic dependencies *R*(*t*) obtained from the direct measurements of optical breakdown images at the late stages of their evolution.

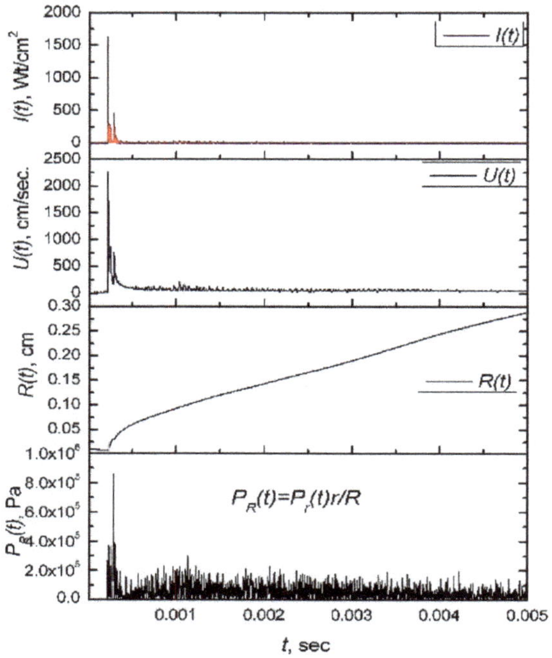

Figure 19. Acoustic emission and bubble dynamics (from **bottom** to **top**): time dependence of pressure *P(t)* in the received acoustic pulse on the optical breakdown region; function *R(t)* calculated by Formula (35) from data for *P(t)*; function *dR(t)/dt*, constructed for calculated function *R(t)*; intensity in the emitted acoustic pulse *I(t)*.

3. Discussion of the Results

The main attention in the article was focused on the description of the experimental methods and theoretical foundations that allow the obtaining and analyzing of the experimental data on the effective parameters of micro-inhomogeneous liquids. Real liquids always contain phase inclusions, among which bubbles are the most significant with regard to the changing of the properties of liquids. Theoretical estimates of the effective parameters of such bubble liquids are carried out in the article. All the calculations were carried out in the framework of a homogeneous approximation, while the calculations of the bubble dynamics were used in the framework of a linear approximation to analyze the complex effective compressibility of a micro-inhomogeneous liquid. The resonance and relaxation (thermal and diffusion) characteristics of the bubbles, as well as their size distribution, were taken into account.

Previously conducted studies of the distribution of bubbles in typical real liquids (including seawater) allowed this article to use the model function of bubble size distribution (28) to calculate the effective compressibility β_e (11) and, on this basis, to obtain estimates for the absorption coefficient α and the speed of sound c_e, Equations (21) and (22), which were valid for sufficiently high concentrations of bubbles. The article did not have the opportunity to discuss the limits of applicability of the approximations made, but the authors, when conducting the analysis, understood the need to take into account the effects of multiple scattering and the overlap of sound scattering cross-sections at high concentrations of bubbles; therefore, all the calculations were carried out in areas where these effects should not be significant.

The effective nonlinear parameter of a micro-inhomogeneous liquid was estimated under the assumption that the main contribution is made by the nonlinearity of radially symmetric bubble pulsations, the contribution of which differs for bubbles with different sizes. In this regard, it was important to take into account the dimensional effects near the monopole resonance and the relaxation near the "second resonance"—the maximum compressibility when the bubble size coincides with the heat wavelength, which led to the general Formula (26). The estimates of the nonlinearity parameter obtained on this basis and measured experimentally by the methods analyzed in Section 2.1 showed the similarity of the results obtained, which is illustrated by the graphs in Figures 6–9. Here, attention should be paid to the practicality of using the nonlinear acoustic probe described in the article when using a parametric emitter in marine research. The probe allows you to obtain a fine structure of water with high spatial resolution.

The calculation of the tensile strength of a liquid is usually carried out by taking into account the theory of thermal heterophase fluctuations and the analysis of the size composition of the nuclei. In the article, we took as a basis Equation (29) obtained in [30] for a pure liquid. However, it was not difficult to also generalize Equation (29) to a micro-inhomogeneous liquid. As a result, it was possible to link various characteristics of a micro-inhomogeneous liquid with simple formulas. Such a phenomenological approach allowed us to obtain Equation (31) for cavitation strength ΔP_k, which at low concentrations gives the tensile strength of a pure liquid and at high concentrations of bubbles $x > x_{**} = \beta\delta/\beta' \approx 10^{-5}$ gives the minimum values of $\Delta P_{k,min} \approx 10^{-5}\Delta P_{k0} \approx 10^4$ Pa. As the concept of cavitation is a rather multilateral phenomenon, which is characterized by various approaches to measuring cavitation strength, this article, in Section 2.4, gives a small overview of the theoretical estimates of the alternative approaches to the cavitation threshold based on the effects of rectified heat transfer and rectified gas diffusion, which in conditions of gas saturation of a liquid can lead to a sharp decrease in cavitation thresholds, which is illustrated in Figures 3 and 4. Later, in Section 2.7, the dependence of the various criteria of the cavitation threshold for detecting a discontinuity in seawater was studied: by the nonlinearity of the radiated power curve at the frequency of the radiated signal; by the second harmonic; and by the total higher harmonics, as well as by the subharmonics. Two cavitation thresholds were identified, of which the first threshold corresponds to the beginning of cavitation, and the second to the beginning of violent cavitation, accompanied by a sharp decrease in acoustic impedance. In the work, from the standpoint of the precise detection of the beginning of cavitation, as the beginning of the discontinuity of the continuity of the liquid and the beginning of the formation of bubbles in the liquid, a threshold was chosen, which can be conditionally considered the cavitation strength of the liquid. Figure 13 shows that for the selected cavitation threshold, the cavitation strength of seawater significantly depends on the depth in the subsurface layer up to 6 m thick; then, the dependence on depth is weakly expressed, which indicates a connection with the presence of gas bubbles, which are always present in the subsurface layer; this is shown clearly in Figure 9.

Unfortunately, it was not possible to measure cavitation strength by acoustic methods below a depth of 14 m due to significant experimental difficulties in creating cavitation at the ultrasound frequencies used. In connection with this circumstance, the optical breakdown in the liquid was studied, and the connection with the cavitation strength was established. The obtained dependences of the sound pressure P at the leading edge and the spectral characteristics of the generated acoustic pulse on the energy E of the laser pulse causing the optical breakdown allowed us to estimate the dynamics of the breakdown region based on acoustic emission data.

By comparing the data obtained for the cavitation strength of water presented in Figures 13 and 19, one can see the fairly close values of the experimental measurements of the cavitation strength obtained by the acoustic method and by using laser breakdown. The lower graph of Figure 19 shows an estimate of the pressure $P_R(t)$ on the wall of the cavity at which a breakdown occurs, which turned out to be close to the value of the cavitation strength in seawater at a depth of 5 m.

Here, we assume that such a coincidence is not accidental, i.e., the cavitation strength of liquids in the case of vapor cavitation, when there are no gas bubbles, will correspond to the boiling temperature threshold of the liquid when it is locally overheated by a laser pulse, provided that the corresponding overheating ΔT_k is associated with the cavitation strength by dependence $\Delta T_k = (dT/dP)_\sigma \Delta P_k$, where $(dT/dP)_\sigma$ corresponds to the slope of the phase equilibrium curve of the liquid–vapor.

This assumption needs to be thoroughly verified, especially at the high hydrostatic pressures corresponding to the great depths of seawater, for which, according to Figure 2, there is a strong dependence of cavitation strength on depth. The use of an independent assessment of the cavitation threshold of a liquid using a laser breakdown of a liquid may be of practical importance if it is necessary to measure cavitation strength at great depths in marine conditions where it is extremely difficult to realize high acoustic pressure values using traditional radiators. The use of optical cavitation would be a way out of this situation due to the use of a completely different mechanism of discontinuity of the medium—induced laser pulse boiling, which corresponds in all parameters to the formation of a critical nuclei of a new phase and its further growth in the supercritical region to macroscopic dimensions. It is this mechanism that is usually implemented during cavitation in the absence of gas bubbles, which significantly facilitate the rupture of the liquid.

4. Conclusions

In the work, within the framework of the homogeneous approximation of a micro-inhomogeneous liquid, the studies of the effective parameter of acoustic nonlinearity and the cavitation strength of a liquid with bubbles were carried out. The analysis used data on the bubble size distribution function in seawater, which were written in the form of a formula that simplified the calculation of the effective parameters of a micro-inhomogeneous liquid.

The relationship of the acoustic nonlinearity parameter, cavitation strength, absorption coefficient, and velocity dispersion with the bubble size distribution function was established, and the typical values of these parameters were calculated for different gas contents in water in the form of bubbles.

Experimental measurements of the acoustic nonlinearity parameter and the cavitation strength of seawater at various depths have been carried out, which are consistent with theoretical estimates. The interrelation of these characteristics for seawater is shown.

Due to the difficulties of measuring cavitation strength by acoustic methods, other methods are being sought. It is proposed to use an optoacoustic method associated with the use of laser radiation, which causes the optical breakdown of optical cavitation, accompanied by a strong sound generation effect. The connection between optical breakdown thresholds and acoustic cavitation thresholds has been established and can later be used to identify cavitation thresholds at high static pressures (at great depths in the sea), where the use of acoustic emitters is extremely difficult.

Author Contributions: Conceptualization, A.V.B. and V.A.B.; methodology, A.V.B., V.A.B. and I.V.K.; software, A.V.B. and E.V.S.; validation, A.V.B., V.A.B. and I.V.K.; formal analysis, A.V.B. and V.A.B.; investigation, A.V.B., V.A.B. and I.V.K.; resources, V.A.B. and A.V.B.; data curation, A.V.B., V.A.B. and E.V.S.; writing—original draft preparation, A.V.B., V.A.B. and E.V.S.; writing—review and editing, A.V.B. and V.A.B.; visualization, A.V.B., V.A.B. and E.V.S.; supervision, A.V.B. and V.A.B.; project administration, A.V.B.; funding acquisition, V.A.B. All authors have read and agreed to the published version of the manuscript.

Funding: The work was carried out within the framework of the state projects: 0211-2021-0002 (reg. No. AAAAA-A20-120031890011-8), 0211-2021-0007 (reg. No. 1210215000054-3) and FFMM-2024-0032.

Data Availability Statement: Data are contained within the article.

Conflicts of Interest: The authors declare no conflict of interest.

References

1. Hovem, J.M. *Marine Acoustics: The Physics of Sound in Underwater Environments*; Peninsula Publishing: Newport Beach, CA, USA, 2012; 656p.
2. Beyer, R.T. The parameter B/A. In *Nonlinear Acoustics*; Hamilton, M.F., Blackstock, D.T., Eds.; Academic Press: Cambridge, MA, USA, 1997; pp. 25–39.
3. Naugolnykh, K.A.; Ostrovsky, L.A. *Nonlinear Wave Processes in Acoustics*; University Press: Cambridge, UK, 1998; 298p.
4. Leighton, T.G. *The Acoustic Bubblel*; Academic Press: San Diego, CA, USA, 1994.
5. Akulichev, V.A.; Il'ichev, V.I. Thresholds of acoustic cavitation in sea water in various areas of the World ocean. *Phys. Acoust.* **2005**, *51*, 167–179. [CrossRef]
6. Brekhovskikh, L.M.; Lysanov, Y.P. *Fundamentals of Ocean Acoustics*; Springer: Berlin, Germany, 2013; 250p.
7. Karpov, S.; Prosperetti, A.; Ostrovsky, L. Nonlinear wave interactions in bubble layers. *J. Acoust. Soc. Am.* **2003**, *113*, 1304–1316. [CrossRef]
8. Holliday, D.V.; Greenlaw, C.F.; Thistle, D.E.; Rines, J.E.B. A biological source of bubbles in sandy marine sediments. *J. Acoust. Soc. Am.* **2003**, *114*, 2317–2326. [CrossRef]
9. Baschek, B.; Farmer, D.M. Gas Bubbles as Oceanographic Tracers. *J. Atmos. Ocean. Technol.* **2010**, *27*, 241–245. [CrossRef]
10. Vagle, S.; McNeil, C.; Steiner, N. Upper ocean bubble measurements from the NE Pacific and estimates of their role in air-sea gas transfer of the weakly soluble gases nitrogen and oxygen. *J. Geophys. Res.* **2010**, *115*, C12054. [CrossRef]
11. Czerski, H. An inversion of acoustical attenuation measurements to deduce bubble populations. *J. Atmos. Ocean. Technol.* **2012**, *29*, 1139–1148. [CrossRef]
12. Liu, R.; Li, Z. The effects of bubble scattering on sound propagation in shallow water. *J. Mar. Sci. Eng.* **2021**, *9*, 1441. [CrossRef]
13. Czerski, H.; Brooks, I.M.; Gunn, S.; Pascal, R.; Matei, A. Ocean bubbles under high wind conditions—Part 1: Bubble distribution and development. *Ocean Sci.* **2022**, *18*, 565–586. [CrossRef]
14. Novikov, B.K.; Rudenko, O.V.; Timoshenko, V.I. *Nonlinear Underwater Acoustics*; American Institute of Physics: New York, NY, USA, 1987.
15. Grelowska, G.; Kozaczka, E. Nonlinear properties of the Gotland deep—Baltic Sea. *Arch. Acoust.* **2015**, *40*, 595–600. [CrossRef]
16. Neppiras, E.A. Acoustic cavitation. *Phys. Rep.* **1980**, *61*, 159–251. [CrossRef]
17. Caupin, F.; Herbert, E. Cavitation in water: A review. *Comptes Rendus Phys.* **2006**, *7*, 1000–1017. [CrossRef]
18. Akulichev, V.A.; Bulanov, V.A. *Acoustic Study of Small-Scale Heterogeneities in the Marine Environment*; POI FEB RAS: Vladivostok, Russia, 2017; 414p. Available online: https://www.poi.dvo.ru/node/470 (accessed on 19 December 2023).
19. Lauterborn, W.; Kurz, T. Physics of bubble oscillations. *Rep. Prog. Phys.* **2010**, *73*, 106501. [CrossRef]
20. Padilla-Martinez, J.P.; Berrospe-Rodriguez, C.; Aguilar, G.; Ramirez-San-Juan, J.C.; Ramos-Garcia, R. Optic cavitation with CW lasers: A review. *Phys. Fluids* **2014**, *26*, 122007. [CrossRef]
21. Cremers, D.A.; Radziemski, L.J. *Handbook of Laser–Induced Breakdown Spectroscopy*; John Wiley& Sons: New York, NY, USA, 2006; 282p.
22. Musazzi, S.; Perini, U. *Laser-Induced Breakdown Spectroscopy*; Springer Series in Optical Sciences 182; Springer: Berlin/Heidelberg, Germany, 2014; 575p.
23. Kudryashov, S.I.; Lyon, K.; Allen, S.D. Photoacoustic study of relaxation dynamics in multibubbles systems in laser–superheated water. *Phys. Rev. E* **2006**, *73*, 055301. [CrossRef]
24. Byun, K.T.; Kwak, H.Y.; Karng, S.W. Bubble evolution and radiation mechanism for laser–induced collapsing bubble in water. *Jpn. J. Appl. Phys.* **2004**, *43*, 6364–6370. [CrossRef]
25. Bulanov, V.A.; Korskov, I.V.; Popov, P.N. Measurements of the nonlinear acoustic parameter of sea water via a device using reflected pulses. *Instrum. Exp. Technol.* **2017**, *60*, 414–417. [CrossRef]
26. Farmer, D.M.; Vagle, S.; Booth, D. Reverberation effects in acoustical resonators used for bubble measurements. *J. Acoust. Soc. Am.* **2005**, *118*, 2954–2960. [CrossRef]
27. Garrett, C.; Li, M.; Farmer, D. The Connection between bubble size spectra and energy dissipation rates in the upper ocean. *J. Phys. Oceanogr.* **2000**, *30*, 2163–2171. [CrossRef]
28. Bulanov, V.A.; Bugaeva, L.K.; Storozhenko, A.V. On sound scattering and acoustic properties of the upper layer of the sea with bubble clouds. *J. Mar. Sci. Eng.* **2022**, *10*, 872. [CrossRef]
29. Deane, G.B.; Preisig, J.C.; Lavery, A.C. The suspension of large bubbles near the seasurface by turbulence and their role in absorbing forward-scattered sound. *IEEE J. Ocean. Eng.* **2013**, *38*, 632–641. [CrossRef]
30. Sehgal, C.M. Non-linear ultrasonics to determine molecular properties of pure liquids. *Ultrasonics* **1995**, *33*, 155–161. [CrossRef]
31. Fofonoff, N.P.; Millard, R.C., Jr. Algorithm for computation of fundamental properties of seawater. In *UNESCO Technical Papers in Marine Science*; No. 44; UNESCO: Paris, France, 1983.
32. Bukin, O.A.; Salyuk, P.A.; Maior, A.Y.; Golik, S.S.; Il'in, A.A.; Bulanov, A.V.; Baulo, E.N.; Akmaikin, D.A. The use of laser spectroscopy methods in the Investigation of the carbon cycle in the ocean. *Atmos. Ocean. Opt.* **2010**, *23*, 328–333. [CrossRef]
33. Ilyin, A.A.; Nagorny, I.G.; Bukin, O.A.; Bulanov, A.V.; Shmirko, K.A. Features of development of optical breakdown on an inclined aluminum surface. *Tech. Phys. Lett.* **2012**, *38*, 975–978. [CrossRef]
34. Bulanov, A.V.; Nagorny, I.G.; Sosedko, E.V. A study of the optical and acoustic spectral characteristics by laser breakdown of water in an ultrasonic field. *Tech. Phys. Lett.* **2019**, *45*, 1200–1203. [CrossRef]

35. Bulanov, A.V.; Sosedko, E.V. Opto-acoustic effects by laser breakdown of seawater in an ultrasonic field. *Dokl. Earth Sci.* **2020**, *491*, 183–186. [CrossRef]
36. Bulanov, A.V. Using of ultrasound in automated laser induced breakdown spectroscopy complex for operational study of spectral characteristics of seawater of carbon polygons. *Bull. Russ. Acad. Sci. Phys.* **2022**, *86*, S32–S36. [CrossRef]

fluids

Article

Damage on a Solid–Liquid Interface Induced by the Dynamical Behavior of Injected Gas Bubbles in Flowing Mercury

Hiroyuki Kogawa *, Takashi Wakui and Masatoshi Futakawa

Japan Atomic Energy Agency, Ibaraki 319-1195, Japan
* Correspondence: kogawa.hiroyuki@jaea.go.jp

Abstract: Microbubbles have been applied in various fields. In the mercury targets of spallation neutron sources, where cavitation damage is a crucial issue for life estimation, microbubbles are injected into the mercury to absorb the thermal expansion of the mercury caused by the pulsed proton beam injection and reduce the macroscopic pressure waves, which results in reducing the damage. Recently, when the proton beam power was increased and the number of injected gas bubbles was increased, unique damage morphologies were observed on the solid–liquid interface. Detailed observation and numerical analyses revealed that the microscopic pressure emitted from the gas bubbles contracting is sufficient to form pit damage, i.e., the directions of streak-like defects which are formed by connecting the pit damage coincides with the direction of the gas bubble trajectories, and the distances between the pits was understandable when taking the natural period of gas bubble vibration into account. This indicates that gas microbubbles, used to reduce macroscopic pressure waves, have the potential to be inceptions of cavitation damage due to the microscopic pressure emitted from these gas bubbles. To completely mitigate the damage, we have to consider the two effects of injecting gas bubbles: reducing macroscopic pressure waves and reducing the microscopic pressure due to bubble dynamics.

Keywords: microbubble; mercury target; cavitation damage; pressure wave; solid–liquid interface; impact pressure; gas bubble vibration

1. Introduction

Microbubbles have been applied in various fields. For example, they are used to reduce flow resistance by controlling the boundary layer in ships [1], to clean semiconductors by using microscopic impact pressure emitted during bubble collapse [2], and more, including as contrast agents in ultrasounds in the medical field [3]. In mercury targets, which are high-intensity spallation neutron sources, microbubbles are injected into the mercury to absorb the rapid thermal expansion of mercury. The neutron beams produced in the mercury target are applicable to create innovative research. Neutron beams are produced by spallation reactions when pulsed proton beams are injected into heavy metals. In high-power spallation neutron source facilities such as the SNS in the US [4] and the MLF/J-PARC [5], liquid mercury is used as the target material from the viewpoint of both neutron yield efficiency and cooling performance. Pulsed proton beam injections (pulse duration, 1 μs; repetition cycle, 25 Hz) cause macroscopic pressure waves (hereafter, called MACP) due to the rapid thermal expansion of mercury. Cavitation incidents [6,7] induce the nearby interface between the structure wall and the liquid mercury in the process of MACP propagation in the mercury target vessel. As such, cavitation damage is imposed on the vessel wall [8,9]. This cavitation damage becomes one of crucial issues for keeping the structural integrity and prolonging the lifetime of the target vessels. In order to reduce the MACPs, the gas bubbles were injected into the flowing mercury to absorb the rapid thermal expansion through the mechanism of the bubbles' contraction, and thereby the cavitation damage is expected to be mitigated [10–12]. On the other hand, microscopic pressure (hereafter, called MICP) is emitted by the vibration of injected gas bubbles which is excited by MACPs. The MICPs emitted by gas bubble vibrations are deduced to have two kinds of effects: one is positive, and another is negative. As for the positive effect, the inception of cavitation bubbles may be suppressed by the MICPs, a so called "suppression effect" [13,14]. As for the negative effect, these gas bubbles could become cavitation nuclei, which should be considered carefully.

It was confirmed by measuring the vibration of the vessel wall using a laser Doppler vibration (LDV) system during the operation of mercury targets in MLF that the MACPs were dramatically reduced by injecting gas bubbles, which resulted in decreases in the cavitation damage. Nevertheless, damage, inferred to be caused by the MICPs emitted by the gas bubbles vibrations, was observed. The pits, which are generated by cavitation damage, were much shallower in depth and wider in radius under the gas bubble injection conditions compared to the pits generated without gas bubble injection. And, damage morphology was observed with these features as follows.

(1) Streak-like defects were observed, and the directions roughly coincided with the flow direction of the injected gas bubbles, which was evaluated through computational fluid dynamics (CFD) on the flowing bubble distribution.
(2) Series of beaded pits were observed in the streak-like defects. The distances between each pit were wider in the upper area than in the lower area, which indicated a correlation that bubble size distribution was affected by the buoyancy.

In this paper, the damage morphology observed in the mercury target vessel is described in detail, and the damage generation mechanism by the gas bubbles is discussed qualitatively. The MICP emitted by a single gas bubble vibration, P_L, was estimated based on the bubble dynamics. And, the MICP applied on the solid wall, P_{imp}, was estimated, considering the distance between the gas bubbles and the solid wall; the damage mechanism due to the MICPs emitted by the gas bubbles is discussed.

2. Damage Morphology Observation

2.1. The Effect of Gas Bubbles on the (Macroscopic Pressure Wave)–Materials and Methods

Figure 1 shows the cross-sectional view of the mercury target vessel in J-PARC. The mercury was enclosed with a mercury target vessel. The material of the mercury target vessel was type 316 L stainless steel. The proton beams were injected through its tip (called the beam window). The mercury flowed from the inlet side to the other side of the target vessel. Flow guide vanes were installed in the target vessel so that the mercury flowed across the proton beam injection path at the horizontal center of the vessel. A bubble generator was installed at the mercury inlet side [11]. Since the heat generation in the mercury owing to the proton beam injection was highest near the center of the beam window, and the amplitude of the generated pressure wave (MACP) was also larger there, the specimen for observing the damage was cut out from the center of the beam window after the operation [15]. The bubble generator was about 400 mm away from the beam window. Near the beam window, more bubbles were distributed on the upper area rather than on the lower area in the direction of gravity due to buoyancy. Figure 2 shows the distribution of the gas bubble radius obtained by computational fluid dynamics (CFD) in the upper and lower areas in the mercury near the cut-out specimen [16]. Many of the gas bubbles with larger sizes were distributed in the upper area compared to the lower area.

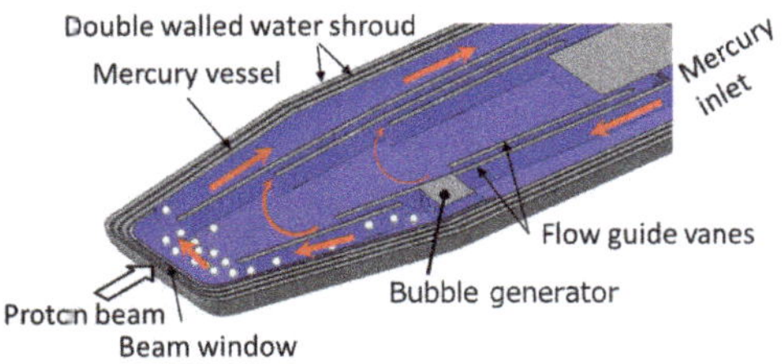

Figure 1. Cross sectional view of a mercury target vessel. Pulsed proton beams were injected at a beam window. The mercury flowed from one side to another side. At the horizontal center, mercury flowed in, perpendicular to the proton beam's injected direction. The bubble generator was set on the inlet side.

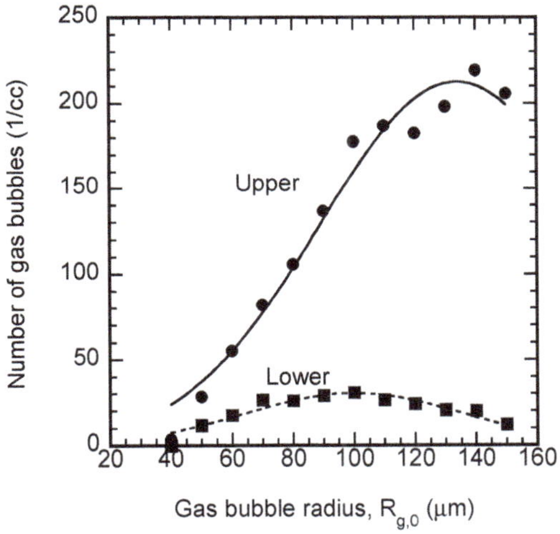

Figure 2. The gas bubble population near the beam window calculated by CFD. Many gas bubbles were distributed in the upper area because of buoyancy. The mean bubble radius was larger in the upper area than that in the lower area.

Figure 3a,b show photographs of the damage observed on the specimens cut out after operations in the cases that the amount of gas supply used to bubble was low and relatively high, respectively. In the case of Figure 3a, gas was not supplied for 30% of the entire operation period because of trouble within the gas supplying system. The proton beam conditions in Figure 3a,b were 530 kW and 570 kW in the average power, and 165 days and 121 days in the operation time, respectively. That is, the accumulated beam powers were almost the same between the cases shown in Figure 3a,b. In Figure 3a, the maximum depth of damage reached 3.3 mm for a wall thickness of 5 mm. On the other hand, in Figure 3b, the maximum depth decreased to 0.4 mm, confirming a damage mitigation effect owing to the reduction in the MACPs due to gas bubble injection. The maximum depth occurred at the center of the gravity direction where the MACP was highest in Figure 3a, while in Figure 3b, it was observed in the lower area. This is thought to be due not only to the reduction in MACPs but also to the suppression effect of gas bubbles [13,14], which suppresses the inception of cavitation in the upper region where more bubbles are distributed, as shown in Figure 2. These results suggest that this number of gas bubbles is effective for the damage mitigation. Furthermore, the fatigue damage combined with cavitation erosion was recognized strongly in the case of Figure 3a. The fatigue phenomenon was also mitigated, owing to the reduction in MACPs by the higher number of gas bubbles as shown in Figure 3b. The fatigue damage was complicated and included uncertainty factors including material degradation due to irradiation and/or LME (liquid metal embrittlement), etc. That is, it is essential from the viewpoint of lifetime estimation to eliminate the fatigue phenomenon caused by the MACPs as well.

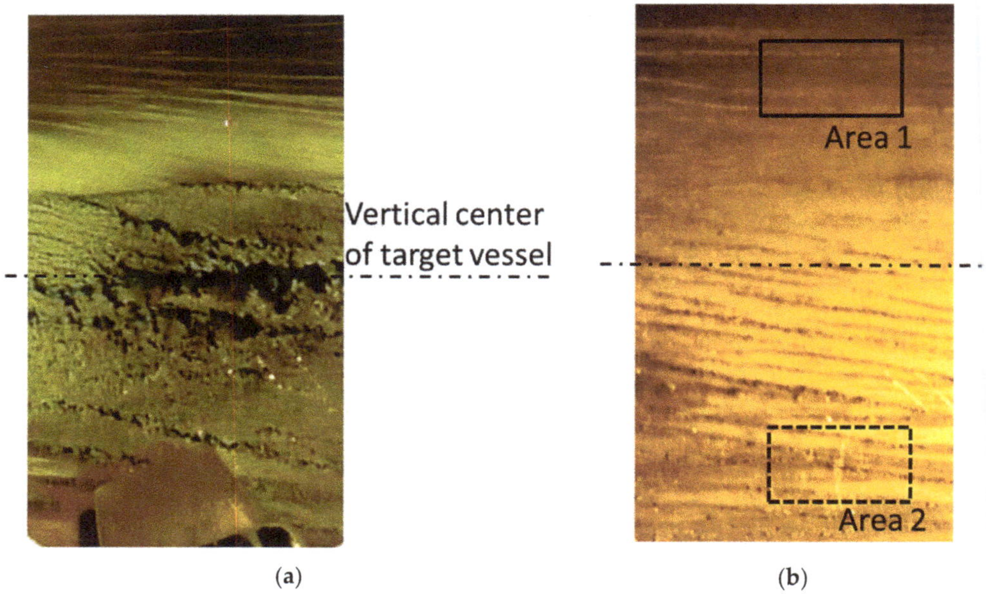

Figure 3. Damage at the beam window in contact with mercury in the cases of (**a**) low gas bubble injection and (**b**) relatively high gas bubble injection. Damage was mitigated in (**b**) but inclined streak-like defects were observed.

2.2. The Effect of MICP (Microscopic Pressure) Emitted from Gas Bubbles

Detailed observation of Figure 3b revealed many streak-like defects tilted downward relative to the horizontal direction, i.e., the primary flow direction of mercury. Since the gas bubbles were injected into the flowing mercury by using a swirling flow into the mercury target [11], the trajectories of the injected gas bubbles may have also been tilted to the horizontal direction. Figure 4 shows the trajectories of the injected gas bubbles on the interface between the mercury and the wall of the mercury target vessel calculated by CFD. It was noted that the darker colors indicated the trajectory of larger size bubbles. Although a mixture of trajectories of large and small gas bubbles was recognized near the vertical center, it was recognized that many large bubbles flowing in the upper section and many small bubbles flowing in the lower section. The trajectories were tilted downward to the direction of the primary flow of mercury. Compared to Figure 3b, the tilted slope of the observed streak-like damage was consistent with the trajectories of the injected gas bubbles. This result suggested that the injected gas bubbles are affecting the damage.

Furthermore, beaded pits were observed in the streak-like defects as shown in Figure 5a,b, which show detailed observations in the streak-like defects in the upper (Area 1 in Figure 3b) and lower (Area 2 in Figure 3b) areas, respectively. The distance to the next pit, ΔD, was relatively wide in the upper area shown in Figure 5a than that in the lower area shown in Figure 5b. These observation results seemed to correlate with the radius of the distributed gas bubbles.

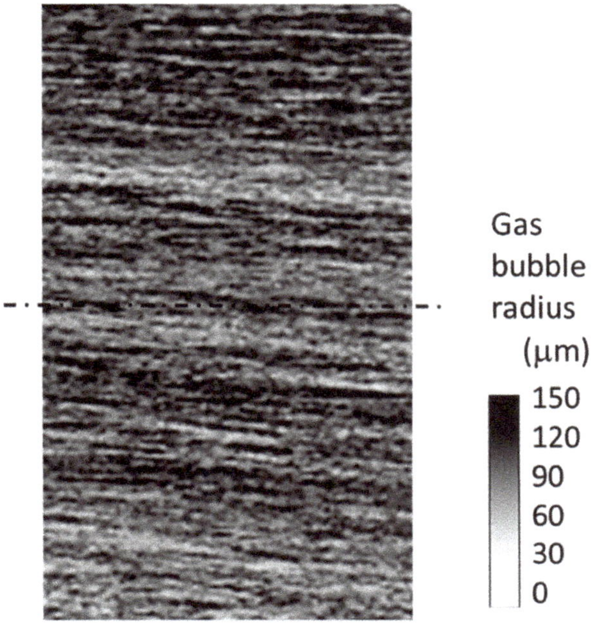

Figure 4. Gas bubble trajectories obtained by CFD analysis. Swirl flow, generated in a bubble generator, affected the slopes of the trajectories, which were similar with those of the streak-like defects observed in Figure 3b.

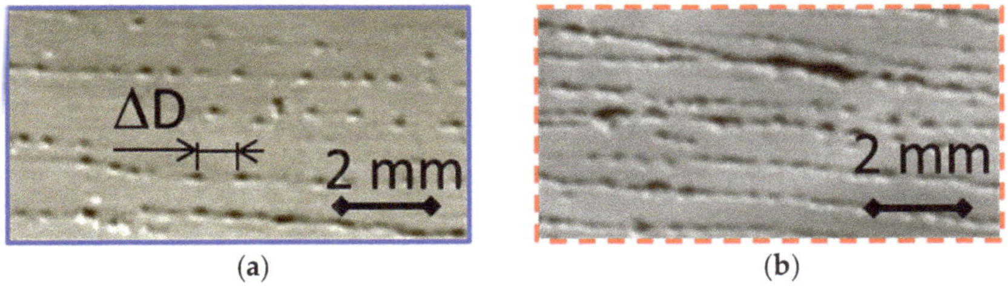

Figure 5. Detailed observation of the streak-like defects in (a) Area 1 and (b) Area 2 in Figure 3b. Beaded pits were observed in the streak-like defects. The distance to the next pit, ΔD, was relatively narrow in the lower area in (b).

Therefore, we considered the possibility that the gas bubbles, injected to absorb thermal expansion and reduce MACPs, could become the inception of damage. We focused on the natural period, T_0, of the gas bubbles and estimated the distance to the next pit, ΔD_{CFD}, with the following assumptions based on the distribution of the gas bubbles shown in Figure 2.

Pits are formed by repeated MICPs due to the repeated expansion and collapse (vibration) of gas bubbles flowing in mercury.

The natural period, T_0, of a bubble estimated by Equation (1) [17] depends on the bubble radius, $R_{g,0}$ (where the larger the bubble, the longer the T_0).

$$T_0(R_{g,0}) = R_{g,0} / \sqrt{\frac{3\gamma}{\rho_L}\left(p_\infty + \frac{2\sigma}{R_{g,0}}\right) - \frac{2\sigma}{\rho_L R_{g,0}}} \quad (1)$$

where $R_{g,0}$ is the initial radius of the gas bubble, γ is the specific heat ratio of gas, ρ_L is the density of liquid, and p_∞ is the liquid pressure at a distance.

The bubble flows in mercury, and the flow speed of the bubble is represented by v_{Hg}. Thus, ΔD_{CFD} at any initial bubble radius can be calculated as

$$\Delta D_{CFD} = v_{Hg} \times T_0(\mathrm{R_{g,0}}) \tag{2}$$

Figure 6a,b show the comparisons of the distributions between ΔD_{CFD} and the observed ΔD for the upper area (Figure 5a) and the lower area (Figure 5b), respectively. The observed distribution of ΔD was similar to the distribution of ΔD_{CFD} based on the estimation. The peak value of ΔD was wider in Area 1 (upper part) where many large bubbles were distributed. The results of these studies indicated that the distribution of the gas bubble radius correlated with the distribution of ΔD, and that the gas bubbles injected to reduce thermal expansion may become cavitation nuclei and cause damage.

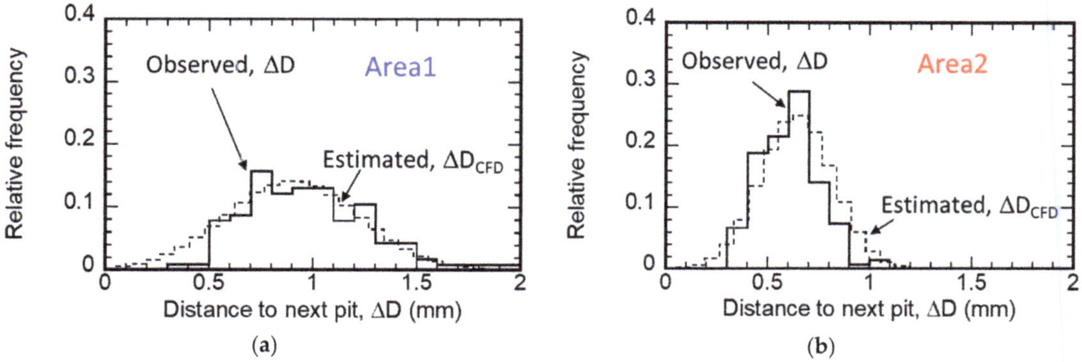

Figure 6. Comparison of the ΔD_{CFD} distribution based on the bubble distribution by CFD and the observed ΔD distribution for the (**a**) upper area (Figure 5a) and (**b**) lower area (Figure 5b). The observed distribution of ΔD was similar to the distribution of ΔD_{CFD}. The correlation between bubble size and damage was recognized.

In the damage observation, not only were beaded pits observed in the streak-like defects, but also isolated pits that were not on the streak-like defects. Morphologies of the isolated pits, which are considered to be the initiation for the formation of streak-like defects, were measured. The pit morphologies for depth, D_{max}, and equivalent radius, R_d, were measured with a One-shot 3D Measurement Macroscope VR-3200 (KEYENCE, Osaka, Japan). The equivalent radius, R_d, was calculated, assuming that the area of the measured pit was a circle. Figure 7 shows the relationship between the maximum pit depth, D_{max}, and the pit radius, R_d, of the isolated pits observed in Area 1 and Area 2 of Figure 3b. D_{max} in Area 1 and Area 2 were distributed with similar variations, but the R_d tended to be larger in Area 1, where larger gas bubbles existed. This suggested that the size of the gas bubbles affects the size of the damage. Furthermore, the isolated pits observed in this study tended to have a larger radius and shallower depth than the ones formed by microjets during cavitation collapse when microbubbles were not injected [7,18]. These results also suggested that the isolated pits observed in this study were formed by a different mechanism than the pits caused by microjets.

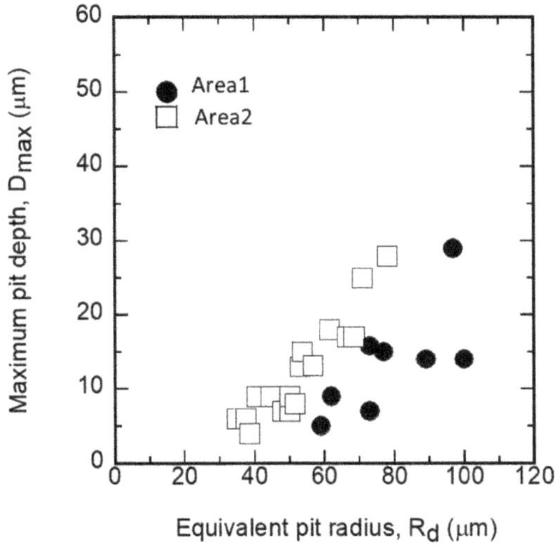

Figure 7. Relationship between the maximum pit depth, D_{max}, and the pit radius, R_d, of the isolated pits observed in Area 1 and Area 2 of Figure 3b. The D_{max} in Area 1 and Area 2 were distributed with similar variations, but the R_d tended to be larger in Area 1, where larger gas bubbles existed.

3. Numerical Analysis

The observed damage results and analyses on the damage indicated that the gas bubbles, injected to reduce MACPs, had the potential to be nuclei of cavitation. As for the isolated pits, those were shallow and wide in Area 1 compared to those in Area 2. The gas bubble sizes near Area 1 and Area 2 were different. In order to investigate the effect of gas bubble size on the formation of pits, the MICP applied on the solid wall was estimated by numerical analysis, and the formation mechanism of the observed pits were considered. At first, the MICP emitted by vibration of a single gas bubble was estimated based on bubble dynamics. Next, the MICP applied on the solid wall was estimated considering the distance between the gas bubble and the solid wall.

3.1. *The Estimation of MICP Emitted by Gas Bubble Vibration*

First, the Keller-Miksis equation in Equation (3) [19] was used to calculate the pressure at the bubble surface, which was the MICP emitted by gas bubble vibrations.

$$\left(1-\frac{1}{c_L}\dot{R}_g\right)R_g\ddot{R}_g+\frac{3}{2}\left(1-\frac{1}{3c_L}\dot{R}_g\right)\dot{R}_g^{\,2}=\frac{1}{\rho_L}\left(1+\frac{1}{c_L}\dot{R}_g\right)\left(p_L-p_\infty\left(t+\frac{R_g}{c_L}\right)\right)+\frac{\dot{p}_L}{\rho_L c_L}R_g \tag{3}$$

where c_L is the sound velocity, p_g is the pressure inside the bubble was estimated by Equation (4), and the pressure at the bubble interface, p_L, was estimated by Equation (5).

$$p_g(R)=\left(p_{\infty 0}-p_v+\frac{2\sigma}{R_{g,0}}\right)\left(\frac{R_{g,0}}{R_g}\right)^{3\gamma} \tag{4}$$

$$p_L=p_g-\frac{1}{R}\left(2\sigma+4\mu\dot{R}_g\right)+p_v \tag{5}$$

The change in MACP around the bubble, p_∞, was given as in Figure 8, simplifying the change in the mercury target, since the duration of the negative pressure in the mercury near the target vessel lasted for a few milliseconds in the pressure wave analysis of the mercury target [12]. The estimation was performed by systematically varying $R_{g,0}$, $p_{\infty,0}$, and t_1 at 40~500 μm, −0.040~0 MPa, and 1~5 ms, respectively.

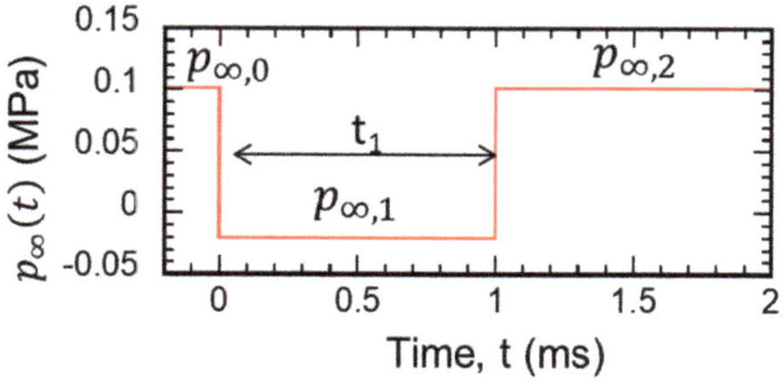

Figure 8. The pressure change around a gas bubble. This time history was used for bubble dynamics calculations in Equation (3).

Figure 9 shows the time behaviors of the gas bubble radius and the pressure at the bubble interface under the condition that the initial gas bubble radius, $R_{g,0}$, was 100 μm, and that the pressure around the gas bubble, $p_{\infty,0}$, decreased stepwise from atmospheric pressure to $p_{\infty,1} = -0.02$ MPa for t_1 = 1 ms then returned to atmospheric pressure. While the pressure, p_∞, was negative, the bubble continued to expand. When the pressure changed to positive, the bubble began to contract and reached a minimum radius. The gas bubble then oscillated through repeated expansion and contraction. As the bubble contracted, the pressure at the bubble interface increased. The increasing pressure at the bubble interface was greatest at the initial contraction, and then local impact pressure at the bubble interface decreased with the number of repetitions. Under the conditions shown in Figure 9, the local impact pressures at the bubble interface exceeded the 0.2% proof stress of the mercury target vessel material up to the fifth repetition. As shown in Equation (3), the density and surface tension of the liquid affect the bubble behavior and the pressure at the bubble interface. The density and surface tension of mercury were about 13.5 and 6.7 times higher than those of water, respectively. These effects on the local impact pressure at the bubble interface, p_L, was estimated. As a result, the pressure, p_L in mercury was 10 times higher than that in water when $R_{g,0}$ = 100 μm, $p_{\infty,1} = -0.02$ MPa, and t_1 = 5 ms.

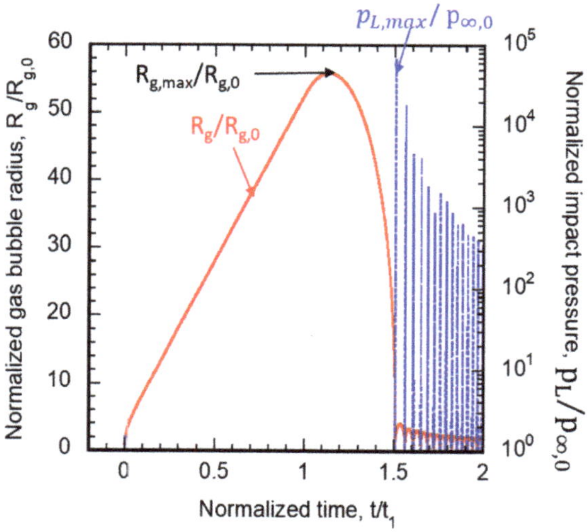

Figure 9. Time behaviors of the gas bubble radius and the pressure at the bubble interface at $R_{g,0} = 100$ μm, $p_{\infty,1} = -0.02$ MPa for t_1 = 5 ms. An impact pressure exceeding the 0.2% proof stress of the wall material was generated, which indicated that the local impact pressure generated from the gas bubble had the potential to form a pit on the wall surface.

3.2. The Estimation of MICP Applied on a Solid Wall

Figure 10 shows a schematic diagram of the analysis for the estimation of MICP applied on a solid wall. Here, for the simple estimation of whether the MICP applied to the solid wall from the gas bubble vibration was enough to generate the damage, the following were assumed:

(a) There is no collapse of the bubbles, which would be accompanied by microjets since the correlation between the observed distance to the next pit, ΔD, and the natural period of the existing gas bubbles was recognized, as shown in Figure 6;
(b) The maximum pressure generated at the bubble interface when a single bubble contracted at an arbitrary distance from the solid wall, h, propagated as a spherical wave [20].

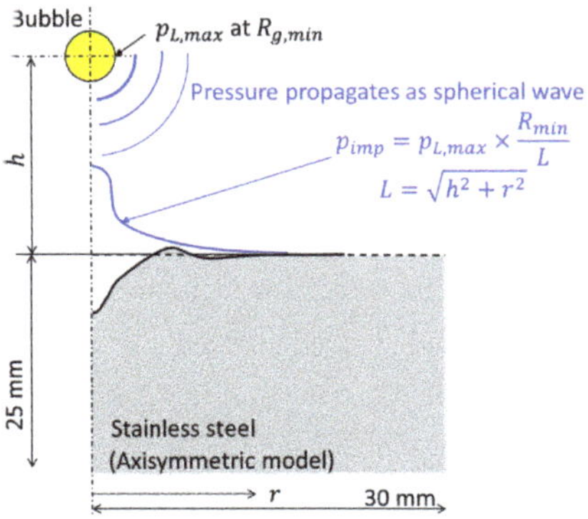

Figure 10. Schematic diagram of the analysis for the calculation of pit morphology. For analysis, the applied pressure on the wall surface was assumed that the local impact pressure, $p_{L,max}$, generated on the bubble surface propagated as a spherical wave. The applied pressure on the solid wall was expressed by $p_{imp}(r)$.

Therefore, the MICP applied on the solid wall, $P_{imp}(r)$, was estimated by using the general equation for the distance decay of pressure for a spherical wave in liquid, as shown in Equation (6).

$$P_{imp}\ (r) = p_{L,max} \times \frac{R_{g,min}}{L} \tag{6}$$

$$L = \sqrt{h^2 + r^2} \tag{7}$$

where r is the distance on the wall surface from the center where the gas bubble exists directly above it, $R_{g,min}$ is the gas bubble radius at which $p_{L,max}$ occurs when the bubble contracts, and L is the distance from the bubble center to any position on the solid wall surface as expressed in Equation (7). The MICP applied on the solid wall varies with not only $p_{L,max}$ but also the distance from the bubble center to the wall surface, h.

Figure 11 shows the relationship between the initial gas bubble radius and the normalized MICP applied on the solid wall, $P^*_{imp,r=0}$, for different distances of the gas bubble to the solid wall, h, for $p_{\infty,1} = -0.02$ MPa and t_1 = 5 ms. MCIP applied on the solid wall, $P^*_{imp,r=0}$, depended on not only the radius of the initial gas bubble but also with the distance between the gas bubble and h. $P^*_{imp,r=0}$ became larger when larger gas bubbles existed near the wall.

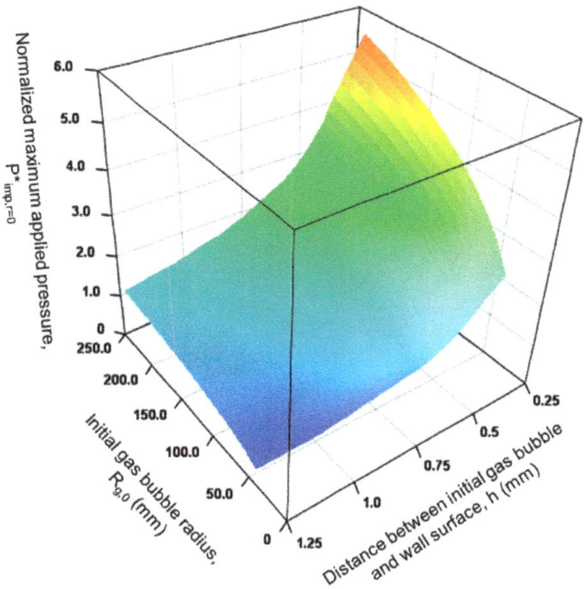

Figure 11. Relationship between the initial gas bubble radius and the maximum applied pressure on the wall surface, $P_{imp,r=0}$, for different distances between the gas bubble and the solid wall, h, in the case of $p_{\infty,1} = -0.02$ MPa and $t_1 = 5$ ms. The maximum applied pressure, $P_{imp,r=0}$, increased with increases in $R_{g,0}$ and decreases in h.

4. Discussion

4.1. The Generation Mechanism of Streak-like Defects with Beaded Pits

As shown in Figure 3, the damage was mitigated by the stable injection of gas bubbles to reduce MACPs. On the other hand, damage that might have been caused by the gas bubbles was observed. In this section, the generation mechanism of the streak-like defects with beaded pits is discussed. The following are possible reasons for the generation of damage caused by the gas bubbles.

(A) Generation of isolated pits

Figure 12 shows a photograph of a series of the isolated pits together with the pressure history emitted from the gas bubble (Enlarged Figure 9). As shown in Figure 12, the gas bubble near the solid wall vibrated by being subjected to the macroscopic pressure changes and the MICP emitted when the gas bubble contracted. The local impact pressure caused the plastic deformation to form a pit on the solid wall. In the photograph of Figure 12, the isolated pits were aligned with a slope consistent with the trajectory of the gas bubbles shown in Figure 4. The radius of the pits decreased with the direction of the gas bubble travel, and the distance between the pits became smaller. On the other hand, in the numerical results, local impact pressure was generated when the bubble contracted, and the local impact pressure decreased with the number of contractions. As such, the time interval between the next pressure generation incident became short. These results reproduce the observed results. That is, a series of isolated pits can be caused by the vibration of a single gas bubble.

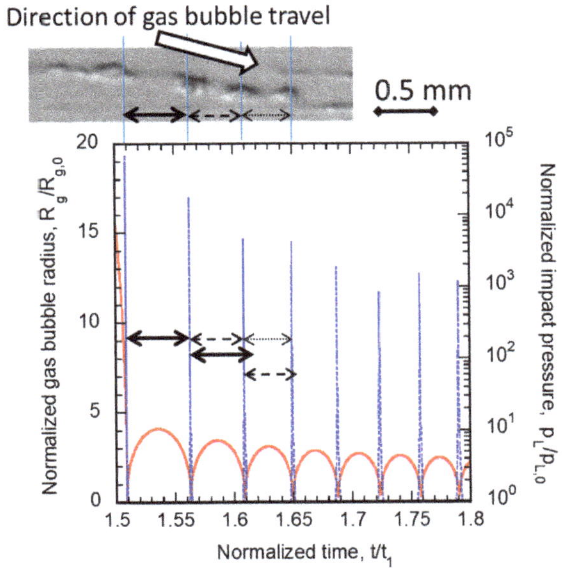

Figure 12. Photograph of a series of the isolated pits together with the pressure history emitted from the gas bubble (Enlarged Figure 9).

(E) Generation of the streak-like defects

In a mercury target, about 10^8 pulsed proton beams were injected during the lifetime of the vessel. It is thought that the continuous pits overlap in the paths where the gas bubbles pass more frequently, as shown in Figure 13, resulting in streak-like defects.

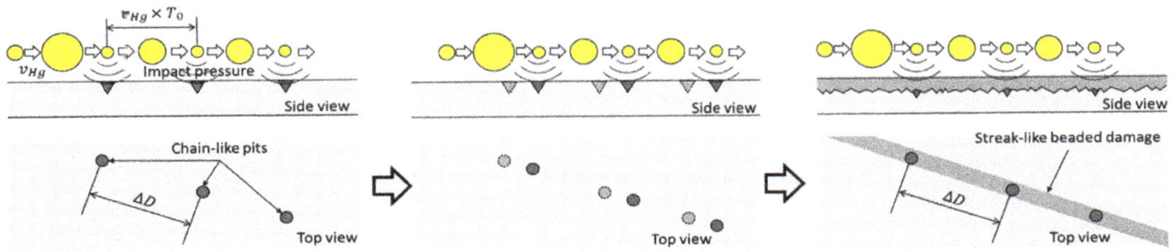

Figure 13. Formation mechanism of the streak-like defect with beaded pits.

(C) Formation of bead-like continuous pits over the streak-like defects

It is thought that gas bubbles pass over the streak defects, and the miniscule impact pressure released by the vibration of the gas bubbles forms a series of beaded pits in the streak-like defects.

4.2. The Effect of the Initial Bubble Radius on the Generation of Isolated Pits

As shown in Figure 11, the MICP applied on the solid wall, $P_{imp}(r)$, depended on the gas bubble diameter, $R_{g,0}$, and the distance between the gas bubble and the wall, h. In the mercury target, larger gas bubbles were distributed in the upper part of the target (Area 1) than in the lower part (Area 2) as shown in Figure 2, and thus larger pits were expected to be observed in the Area 1. However, the observations shown in Figure 7 showed no significant difference in pit depth between Area 1 (upper) and Area 2 (lower), although the pit radius was larger in Area 1. The applied pressure on the wall depends not on the initial bubble radius, $R_{g,0}$, but also on the distance between the bubble and wall, h, as shown in Figure 11. Therefore, we discuss the distance between the gas bubble and the wall surface. Figure 14 showed a distribution in the radial direction of the MICP applied on the solid wall, $P_{imp}(r)$, while varying $R_{g,0}$ and h. Comparing $P_{imp}(r)$ with the same $R_{g,0}$, $P_{imp}(r = 0\ \mu m)$ became larger with decreased of h. However, $P_{imp}(r > 150\ \mu m)$ was almost similar. On the other hand, comparing $P_{imp}(r)$ with same h, $P_{imp}(r = 0\ \mu m)$ was larger when R_g was larger, regardless of r. These results indicated that larger bubbles existing away from the solid wall and small bubbles existing near the solid wall generate pits with similar depth, but the former generates relatively wide pits.

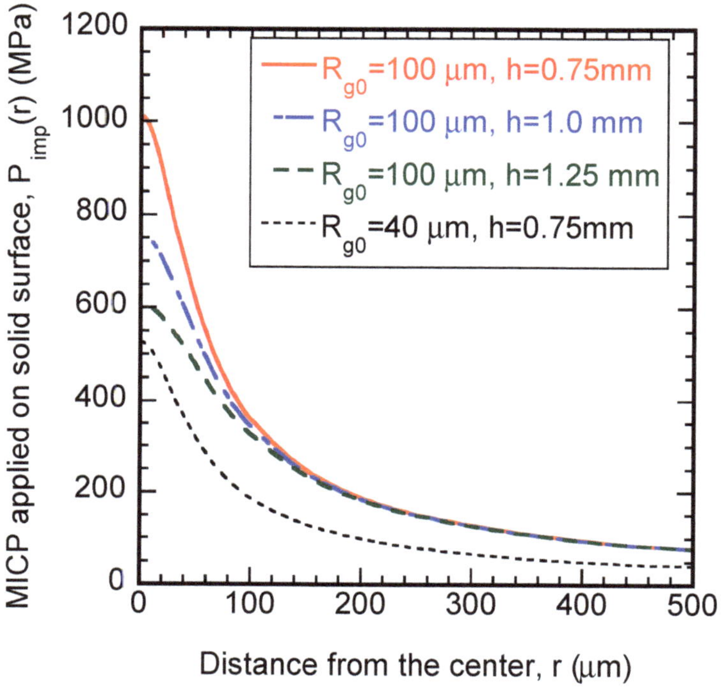

Figure 14. Distribution in the radial direction of the MICP applied on the solid wall, $P_{imp}(r)$, while varying $R_{g,0}$ and h.

Gas bubbles flowing near the wall are subject to centripetal force and drag force due to wall interaction. Near the beam window, bubbles tend to move away from the beam window due to the centripetal force caused by the upstream corners in the mercury target. The mercury flow bends from the entrance side to the beam window. Therefore, centripetal force acts on the bubbles toward the center of the radius of curvature. The centripetal force acting on the bubbles in the flow is generally expressed as shown in Equation (8), which depends on the size of the bubble diameter, like the buoyancy force.

$$F_c = \rho \times a \times \frac{4}{3}\pi R_{g,0}^3 \tag{8}$$

where a is the acceleration of the mercury in the direction of the center of curvature in the corner.

Guan et al. calculated the trajectory of a gas bubble rising near a wall in a static liquid and showed that the gas bubble rises while leaving the wall. And, when the bubble's Reynolds number is high, i.e., the bubble radius is large, the distance between the bubble leaving the wall while rising becomes larger due to the lift generated by the influence of the wall surface [21]. In the mercury target vessel, the large bubbles were considered to be moving away from the wall, resulting in the applied local pressure, $P_{imp}(r)$, equivalent to that of the small bubbles.

5. Concluding Remarks

In the mercury target vessel of a spallation neutron source, it has been recognized that gas microbubbles reduce the MACPs caused by the thermal expansion of mercury due to proton beam injection. On the other hand, the mechanism of damage formation by the MICP emitted by the gas bubbles' vibration was considered in the formation of isolated pits, streak-like defects, and the beaded pits in these defects, which had not been observed under low proton beam power. The knowledge obtained in this study is as follows:

(1) As a result of operating the mercury target at a high proton beam power with sufficient injection of gas microbubbles, streak-like defects with beaded pits were observed in the mercury target vessel. The direction of the streak-like defects were almost consistent with the flow direction of the gas bubbles estimated by CFD analysis. The distance between the continuous beaded pits observed on the streak-like defects was wider in the upper area than in the lower area in the gravity direction. The distance to the next pit on the beaded pits was estimated assuming that the gas bubbles moved with mercury flow velocity and pits were formed by the natural period of bubble vibration, based on the bubble distribution results from the CFD analysis. The distance was wider in the upper area where more large bubbles were distributed, which was consistent with the observation results;

(2) In the estimation based on the Keller-Miksis equation, the gas bubbles expanded, then contracted, and repeated the expansion and contraction again. The MICP was emitted in each contraction. The MICP emitted by the gas bubble vibrations decreased with the number of repetitions of the vibration and its period became shorter. In the observed series of isolated pits, the damage size decreased and the distance between the pits became shorter in the flow direction of the gas bubble. These results indicated a correlation between the motion of the gas bubbles and the damage;

(3) The MICP applied on the solid wall, which were generated in the contractions of gas bubbles, could be sufficient to generate the plastic deformation and pits in the wall material. The applied pressure on the solid wall was higher when the initial radius of the gas bubble was larger and when the gas bubble existed closer to the solid wall;

(4) In the observed damage, in the upper area of the gravity direction where many of the gas bubbles with a large initial radius existed, the pit radius tended to be large, although the maximum pit depth was similar to those in the lower area. This was considered that the large gas bubbles were further away from the wall surface than the small gas bubbles and the maximum MICP applied on the solid wall surface was similar in the upper and lower areas.

In this paper, we showed the potential of gas microbubbles in causing damage on solid wall surfaces. In the mercury target, gas microbubbles reduce MACPs, resulting in suppressing cavitation damage. On the other hand, the gas microbubbles emit MICP, which causes the damage. Increasing the number of gas microbubbles is effective, but the gas bubbles may become cavitation nuclei which emit MICP. It will be important to optimize the number of gas bubbles to be injected, taking into account the interaction of multiple gas bubbles for the operation with a higher proton beam power and longer lifetime of the mercury target vessel. In addition, for various application techniques using gas microbubbles, especially when damage is a concern, it is necessary to consider the local impact pressure emitted from the gas bubbles themselves as shown in this paper, not only the microjets generated in the bubbles collapsing near the solid wall.

Author Contributions: Conceptualization, M.F. and H.K.; methodology, H.K. and M.F.; validation, M.F., T.W., and H.K.; formal analysis, H.K. and T.W.; investigation, H.K. and T.W.; data curation, H.K. and T.W.; writing—original draft preparation, H.K.; writing—review and editing, T.W. and M.F.;

visualization, H.K. and M.F.; project administration, M.F. and H.K. All authors have read and agreed to the published version of the manuscript.

Funding: This research received no external funding.

Data Availability Statement: The raw data supporting the conclusions of this article will be made available by the authors on request.

Acknowledgments: In compiling this paper, we obtained co-operation from the members of the Neutron Source Section, Materials and Life Science Division, JAEA, for their assistance in observing the damage to the mercury target vessel. We hereby express our acknowledgments.

Conflicts of Interest: The funders had no role in the design of this study; in the collection, analyses, or interpretation of data; in the writing of the manuscript; or in the decision to publish the results.

References

1. Hashim, A.; Yaakob, O.B.; Koh, K.K.; Ismail, N.; Ahmed, Y.M. Review of Micro-bubble Ship Resistance Reduction Methods and the Mechanisms that Affect the Skin Friction on Drag Reduction from 1999 to 2015. *J. Teknol.* **2015**, *74*, 105–114. [CrossRef]
2. Takahashi, M. Microbubble: Fundamental Properties and Application for Semiconductor Cleaning. *J. Jpn. Soc. Precis. Eng.* **2017**, *83*, 636–640. [CrossRef]
3. Chomas, J.E.; Dayton, P.; Allen, J.; Morgan, K.; Ferrara, K.W. Mechanisms of contrast agent destruction. *Trans. Ultrason. Ferroelectr. Freq. Control* **2001**, *48*, 232–248. [CrossRef] [PubMed]
4. Available online: https://neutrons.ornl.gov/ (accessed on 20 September 2024).
5. Available online: https://j-parc.jp/c/en/index.html (accessed on 20 September 2024).
6. Robert TKnapp, R.T. Recent Investigations of the Mechanics of Cavitation and Cavitation Damage. *Trans. Am. Soc. Mech. Eng.* **1955**, *77*, 1045–1054.
7. Plesset, M.S.; Chapman, R.B. Collapse of an Initially Spherical Vapour Cavity in the Neighbourhood of a Solid Boundary. *J. Fluid Mech.* **1971**, *47*, 283–290. [CrossRef]
8. Futakawa, M.; Kogawa, H.; Hino, R.; Date, H.; Takeishi, H. Erosion damage on solid boundaries in contact with liquid metals by impulsive pressure injection. *Int. J. Impact Eng.* **2003**, *28*, 123–135. [CrossRef]
9. Kogawa, H.; Futakawa, M.; Ishikura, S. Dynamic Response of Mercury Subjected to Pressure Wave. *J. Nucl. Sci. Technol.* **2007**, *44*, 523–529. [CrossRef]
10. Futakawa, M.; Kogawa, H.; Hasegawa, S.; Naoe, T.; Ida, M.; Haga, K.; Wakui, T.; Tanaka, N.; Matsumoto, Y.; Ikeda, Y. Mitigation technologies for damage induced by pressure waves in high-power mercury spallation neutron sources (II)—Bubbling E. ect to Reduce Pressure Wave—. *J. Nucl. Sci. Technol.* **2008**, *45*, 1041–1048. [CrossRef]
11. Kogawa, H.; Naoe, T.; Kyotoh, H.; Haga, K.; Kinoshita, H.; Futakawa, M. Development of microbubble generator for suppression of pressure waves in mercury target of spallation source. *J. Nucl. Sci. Technol.* **2015**, *52*, 1461–1469. [CrossRef]
12. Kogawa, H.; Naoe, T.; Futakawa, M.; Haga, K.; Wakui, T.; Harada, M.; Takada, H. Mitigation technologies for damage induced by pressure waves in high-power mercury spallation neutron sources (IV)—Measurement of pressure wave response and microbubble effect on mitigation in mercury target at J-PARC—. *J. Nucl. Sci. Technol.* **2017**, *54*, 733–741. [CrossRef]
13. Ida, M.; Naoe, T.; Futakawa, M. Suppression of cavitation inception by gas bubble injection: A numerical study focusing on bubble-bubble interaction. *Phys. Rev. E* **2007**, *76*, 046309. [CrossRef] [PubMed]
14. Ida, M.; Naoe, T.; Futakawa, M. On the effect of microbubble injection on cavitation bubble dynamics in liquid mercury. *Nucl. Instrum. Methods Phys. Res. Sect. A Accel. Spectrometers Detect. Assoc. Equip.* **2009**, *600*, 367–375. [CrossRef]
15. Naoe, T.; Kinoshita, H.; Wakui, T.; Kogawa, H.; Haga, K. *Technical Note for the Cavitation Damage Inspection for Interior Surface of the Mercury Target Vessel (1) Development of Specimen Cutting Machine for Remote Handling. JAEA-Technology 2022-018*; Japan Atomic Energy Agency: Ibaraki, Japan, 2022. (In Japanese)
16. Kogawa, H.; Haga, K.; Futakawa, M.; Tsuzuki, T.; Murai, T. *Optimization of Mercury Flow with Microbubbles in the Target-Vessel Design by Means of Machine Learning. JAEA-Technology 2022-023*; Japan Atomic Energy Agency: Ibaraki, Japan, 2022. (In Japanese)
17. Kameda, M.; Yamada, M.; Matsumoto, Y. Motion of a Spherical Bubble in an Oscillatory Pressure Field. *Trans. JSME (B)* **1998**, *64*, 1650–1658. (In Japanese) [CrossRef]
18. Futakawa, M.; Naoe, T.; Kogawa, H.; Date, H.; Ikeda, Y. Micro-Impact Damage Caused by Mercury Bubble Collapse. *JSME Int. J. Ser. A* **2005**, *48*, 234–239. [CrossRef]
19. Keller, J.B.; Miksis, M. Bubble oscillations of large amplitude. *J. Acoust. Soc. Am.* **1980**, *68*, 628–633. [CrossRef]
20. Kandula, M.; Freeman, R. On the Interaction and Coalescence of Spherical Blast Waves. *Shock Waves* **2008**, *18*, 21–33. [CrossRef]
21. Guan, C.; Yanase, S.; Matsuura, K.; Kouchi, T.; Nagata, Y. Numerical Study on Motion of a Single Spherical Bubble near Vertical Wall in the quiescent Fluid using the Modified Force-Coupling Method. *Nagare* **2018**, *37*, 281–289. (In Japanese)

fluids

Article

Simplified Approach to Evaluate Cavitation Intensity Based on Time Information on Imposed Pressure in Liquid

Hiroyuki Kawashima [1,*], Hiroyuki Kogawa [2], Masatoshi Futakawa [2] and Nobuatsu Tanaka [1]

[1] Graduate School of Science and Technology, Ibaraki University, Mito 316-8511, Ibaraki, Japan; nobuatsu.tanaka.07@vc.ibaraki.ac.jp
[2] J-PARC Center, Japan Atomic Energy Agency, Tokai 319-1195, Ibaraki, Japan; kogawa.hiroyuki@jaea.go.jp (H.K.); futakawa.masatoshi@jaea.go.jp (M.F.)
* Correspondence: 22nd252g@vc.ibaraki.ac.jp

Abstract: Cavitation damage is an important research topic in fluid–structure interactions, such as those being studied using the mercury target for the pulsed neutron source at the Materials Life Science Experimental Facility/Japan Proton Accelerator Complex. Hence, the estimation of cavitation damage (cavitation intensity) is required from the perspective of structural integrity. The results of previous studies suggest that the maximum radii of cavitation bubbles immediately prior to collapse are related to cavitation intensity. Therefore, we propose a method for estimating the maximum radius from the time information by measuring the vibrations of structure walls that are induced by collapsing cavitation bubbles in a confined liquid. In this study, we used a magnetic impact testing machine to experimentally investigate the cavitation bubble dynamics, directly observe the bubble collapsing behavior, and measure the induced vibration. We experimentally confirmed that the time information is useful in the estimation of the maximum radii of bubbles. Moreover, we theoretically derived a simple evaluation formula to estimate the maximum radius from the time responses of the imposed pressure in a confined liquid in a structure.

Keywords: cavitation; bubble behavior; mercury; high-speed video camera; Keller–Miksis equation

1. Introduction

From the perspective of structural integrity, cavitation is a critical issue in fluid–structure interfaces, e.g., pumps, ship propellers, and pipelines [1–3]. Hence, the quantitative evaluation of the impacts induced by cavitation bubbles has become a key issue from the perspective of the structural integrity of fluid machines. In cases of high-intensity spallation neutron sources using liquid heavy metals, such as the mercury targets in the spallation neutron sources at Oak Ridge National Laboratory (ORNL) [4] and Materials Life Science Experimental Facility/Japan Proton Accelerator Complex (MLF/J-PARC) [5], local impact evaluation is a critical issue in estimating lifetime and improving durability. The liquid mercury target of a spallation pulsed neutron source at the J-PARC, in which rapid pressure fluctuation is induced by the thermal expansion of mercury owing to high-intensity pulsed proton beam injection. Negative pressure is generated by the interaction between the structure and confined liquid mercury, and cavitation is induced [6,7]. Cavitation bubbles impose localized impact forces on the interface, causing severe cavitation damage to structural solid walls. Cavitation damage is an issue related to the structural integrity of mercury targets and the long-term stable operation of neutron sources. Therefore, it is important to develop a technique for estimating cavitation damage, that is, cavitation intensity, near the vessel walls of mercury targets [8].

In recent years, numerous studies have been reported on cavitation dynamics causing wall damage. Reuter et al. [9] followed the flow field around a collapsing bubble near the wall using the PTV method, even after the collapse. Huang et al. [10] showed the formation of two vortex flow patterns caused by the collapse of cavitation bubbles near a rigid wall. Gonzalez-Avila et al. [11] studied a biomimetic gas-entrapping microstructure surface (GEMS) and found that the entrapment of air inside the cavities repels cavitation bubbles away from the surface, thereby preventing cavitation damage. Sun et al. [12] investigated the effect of rigid walls with gas capture holes on the dynamic behavior of laser-induced bubble collapse. They showed that gas capture holes affected the jet morphology and bubble migration direction, providing new insight into how to prevent cavitation from occurring.

Among various studies on cavitation dynamics, we are most interested in the correlation between cavitation intensity and maximum bubble radius. Isselin et al. [13] and Soyama [14] showed that the impact pressure caused by cavitation collapse has a linear relationship with the maximum bubble radius R_{max} and time T_D (time duration between bubble generation and collapse). Philipp and Lauterborn [15] showed that the area of cavitation damage on a wall is linearly correlated with the maximum radius of the cavitation bubble. Wakui et al. [16] applied these relationships to gain new insight into the relationship between power and cavitation damage evolution in the J-PARC mercury target vessel. The practical application of these relationships to mercury target vessels requires the estimation of the size of cavitation bubbles in the mercury target vessel. Therefore, we propose a method by which to extract the maximum bubble volume from the vibration information of a vessel wall exposed to cavitation phenomena. That is, the vibration waveform data correspond to the dynamic behavior of the bubbles. The relationship between the wall vibration waveform and bubble behavior has been investigated in underwater spark discharge and laser irradiation experiments, and there is a proportional correlation between the maximum bubble radius R_{max} and time T_D (time duration between bubble generation and collapse) of the wall vibration that occurs during bubble collapse [14,17]. This relationship between T_D and R_{max} is only valid for time-symmetric bubble behavior in which the bubble growth time T_α (time duration between bubble generation and maximum) and bubble collapse time T_β (time duration between bubble maximum and collapse) are equal, such as observations made during spark discharge or laser irradiation.

However, cavitation bubbles in mercury target vessels are estimated to behave asymmetrically with respect to the time axis when $T_\alpha \neq T_\beta$ [18]. In such cases, the relationship between R_{max} and T_D cannot be simply applied. In our previous report, we focused on T_β as the time parameter for predicting the bubble radius and found that T_β has a proportional correlation with R_{max}. This general relationship can be explained by bubble dynamics and is applicable to the prediction of the bubble radius in mercury target vessels as well as bubbles caused by spark discharges and laser irradiation [19]. Recently, a non-contact vibration measurement system using laser Doppler in a high-radiation field was realized in the J-PARC mercury target vessel [20], and the data are being accumulated. This study provides fundamental insights for estimating the amount of damage to the J-PARC mercury target vessel.

Hence, in this study, we first investigated the correlation between the behavior of bubbles generated in mercury and pressure fluctuations using a magnetic impact testing machine (MIMTM) that reproduces pressure waves generated inside a mercury target vessel [6]. Because it is difficult to directly observe the behavior of bubbles in mercury, a high-speed video camera was used to observe the behavior of bubbles adhering to the glass wall of the vessel, which are likely to be precisely different from the bubbles influenced by the wall boundary. Simultaneously, T_β was measured from the correlation between the time history data on the acceleration vibration waveform of the vessel, pressure fluctuation in the mercury, and visualized bubble behavior, considering wettability. As a result, we propose a simplified formula that allows T_β to be estimated from T_D based on the imposed pressure responses into the confined liquid. This demonstrates the possibility of a simple and rapid evaluation of R_{max} (cavitation intensity) via the in situ measurement of cavitation-derived vibration waveforms of real structures.

2. Experimental Setup

Figure 1 shows the mercury chamber and specimen in the MIMTM. The inner diameter and height were 100 and 15 mm, respectively. Impulsive pressure was imposed on the mercury through a disk plate specimen driven by a striker, which was controlled by an electromagnetic force associated with the input electric power. The magnitude of pressure varied according to the power imposed on the MIMTM [6]. The pressure in the mercury was measured using a pressure gauge (Entran®, EPX, Fairfield, NJ, USA) that was installed into the mercury through the lid of the mercury chamber, as illustrated in Figure 1. Figure 2 shows a schematic diagram of the cavitation visualization setup. A high-speed CCD video camera (NAC, MEMRECAM RX-6, Tokyo, Japan) with a frame rate of up to 2×10^4 f/s was used to visualize the cavitation bubbles. The observed bubbles were generated at the interface between the mercury and a glass window installed in the lid of the mercury chamber. A trigger signal was precisely controlled and input from the MIMTM to the camera at the onset of the striker driving to investigate the relationship between the time response of the imposed pressure in the mercury, the acceleration measured at the striker, and the optical images of the bubble behavior. Acoustic vibrations were measured using an accelerometer (RION, PV-90B, Tokyo, Japan) fixed on the striker.

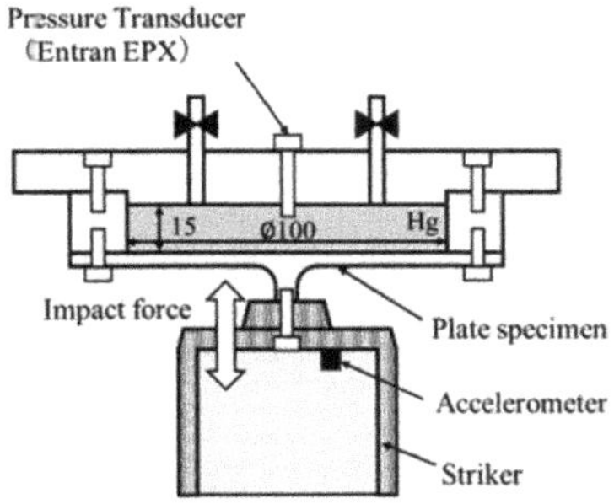

Figure 1. Schematic of the experimental setup.

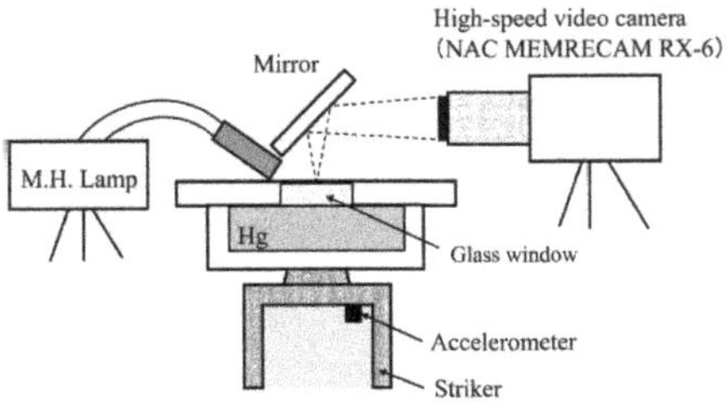

Figure 2. Schematic of the cavitation visualization setup.

3. Theoretical Investigation

In this study, a one-dimensional bubble dynamics model was applied to rationally discuss the dependence of bubble behavior on physical properties and pressure fluctuations. Therefore, a theoretical investigation was conducted using the Keller–Miksis equation [21], considering the compressibility and acoustic radiation in the surrounding liquid, as given by Equation (1):

$$\left(1-\frac{\dot{R}}{S}\right)R\ddot{R}+\frac{3}{2}\left(1-\frac{\dot{R}}{3S}\right)\dot{R}^2=\left(1+\frac{\dot{R}}{S}\right)\frac{p_b(R,t)-p_\infty\left(t+\frac{R}{S}\right)}{\rho}+\frac{R}{\rho S}\frac{dp_b(R,t)}{dt}, \quad (1)$$

where R is the bubble radius, $\dot{R}$ is the first-order derivative of the bubble radius over time, $\ddot{R}$ is the second-order derivative of the bubble radius, ρ is the density of the surrounding liquid, p_∞ is the liquid pressure at infinity, S is the sound speed of the liquid, and p_b is the pressure at the bubble surface, which is expressed as follows:

$$p_b(R,t) = p_g(R) - \frac{1}{R}\left(2\sigma + 4\mu\dot{R}\right), \quad (2)$$

where μ is the liquid surface tension and p_g is the pressure inside the bubbles. The polytropic change, denoted by p_g is given by Equation (3):

$$p_g(R) = \left(p_{\infty 0} - p_v + \frac{2\sigma}{R_0}\right)\left(\frac{R_0}{R}\right)^{3\gamma} + p_v, \quad (3)$$

where $p_{\infty 0}$ is the initial liquid pressure at infinity and γ is the specific heat ratio of the gas in the bubble. The time evolution methods used in this analysis were the Richardson extrapolation and the Bulirsch–Stoer method [22]. Table 1 lists the physical properties of the liquid mercury used in the analysis. The specific heat ratio γ of the gas was 1.41, assuming the presence of air. The initial radius of the bubbles was set to 50 μm.

Table 1. Physical properties of mercury.

	Mercury (298 K)
Density (kg/m^3)	13,500
Saturated vapor pressure (Pa)	0.16
Surface tension (N/m)	0.4855
Sound speed (m/s)	1400
Viscosity (Pa · s)	1.49×10^{-3}

4. Results

4.1. Visualization Results of Cavitation Bubbles

Figure 3 shows the typical cavitation bubble behavior of mercury. This behavior was visualized via the complex reflection of the insides of bubbles that had adhered to the glass wall under illumination. The bubble generation recognized as the reflected portion of the illumination gradually increased in size from t = 0.6 ms. By t = 0.8 ms, the contours of the bubbles were clearly visible. Thereafter, the bubble gradually grew and reached its maximum radius at t = 1.8 ms. The bubble then began to contract at approximately t = 2.0 ms and disappeared only 0.2 ms later (i.e., at approximately 2.2 ms). The bubble took 1.8 ms to grow to its maximum radius and 2.0 ms to disappear. These results indicate that bubble behavior in the MIMTM differs from that of bubble generation via underwater spark discharge or the laser irradiation method. That is, the growth and collapse times (T_α and T_β) are different. Hence, the behavior is asymmetric with respect to the time axis.

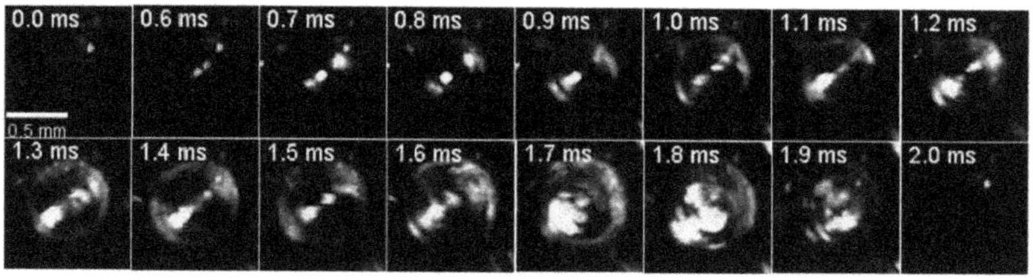

Figure 3. Visualization results of cavitation bubble behavior in the MIMTM.

Figure 4 shows (a) the radius of bubbles adhered to the glass wall, (b) the vibration acceleration measured by the striker, and (c) the time variation in the pressure transducer measurement. Panel (a) shows the time evolution of the radii of three bubbles generated by the same impulsive pressure. The radius of the adherend, denoted by c, was derived as an ellipse with a major axis of a and a minor axis of b. The equivalent radius $c = \sqrt{ab}$ was obtained from the area. The measured vibration acceleration was subjected to a high-pass filter to remove low-frequency components owing to the striker motion. Here, t = 0 ms represents the time when the trigger is issued, that is, the time when the striker begins to move. First, bubbles were observed at t = 0.7 ms. That is, 0.7 ms elapsed before bubbles adhered to and grew on the glass wall and became observable. Thereafter, the bubbles gradually grew until they reached their maximum radii at approximately t = 1.6 ms (T_α). At t = 1.6 ms, the mercury pressure began to increase, and the bubbles simultaneously began to contract. The bubbles completely collapsed at t = 2.0 ms (T_D). For all bubbles, the difference between the growth time T_α and collapse time $T_\beta = T_D - T_\alpha$ was large, showing asymmetric bubble growth and collapse behavior with respect to the time axis.

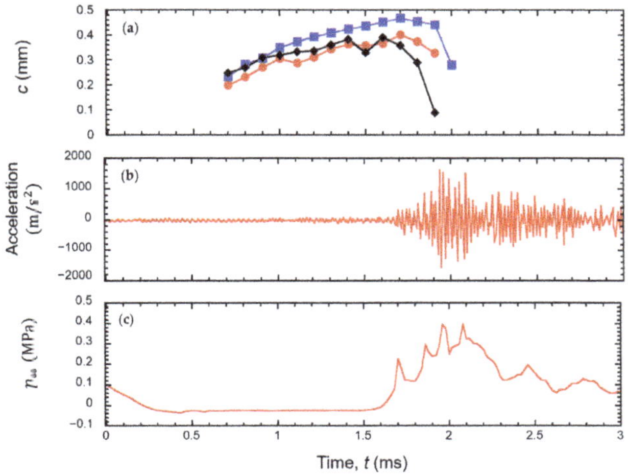

Figure 4. Measurements of (**a**) bubble radius, (**b**) striker acceleration response, and (**c**) mercury pressure response.

In the acceleration waveform measured in the chamber, high-frequency components began appearing at approximately 1.7 ms, immediately before the bubble collapse, as shown in Figure 3. The maximum value was reached at approximately 2.0 ms. Subsequently, the attenuation of the amplitude of the high-frequency components was observed. This attenuation may be due to the reflection of pressure waves inside the chamber during bubble collapse and the collapse of multiple bubbles with a time delay. In the pressure waveform, the pressure gradually decreased from t = 0.0 to t = 0.3 ms and remained constant at approximately −0.026 MPa. Thereafter, the pressure increased sharply. At t = 2.0 ms (T_D), the local pressure derived from the bubble collapse was superimposed, and the maximum pressure was reached. Hence, the maximum pressure can be attributed to the generation of localized impact forces caused by the bubble collapse. The superimposing of the periodical peak pressures with approximately 0.1 ms on the increasing pressure might be due to the repeated reflection of the pressure wave generated by the striker in the chamber.

These results indicate that the onset times of the accelerometer and pressure transducer oscillation peaks correspond to the T_D of bubble growth and collapse. Hence, it is possible to extract time information regarding the maximum radius of the cavitation bubble from the vessel vibration as well as the imposed pressure in the liquid. Moreover, these results suggest that an understanding of the pressure fluctuation can be used to extract parameters related to the impacts of bubble collapse on the wall surface. In Section 5, using the simple one-dimensional bubble dynamics model presented in Section 2, we discuss in detail the relationship between pressure fluctuations and bubble behavior as well as methods for extracting impact-related parameters.

4.2. Comparison of Experimental and Theoretical Results

Figure 5 compares the experimental and theoretical results for the time evolution of the bubble radius. The vertical axis is the bubble radius, normalized by the maximum bubble radius, and the horizontal axis is the time, normalized by the bubble growth time T_α. In the region of $t/T_\alpha < 1$ during the bubble growth process, discrepancies are recognized between the experimental and theoretical results. These discrepancies are particularly pronounced for small values of t/T_α, e.g., an approximately twofold difference was observed when $t/T_\alpha = 0.4$. The discrepancy between the experimentally observed and theoretically obtained bubble radii are attributable to the physical difference between the bubble radii on the glass surface and those in the theoretical investigation. Section 5 discusses the physical factors related to this discrepancy. However, the differences between the experimental and theoretical results are smaller for the bubble collapse process ($t/T_\alpha > 1$) than for the growth process ($t/T_\alpha < 1$), and the results tend to be in general agreement. These findings confirm that the relationship between the collapse time T_β and R_{max} predicted in our previous study is valid for asymmetric bubble behavior with respect to the time axis.

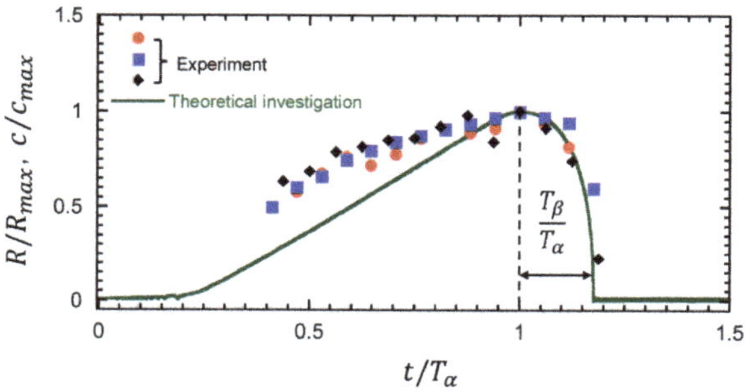

Figure 5. Comparison of experimental and theoretical results.

5. Discussion

5.1. Differences Between Experimental and Theoretical Results for $T/T_\alpha < 1$

Here, we discuss the differences between the experimental and theoretical results shown in Figure 5. During the bubble growth process, the theoretically estimated dimensionless bubble radius tended to be underestimated, compared with the experimental results. This discrepancy arises from the differences in the locations of the measured radii when evaluating the sizes of cavitation bubbles. This is because the theoretical investigation determines the exact radius of the sphere, whereas the experiment measures the radius of the contact area with the glass, that is, the radius of the bottom of the spherical cap. Therefore, it is desirable to measure the exact spherical radius of the cavitation bubble in the experiment. However, this is difficult because mercury is opaque. Therefore, we compared the results more accurately by mapping the theoretical investigation to spherical caps.

The bubbles measured in the experiment were deformed into spherical caps via contact with the wall, and the contact lines on the interface between the spherical cap and glass window were measured as the radius of the bubble. The shape of this spherical cap is affected by the interaction with the wall, considering wettability. Moreover, in the case of a static bubble, the static contact angle θ_s between the bubble and wall can be derived using Young's equation, as follows:

$$\theta_s = \cos^{-1}\left(\frac{\gamma_S - \gamma_{SL}}{\gamma_L}\right) \tag{4}$$

where γ_L is the surface tension of the liquid, γ_S is the surface tension of the solid, and γ_{SL} is the surface tension between the liquid and solid. The behavior of a spherical cap attached to a wall requires the consideration of friction with the wall and changes in surface tension; however, we assume that the pressures outside and inside the bubble are the same for a sphere and the corresponding spherical cap. Boyle's law states that the bubble volume can be considered to be identical for a sphere and spherical cap under an arbitrary internal pressure, and the relationship between the two shapes can be easily derived. The relationship between the radius of the bottom of the spherical cap, denoted by c, and the radius of the sphere, denoted by R, was obtained using Equation (5):

$$c = 2 \times \sqrt[3]{\frac{1}{z(z^2+3)}}R; \quad z \equiv \frac{1+\cos\theta}{\sin\theta}. \tag{5}$$

Using Equation (5), the theoretical results in Figure 5 can be converted to the radius of the contact area, c. However, the derivation of θ_s in Equation (4) is based on static conditions and is not applicable to the experimentally measured dynamic behaviors of cavitation bubbles. Moreover, the difference between the experimental and theoretical results in Figure 4 shows a gradual decrease at $T/T_\alpha < 1$, and this is followed by a collapse with consistent behavior. These results suggest that the contact angle changes dynamically from the bubble generation phase to the collapse phase. To reduce this difference, it is necessary to consider the dynamic contact angle θ_d [23].

In this study, we consider modeling the time evolution of θ_d using a simple function. This function is represented by the following exponential function:

$$\theta_d(t) = AR(t)^B + \theta_0, \tag{6}$$

where A and B are the model coefficients and θ_0 is the initial contact angle. These values were evaluated as inverse values to adequately fit the experimental results. If we consider the initial contact angle to be the static contact angle between the microbubbles and the wall, which is the starting point of a cavitation bubble, then the static contact angle θ_s can be uniquely derived from Equation (4). For bubbles between mercury and glass, $\theta_s \approx 130°$ [24,25]. The coefficients A and B must be used to derive the best value to fit the experiment. Hence, the method of least squares was used to quantitatively evaluate the fitting values for the experiment. The coefficients A and B are determined such that the residual sum of squares, denoted by RSE, as shown in Equation (7), is minimized.

$$RSE = \sum_{i=1}^{N} (c_{M,i} - c_{E,i})^2, \tag{7}$$

where N is the number of data points, c_M is the value obtained from the theoretical investigation, and c_E is the experimental value of the target. The relationships between the coefficients A and B and the residual sum of squares RSE was found to be minimal when $A = -2400$ and $B = 0.5$. Using these values, the bubble radius R obtained from the theoretical investigation can be converted into the radius of the contact area c, which is in close agreement with the experimental results.

Figure 6 shows a comparison of the time evolution of c between the experimental and theoretical results using $A = -2400$ and $B = 0.5$. Panel (a) shows the time evolution of the radius of the glass contact area c, panel (b) shows the time evolution of the dynamic contact angle, and panel (c) shows a schematic of the expected bubble behavior. According to Figure 6a, the change in bubble radius obtained from the theoretical investigation using $A = -2400$ and $B = 0.5$ adequately describes the experimental results. In the time evolution of the contact angle shown in Figure 6b, the contact angle gradually decreased from the initial contact angle θ_0 to a minimum angle of approximately 60° in the process from A to C. Figure 6c schematically shows the shape of the bubble, where A–D correspond to the positions in Figure 6b. From these results, it was expected that the bubbles would first grow away from the wall and then collapse toward the wall. However, it is difficult to verify the experimental validity of dynamic contact angles obtained using an inverse problematic approach. This verification will be a subject of future research.

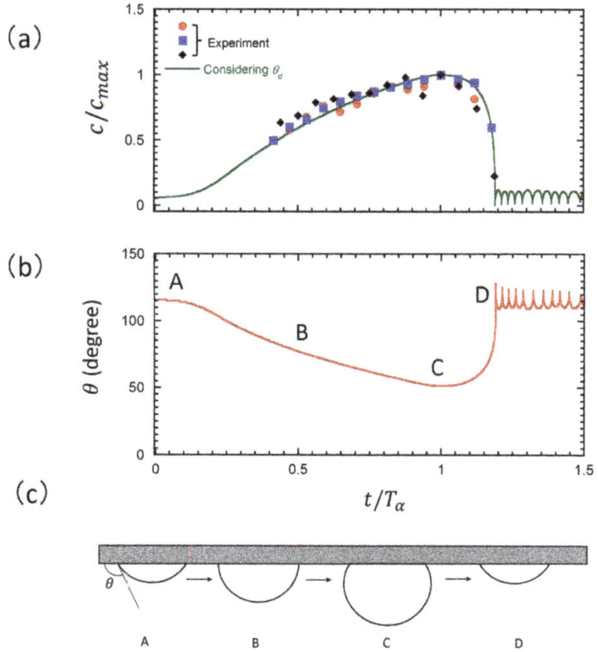

Figure 6. Results of the theoretical investigation of bubble behavior, considering a dynamic contact angle: (**a**) time evolution of the bubble radius, (**b**) time evolution of the contact angle, and (**c**) schematic of the predicted bubble behavior.

5.2. Method for Extracting T_β from T_D

5.2.1. Effect of Pressure Fluctuations on T_D

As shown in the experimental results in Figure 4, the time information relating to cavitation bubble dynamics that can be extracted from the wall vibration is attributable to the timing of the bubble collapse, $T_D = T_\alpha + T_\beta$. The bubble maximum to collapse time T_β is proportional to the maximum radius of the cavitation bubble. Since the prediction of cavitation intensity, which is the objective of this study, requires knowledge of the maximum radius of the bubble, a method by which to accurately extract T_β from T_D is required. The ratio of T_β to T_D is thought to be influenced by the pressure fluctuations around the bubble. For example, in the cases of spark discharge and laser irradiation cavitation, it was experimentally found that $T_\beta/T_D = 0.5$. However, cavitation bubbles inside a mercury target vessel grow gradually due to the negative pressure generated at the interface between the wall and the mercury, resulting in asymmetric behavior with $T_\alpha \neq T_\beta$ [18]. In such cases, T_β/T_D cannot be easily determined. This asymmetry in the bubble behavior was also observed in the experimental results described in Section 4. To investigate the change in the bubble behavior owing to pressure fluctuations, the experimentally measured pressure fluctuations (Figure 4) were approximated using a trapezoidal pressure fluctuation model, as shown in Figure 7, where p_0 is the initial pressure, p_1 is the pressure during bubble growth, p_2 is the pressure during bubble collapse, t_1 is the applied time, and w is the temporal weight of the gradient region. Naturally, $w = 0$ indicates a rectangular pressure fluctuation, and $w = 1$ indicates a triangular pressure fluctuation. This pressure fluctuation model is a prerequisite for the derivation of the simple evaluation formula for T_β/T_D, which is discussed in Section 5.2.2.

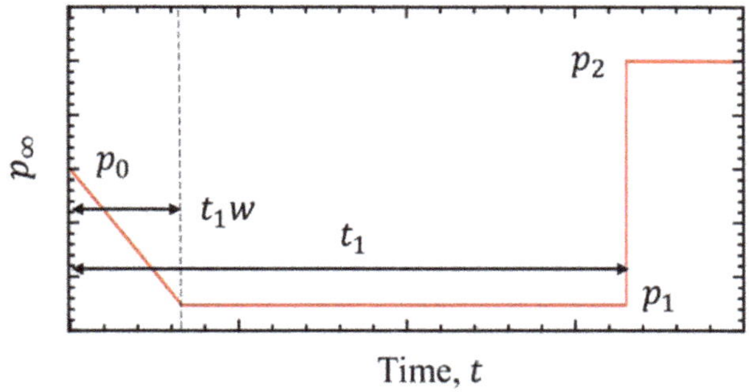

Figure 7. Trapezoidal pressure profile model used in the theoretical investigation.

To validate this trapezoidal pressure fluctuation model, a comparison was made with experimentally measured pressure fluctuations. Figure 8 shows a comparison of the trapezoidal pressure fluctuation with parameters set to fit the experimental pressure fluctuation, where $p_1 = -260$ kPa, $t_1 = 1.65$ ms, and $w = 0.2$. However, it was difficult to fit the pressure fluctuation in the p_2 region perfectly with the experimental pressure fluctuation. Therefore, three pressure values of 0.1, 0.2, and 0.3 MPa were assumed to theoretically represent the bubble behavior. Figure 9 shows the time evolution of the bubble radius analyzed using these trapezoidal pressure fluctuation models. The bubble radius increased gradually during the application of negative pressure and then collapsed within a short period, resulting in an asymmetric bubble behavior with respect to the time axis. Compared with the theoretical results of the bubble behavior based on the experimentally measured pressure fluctuations, the bubble behavior was generally reproduced throughout the entire process. However, slight differences were observed, including immediately before the bubbles reached their maximum radii. These differences are thought to be due to the switch from p_1 to p_2 being a step-like change, unlike that in the experiment. However, from t_1 onward, a change in p_2 produced a clear difference in the value of T_D. The value of p_2 was set to approximately 0.2 MPa to represent the bubble behavior in the experiment. These results confirm that the trapezoidal pressure fluctuation model can adequately represent the behavior of bubbles in mercury by setting the appropriate parameters.

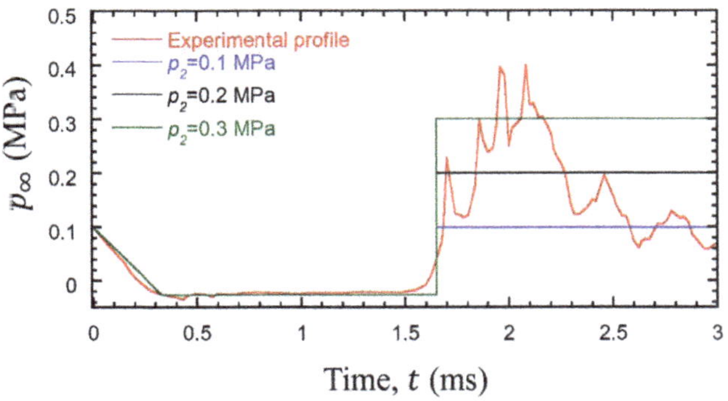

Figure 8. Comparison of the experimental and trapezoidal pressure distribution models.

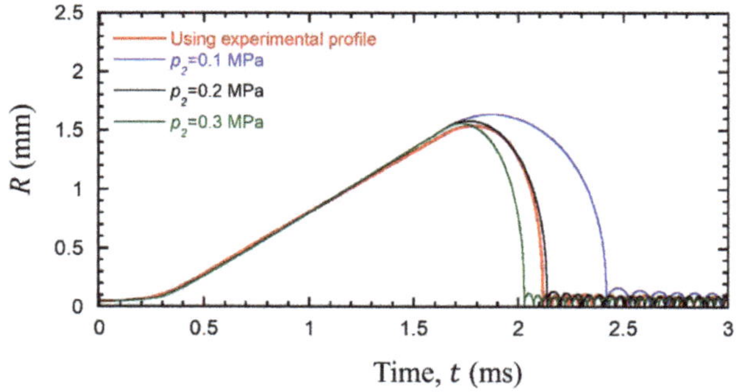

Figure 9. Comparison of bubble behavior between the experimental and trapezoidal pressure profiles.

5.2.2. Estimation of T_β from T_D

Section 5.2.1 demonstrated that trapezoidal pressure fluctuations could simulate the behavior of the bubbles evaluated in the experiment. In general, it is difficult to separate both T_α and T_β from T_D by measuring structural vibration, acceleration, velocity, etc. Therefore, we derived an approximate formula for the relationship between T_D and T_β, based on the pressure fluctuations indicated in Section 5.2.1. The discussion is based on the simple Rayleigh–Plesset equation without considering the viscosity, surface tension, and specific heat ratio, as given by Equation (8):

$$R\ddot{R} + \frac{3}{2}\dot{R}^2 = \frac{p_v - p_\infty}{\rho}, \tag{8}$$

where $\ddot{R}$ is the second-order time derivative of the bubble radius, $\dot{R}$ is the first-order time derivative of the bubble radius, p_v is the vapor pressure, p_∞ is the pressure at infinity, and ρ is the liquid density. Assuming further that p_∞ is constant, Equation (8) can be integrated and derived for $\dot{R}$ [20]:

$$\left[\dot{R}(t)\right]^2 = \frac{2}{3}\frac{p_\infty - p_v}{\rho}\left[\left(\frac{R_0}{R(t)}\right)^3 - 1\right], \tag{9}$$

If the bubble grows sufficiently and $R_0 \ll R$, then Equation (9) can be approximated using Equation (10), which is the governing equation for the bubble surface velocity $\dot{R}$ at a constant pressure after t_1w in Figure 7:

$$\left[\dot{R}(t)\right]^2 = \frac{2}{3}\frac{p_v - p_\infty}{\rho}, \tag{10}$$

The next step was to derive the governing equation for the bubble surface velocity in the gradient region of the trapezoidal pressure fluctuation model, i.e., until t_1w in Figure 7. The pressure fluctuation in this region is expressed as a function of $p_\infty(t) = (p_1 - p_0)t/t_1w + p_0$, which is substituted into Equation (8) and integrated. Thus, the governing equation for the bubble surface velocity is derived as follows:

$$\left[\dot{R}(t)\right]^2 = \frac{2}{3}\frac{p_v - p_\infty(t)}{\rho} + \frac{p_1 - p_0}{6\rho t_1 w}R(t), \tag{11}$$

In the case of pressure fluctuations in the MIMTM, the second term on the right side hardly contributes to the bubble surface velocity, owing to the short duration of the gradient region and the high density of the liquid. Therefore, the second term on the right side can be ignored, and the bubble radius can be derived by combining Equations (10) and (11). By setting $t = t_1$, the bubble radius at the end of the negative pressure, denoted by R_2, can be approximately derived:

$$R_2 = \sqrt{\frac{2}{3}\frac{p_v - p_1}{\rho}}\left[(1-w) + \frac{2}{3}\frac{p_v - p_1}{p_0 - p_1}w\right]t_1. \tag{12}$$

After negative pressure was reached, the bubble surface velocity was governed by Equation (9), as the initial bubble radius $R_0 = R_2$. This equation has $\dot{R} = 0$. That is, the left side is equal to zero at the time of the maximum bubble radius. Hence, R_{max} and R_2 can be derived:

$$\left(\frac{R_2}{R_{max}}\right)^3 + \frac{p_v - p_1}{p_2 - p_v} = 1. \tag{13}$$

Equation (12) shows that R_2 is a function of t_1. Furthermore, when the pressure fluctuation during bubble collapse behavior is constant, the relationship between R_{max} and T_β is as follows [26]:

$$R_{max} = c\sqrt{\frac{p_2 - p_v}{\rho}}T_\beta; \quad c = \frac{\sqrt{6}\,\Gamma\left(\frac{4}{3}\right)}{\Gamma\left(\frac{5}{6}\right)\Gamma\left(\frac{1}{2}\right)} \approx 1.093. \tag{14}$$

Substituting Equations (12) and (14) into Equation (13), we obtain Equation (15) for the relationship between t_1 and T_β:

$$\frac{t_1}{T_\beta} = 1.34P_A^*\sqrt{\frac{1}{P_B^*}}\sqrt[3]{P_B^* + 1}; \quad P_A^* \equiv \frac{3(p_0 - p_1)}{(2p_v - 3p_0 + p_1)w + 3(p_0 - p_1)}; \quad P_B^* \equiv \frac{p_v - p_1}{p_2 - p_v}, \tag{15}$$

where P_A^* is a dimensionless value representing the weight of the gradient region at total negative pressure, $P_A^* \geq 1$, and P_B^* is the ratio of the magnitude of the negative pressure to the positive pressure during the bubble collapse, where $P_B^* > 0$. In Equation (13), the second term on the left-hand side is P_B^*, and if $P_B^* \ll 1$, then R_2/R_{max} can be regarded as approximately 1. That is, the time t_1 required for R_2 and the time T_α to reach the maximum bubble radius are approximately equal. Therefore, if $P_B^* \ll 1$, then T_α/T_β and T_β/T_D can be obtained from Equation (16):

$$\frac{T_\alpha}{T_\beta} \approx 1.34P_A^*\sqrt{\frac{1}{P_B^*}}\sqrt[3]{P_B^* + 1}; \quad \frac{T_\beta}{T_D} = \frac{1}{T_\alpha/T_\beta + 1}; \tag{16}$$

Equation (15) shows that T_β/T_D depends only on P_B^* when there is no gradient region in the pressure fluctuation ($P_A^* = 1.0$). In addition, for $P_A^* > 1.0$, T_β/T_D is determined only by the values of p_0, p_1, p_2, and p_v. Hence, it is possible to predict T_β/T_D using Equation (16).

To verify its accuracy, Equation (16) was compared with the results of a theoretical investigation using Keller's equation, as presented in Equation (1) and shown in Figure 10. There was a tendency for T_β/T_D to increase with an increasing value of P_B^*. This P_B^* dependence was observed in both the theoretical results and Equation (16). Moreover, the difference between the two values increases with an increasing value of P_B^* owing to the approximation used to derive Equation (16). That is, the larger P_B^* becomes, the more difficult it becomes to consider $t_1 \neq T_\alpha$. Furthermore, T_β/T_D tended to decrease with an increasing value of P_A^*. Furthermore, the difference between Equation (16) and the theoretical results both tended to increase as P_A^* increased owing to the increasing influence of the second term on the right side of Equation (10). Therefore, to accurately determine T_β/T_D from Equation (16), both P_A^* and P_B^* should be small. Hence, Figure 10 shows that the prediction is likely to be acceptable as a simplified approach within the ranges of $0.01 < P_B^* < 1.0$ and $1.0 < P_A^* < 2.0$. The actual T_β/T_D measured in this experiment is plotted using black dots in Figure 10. In the experiment, P_A^* and p_2 were set to 1.20 and 0.2 MPa, respectively, based on the results of the analysis in Figures 8 and 9. Slight differences were observed between the experimental values and those predicted using Equation (16) owing to the complex variations in the pressure fluctuations resulting from the interactions between the wall and bubbles as well as interactions among the surrounding bubbles. Therefore, these effects should be investigated in future studies.

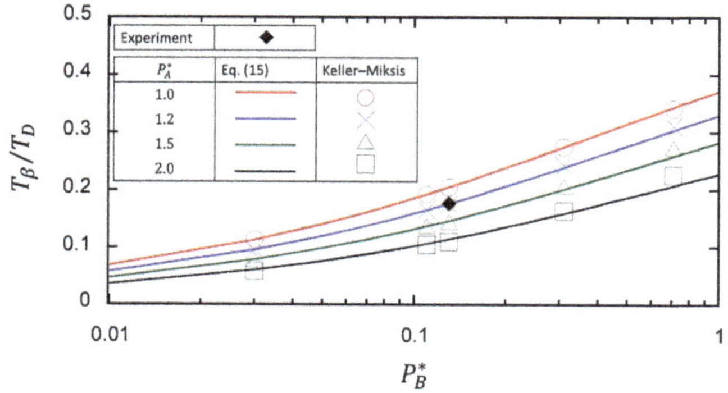

Figure 10. Relationships between P_A^*, P_B^*, and T_β/T_D.

Next, we considered the case in which Equation (16) is applied to an actual mercury target vessel. The value of p_1 is important to uniquely determine P_B^*, as $p_2 > 0.1$ MPa is estimated through a numerical simulation that considers the operational conditions in the target [27]. The value of p_1 was found to be approximately -0.03 MPa for the MIMTM used in this experiment [28], and the numerically estimated value on the mercury target p_1 was -0.025 MPa. These values predict that $P_B^* < 0.3$ for the mercury target vessel. Moreover, P_A^* is expected to be $P_A^* < 1.2$, with a smaller gradient region in the mercury target vessel than in the MIMTM [29]. These two facts suggest that the cavitation bubble behavior in a mercury target vessel can be evaluated quickly and easily via the in situ measurement of the time response waveform of the vibration using Equation (16).

6. Conclusions

The estimation of cavitation damage in a fluid–structure system, for example, the J-PARC mercury target for a neutron source, is important from the perspectives of structural integrity and long-term stable operation. In this study, we focused on the maximum radius of cavitation bubbles associated with the aggression of cavitation. We conducted a basic study using a simplified approach to evaluate the maximum radius using time information regarding the pressure imposed on the liquid. The following conclusions were drawn.

1. The correlation between the measured pressure fluctuations and vibration acceleration waveform of the vessel was experimentally determined by observing the mercury bubble behavior, bubble growth, and collapse behavior.
2. The dynamic contact angle is an important parameter for understanding the dynamics of bubbles with spherical cap shapes in contact with a solid wall. The experimentally measured behavior of bubbles formed at an interface between a liquid and a solid was investigated based on the Keller–Miksis equation, considering wettability. The derivation of the dynamic contact angle and its time variation were discussed as inverse problems.
3. The size of the cavitation bubbles is an indicator that can be used to define the aggressivity, as it is associated with the time T_β. The simple approach was proposed to quantitatively estimate T_β based on the Keller–Miksis equation combined with the experimentally measured or numerically predicted pressure fluctuations.

These facts suggest that the cavitation bubble behavior in a mercury target vessel can be evaluated quickly and easily via the in situ measurement of the time response waveform of the vibration. First, time information T_D is obtained from container vibration and acoustic data, etc. Next, T_β is estimated from T_D using the relationship in Equation (16). Finally, the maximum bubble radius R_{max} is estimated from T_β using a linear relationship. This method is expected to be useful with global applicability for evaluating cavitation damage and intensity.

Author Contributions: Conceptualization, H.K. (Hiroyuki Kawashima), H.K. (Hiroyuki Kogawa) and M.F.; methodology, H.K. (Hiroyuki Kawashima); software, H.K. (Hiroyuki Kawashima); validation, H.K. (Hiroyuki Kawashima), H.K. (Hiroyuki Kogawa) and M.F.; formal analysis, H.K. (Hiroyuki Kawashima); investigation, H.K. (Hiroyuki Kawashima), H.K. (Hiroyuki Kogawa) and M.F.; resources, M.F. and N.T.; data curation, H.K. (Hiroyuki Kawashima); writing—original draft preparation, H.K. (Hiroyuki Kawashima); writing—review and editing, M.F. and N.T.; visualization, H.K. (Hiroyuki Kawashima); supervision, H.K. (Hiroyuki Kawashima); project administration, H.K. (Hiroyuki Kawashima); funding acquisition, H.K. (Hiroyuki Kawashima) and N.T. All authors have read and agreed to the published version of the manuscript.

Funding: This work was supported by the Japan Science and Technology Agency and the establishment of university fellowships for the creation of science and technology innovation (Grant Number JPMJFS2105).

Data Availability Statement: Data are contained within the article.

Conflicts of Interest: The funders had no role in the design of the study; in the collection, analyses, or interpretation of data; in the writing of the manuscript; or in the decision to publish the results.

References

1. Jian, W.; Petkovšek, M.; Houlin, L.; Širok, B.; Dular, M. Combined Numerical and Experimental Investigation of the Cavitation Erosion Process. *J. Fluids Eng.* **2015**, *137*, 051302. [CrossRef]
2. Zheng, R.; Zhao, X.; Dong, L.; Liu, G.; Huang, Y.; Xu, Y. On the Cavitation Erosion-Corrosion of Pipeline Steel at Different Locations of Venturi Pipe. *Eng. Fail. Anal.* **2022**, *138*, 106333. [CrossRef]
3. Zharkovskii, A.; Svoboda, D.; Borshchev, I.; Klyuyev, A.; Ivanov, E.; Shutsky, S. Axial-Flow Pump with Enhanced Cavitation Erosion Resistance. *Energies* **2023**, *16*, 1344. [CrossRef]
4. Spallation Neutron Source | Neutron Science at ORNL. Available online: https://neutrons.ornl.gov/sns (accessed on 16 October 2024).
5. J-PARC MLF (Materials and Life Science Experimental Facility). Available online: https://mlfinfo.jp/en/ (accessed on 16 October 2024).
6. Futakawa, M.; Naoe, T.; Kogawa, H.; Tsai, C.-C.; Ikeda, Y. Pitting Damage Formation up to over 10 Million Cycles: Off-Line Test by MIMTM. *J. Nucl. Sci. Technol.* **2003**, *40*, 895–904. [CrossRef]
7. McClintock, D.A.; Riemer, B.W.; Ferguson, P.D.; Carroll, A.J.; Dayton, M.J. Initial Observations of Cavitation-Induced Erosion of Liquid Metal Spallation Target Vessels at the Spallation Neutron Source. *J. Nucl. Mater.* **2012**, *431*, 147–159. [CrossRef]
8. Futakawa, M.; Naoe, T.; Tsai, C.C.; Kogawa, H.; Ishikura, S.; Ikeda, Y.; Soyama, H.; Date, H. Pitting Damage by Pressure Waves in a Mercury Target. *J. Nucl. Mater.* **2005**, *343*, 70–80. [CrossRef]
9. Reuter, F.; Gonzalez-Avila, S.R.; Mettin, R.; Ohl, C.-D. Flow Fields and Vortex Dynamics of Bubbles Collapsing near a Solid Boundary. *Phys. Rev. Fluids* **2017**, *2*, 064202. [CrossRef]
10. Huang, J.; Wang, J.; Huang, J.; Lv, P.; Li, H.; Wang, Y. Effects of Wall Wettability on Vortex Flows Induced by Collapses of Cavitation Bubbles: A Numerical Study. *Phys. Fluids* **2023**, *35*, 087122. [CrossRef]
11. Gonzalez-Avila, S.R.; Nguyen, D.M.; Arunachalam, S.; Domingues, E.M.; Mishra, H.; Ohl, C.-D. Mitigating Cavitation Erosion Using Biomimetic Gas-Entrapping Microtextured Surfaces (GEMS). *Sci. Adv.* **2020**, *6*, eaax6192. [CrossRef]
12. Sun, Y.; Yao, Z.; Wen, H.; Zhong, Q.; Wang, F. Cavitation Bubble Collapse in a Vicinity of a Rigid Wall with a Gas Entrapping Hole. *Phys. Fluids* **2022**, *34*, 073314. [CrossRef]
13. Isselin, J.-C.; Alloncle, A.-P.; Autric, M. On Laser Induced Single Bubble near a Solid Boundary: Contribution to the Understanding of Erosion Phenomena. *J. Appl. Phys.* **1998**, *84*, 5766–5771. [CrossRef]
14. Soyama, H. Laser Cavitation Peening and Its Application for Improving the Fatigue Strength of Welded Parts. *Metals* **2021**, *11*, 531. [CrossRef]
15. Philipp, A.; Lauterborn, W. Cavitation Erosion by Single Laser-Produced Bubbles. *J. Fluid Mech.* **1998**, *361*, 75–116. [CrossRef]
16. Wakui, T.; Takagishi, Y.; Futakawa, M. Study on Cavitation Damage Evolution influenced by Input Power. *J. Soc. Mater. Sci. Jpn.* **2024**, *73*, 520–526. [CrossRef]

17. Shima, A.; Takayama, K.; Tomita, Y.; Miura, N. An Experimental Study on Effects of a Solid Wall on the Motion of Bubbles and Shock Waves in Bubble Collapse. *Acustica* **1981**, *48*, 293–301.
18. Kogawa, H.; Naoe, T.; Kyotoh, H.; Haga, K.; Kinoshita, H.; Futakawa, M. Development of Microbubble Generator for Suppression of Pressure Waves in Mercury Target of Spallation Source. *J. Nucl. Sci. Technol.* **2015**, *52*, 1461–1469. [CrossRef]
19. Kawashima, H.; Kogawa, H.; Futakawa, M.; Tanaka, N. Local Impact Estimation Method Based on Time Information of the Bubble Growth and Collapse Behavior. *J. Jpn. Soc. Exp. Mech.* **2024**, *24*, 134–140. (In Japanese)
20. Naoe, T.; Kogawa, H.; Wakui, T.; Teshigawara, M.; Haga, K.; Futakawa, M. Pressure Wave Induced Sound Measurement for Diagnosing the Operation Status of the J-PARC Pulsed Spallation Neutron Source. *Nucl. Instrum. Methods Phys. Res. Sect. A Accel Spectrometers Detect. Assoc. Equip.* **2020**, *982*, 164566. [CrossRef]
21. Keller, J.B.; Kolodner, I.I. Damping of Underwater Explosion Bubble Oscillations. *J. Appl. Phys.* **1956**, *27*, 1152–1161. [CrossRef]
22. Stoer, J.; Bulirsch, R. *Introduction to Numerical Analysis*; Texts in Applied Mathematics; Springer: New York, NY, USA, 2002; Volume 12, ISBN 978-1-4419-3006-4.
23. Ardron, K.H.; Giustini, G.; Walker, S.P. Prediction of Dynamic Contact Angles and Bubble Departure Diameters in Pool Boiling Using Equilibrium Thermodynamics. *Int. J. Heat Mass Transf.* **2017**, *114*, 1274–1294. [CrossRef]
24. Fowkes, F.M. Additivity of Intermolecular Forces at Interfaces. I. Determination of The Contribution to Surface and Interfacial Tensions of Dispersion Forces in Various Liquids. *J. Phys. Chem.* **1963**, *67*, 2538–2541. [CrossRef]
25. Ellison, A.H.; Klemm, R.B.; Schwartz, A.M.; Grubb, L.S.; Petrash, D.A. Contact Angles of Mercury on Various Surfaces and the Effect of Temperature. *J. Chem. Eng. Data* **1967**, *12*, 607–609. [CrossRef]
26. Rayleigh, L. On the Pressure Developed in a Liquid during the Collapse of a Spherical Cavity. *Philos. Mag. Series 6* **1917**, *34*, 94–98. [CrossRef]
27. Kogawa, H.; Naoe, T.; Futakawa, M.; Haga, K.; Wakui, T.; Harada, M.; Takada, H. Mitigation Technologies for Damage Induced by Pressure Waves in High-Power Mercury Spallation Neutron Sources (IV)—Measurement of Pressure Wave Response and Microbubble Effect on Mitigation in Mercury Target at J-PARC. *J. Nucl. Sci. Technol.* **2017**, *54*, 733–741. [CrossRef]
28. Ida, M.; Naoe, T.; Futakawa, M. Suppression of Cavitation Inception by Gas Bubble Injection: A Numerical Study Focusing on Bubble-Bubble Interaction. *Phys. Rev. E* **2007**, *76*, 046309. [CrossRef]
29. Futakawa, M.; Wakui, T.; Kogawa, H.; Ikeda, Y. Pressure Wave Issues in High Power Mercury Target. *Nucl. Instrum. Methods Phys. Res. Sect. A Accel. Spectrometers Detect. Assoc. Equip.* **2006**, *562*, 676–679. [CrossRef]

fluids

MDPI

Article

Rheological Features of Aqueous Polymer Solutions Tailored by Hydrodynamic Cavitation

Santiago Nicolás Fleite [1,2,3], María del Pilar Balbi [1,2], María Alejandra Ayude [4,5] and Miryan Cassanello [1,2,*]

1 Instituto de Tecnología de Alimentos y Procesos Químicos (ITAPROQ), CONICET, Universidad de Buenos Aires, Buenos Aires C1428BGA, Argentina; fleite@di.fcen.uba.ar or fleite@agro.uba.ar (S.N.F.); mpbalbi@fi.uba.ar (M.d.P.B.)

2 Laboratorio de Reactores y Sistemas para la Industria (LARSI), Departamento de Industrias, FCEyN, Universidad de Buenos Aires, Buenos Aires C1428BGA, Argentina

3 Cátedra de Química Inorgánica y Analítica, Facultad de Agronomía, Universidad de Buenos Aires, Buenos Aires C1120AAQ, Argentina

4 Departamento de Ingeniería Química y de Alimentos, Universidad Nacional de Mar del Plata (UNMdP), Mar del Plata B7600FDQ, Argentina; mayude@fi.mdp.edu.ar

5 División Catalizadores y Superficies, Instituto de Investigaciones en Ciencia y Tecnología de Materiales (INTEMA), CONICET, Universidad Nacional de Mar del Plata (UNMdP), Mar del Plata B7600FDQ, Argentina

* Correspondence: miryan@di.fcen.uba.ar; Tel.: +54-11-52859029

Abstract

Hydrodynamic cavitation (HC) has emerged as a versatile method for modifying the rheological properties of polymer solutions, offering advantages such as scalability and operational simplicity. This work investigates the effect of HC on aqueous polyacrylamide (PAM) solutions, focusing on viscosity and viscoelasticity changes as a function of the number of passes through a vortex-type HC device and the presence of dissolved salts ($CaCl_2$ or KCl). Viscosity measurements were modeled using the power law equation, while oscillatory tests were used to determine storage and loss moduli. The results show that HC substantially reduced viscosity and elastic behavior, with the degree of modification strongly influenced by the number of passes. A critical molecular size limit was suggested, below which further degradation becomes limited. Salt addition enhanced depolymerization, likely due to charge screening, hydrodynamic radius reduction, and the increased solubility and mobility of polymer chains within cavitation bubbles. HC eliminated elasticity in all cases, yielding solutions with near-Newtonian behavior. The transformation is attributed to molecular weight reduction and changes in molecular size distribution. These findings support the use of HC as a practical approach to tailor the flow properties of PAM solutions, while highlighting intrinsic limitations imposed by cavitation dynamics and polymer chain dimensions.

Keywords: hydrodynamic cavitation; polyacrylamide aqueous solution; viscosity; rheological behavior

1. Introduction

Water-soluble, biodegradable, and biocompatible polymers are widely used for diverse industrial applications. They are typically employed as stabilizers and thickeners for medicines, cosmetics, and coatings, for assisting petroleum extraction, and for wastewater remediation. Although available with different grades, their high molecular weight and broad molar mass distribution can pose handling and performance challenges. Hence, different methods have been proposed and studied to partially reduce the polymer chains after their synthesis, to fit desired conditions [1]. Moreover, the increasing use of these polymers has led to large amounts of polymer-bearing wastewater, which is difficult to process due to its high viscosity. This highlights the need for viscosity reduction, typically achieved through depolymerization by thermal, chemical, mechanical, and enzymatic or biological methods [2,3].

Among the mechanical methods used to post-process or treat polymer aqueous solutions, hydrodynamic cavitation appeared as a promising alternative due to its simple application and ease of scalability [4,5]. Cavitation involves the formation of small bubbles, called cavities, which initially grow and then are compressed and finally collapse, inducing local extreme conditions [6–8]. Apart from high local temperatures, the released energy can result in mechanical effects related to shock waves, microjets, and high shear stress. In addition, the implosion generates highly reactive species, like hydroxyl radicals, which can act at the molecular level to modify polymer chain length. Yan et al. [9] evidenced that mechanical forces like those exerted by cavitation generally lead to cleavage of polymer chains around the center, generating molecules of significantly lower molar mass with a narrow distribution. On the other hand, highly active species, which can also be formed by cavitation, cleave the polymer chains randomly, also leading to smaller molecules but with a broader molar mass distribution.

Examples of polymer aqueous solutions that have been modified through cavitation are carboxymethyl cellulose (CMC) and hydroxypropyl methyl cellulose (HPMC). It was shown that the use of acoustic or hydrodynamic cavitation can successfully cleave the cellulose glycosidic bonds of CMC and HPMC [1,10]. CMC is extensively used for diverse applications in the food and cosmetic industries. HPMC is a non-ionic, semi-synthetic derivative of cellulose commonly applied in medicine for controlled drug delivery and in the food industry as a stabilizer. Another polymer that has been subjected to ultrasonic (US) cavitation for tuning the aqueous solution features is Konjac glucomannan (KGM) [11]. KGM is a polysaccharide used for pharmaceutical and biotechnological applications. However, modifications of the naturally extracted polymer are required to meet the necessary conditions (solubility, viscosity, etc.). US cavitation was found to be appropriate for reducing the molecular weight and viscosity without occasioning changes in the chemical nature of this polymer [11]. The authors found that the initially shear-thinning behavior of a 0.25% (w/v) KGM aqueous solution became Newtonian by applying US cavitation for more than 4 min.

Polyacrylamide (PAM) is also a polymer that has been extensively used in the industry, along with its derivatives. Its main applications are found in water treatment, the textile industry, and especially in oil recovery [12–14]. Aqueous PAM solutions are frequently employed to enhance the sweep capability of injection fluids in oil recovery, as their high viscosity improves oil displacement. However, the synthesis of PAM often results in molar masses and distributions that do not fully meet the application requirements. Hence, post-processing of commercial PAM is used to tailor the molecular weight by partially cleaving the polymer chains.

Thermal, chemical, and enzymatic methods have been proposed to modify the molar mass and distribution of PAM [3], but these treatments are frequently complex and expensive. In contrast, cavitation, either ultrasonic or hydrodynamic, and its combination with other advanced oxidation processes, provides a simple, cleaner, and faster route to reduce the molar mass of PAM chains to desirable levels. Prajapat and Gogate [13] demonstrated that the conditions of hydrodynamic cavitation (HC), applied using a slit venturi device, significantly influence the viscosity reduction in the polymer solution. These authors also combined HC with other advanced oxidation processes to further enhance polymer degradation. HC offers strong potential for scale-up with relatively low investment and operating costs [15,16]. It arises from a perturbation in the flowing fluid that moves the pressure below the solvent (water) vapor pressure, thus inducing the formation of vapor cavities. As the fluid continues downstream, pressure increases and bubbles are compressed and finally collapse violently, producing effects comparable to those of the ultrasound cavitation.

Recently, Zhang et al. [14] used simulations to analyze the impact of bubble collapse on PAM degradation efficiency. They concluded that the local shear after collapse induces stretching of C–C bonds and expansion of C–C–C bond angles in the PAM main chain, leading to fragmentation of the molecules. This mechanical degradation proceeds without significant change in the chemical structure, and the bond breaking occurs in the central region of the collapsing bubble.

In this context, the aim of the present work is to examine the suitability of HC as a method for tailoring the rheological properties of polymeric solutions. PAM is used as a model polymer due to its widespread application in oil recovery. Aqueous PAM solutions at different concentrations are subjected to HC using a vortex-type cavitation device, which achieves cavitation inception at lower pressure drops [17,18]. This study focuses particularly on the effects of the number of passes through the cavitation device and the presence of dissolved salts in the solution.

2. Materials and Methods

2.1. Experimental Setup

The setup for subjecting the polymer solutions to hydrodynamic cavitation (HC) is schematized in Figure 1. The cavitation device was a vortex diode type with a throat diameter of 6 mm, procured from VIVIRA Process Technology Pvt. Ltd. (100 NCL Innovation Park, Dr Homi Bhabha Rd, Pune, Maharashtra 411008, India). A batch assembly powered by a 1 HP centrifugal pump was used. It contained regulating valves to control the flow rate and a cryostat to keep the temperature at 20 ± 1 °C.

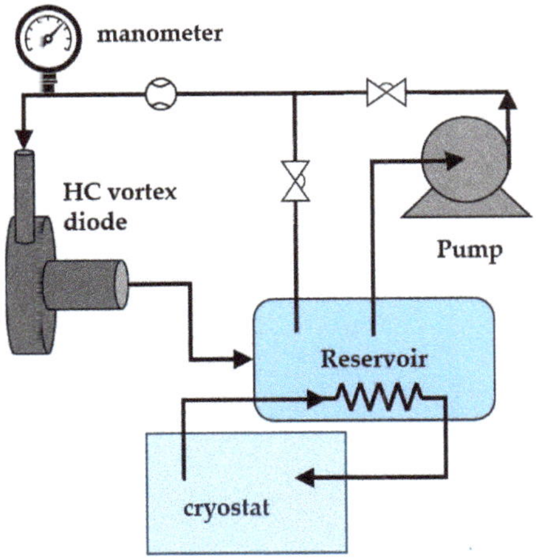

Figure 1. Experimental setup.

The pressure drop through the cavitation device was varied between 2 and 3.5 bar, resulting in liquid flow rates in the range of 3 to 6 L/min. Experiments were carried out for different number of passes through the HC device to analyze the effect on the rheological characteristics of the aqueous polymer solutions. The number of passes (n_p) roughly gives how many times the whole liquid passes through the cavitation device [16]. The time (t) needed for a given number of passes (n_p) depends on the liquid flow rate through the cavitation device (Q) and the volume of the liquid in the reservoir tank (V), as expressed in Equation (1).

$$t = n_p\ V/Q \tag{1}$$

The vortex-type cavitation device consists in a bidirectional fluid flow system, in which the inlet flow enters a swirl chamber where high-speed rotation is achieved. The fluid exits only through a centered outlet channel perpendicular to the base of the swirl chamber, with a narrowing point called the throat [17,18]. The high-speed rotation induces the formation of cavitation bubbles due to the sharp decrease in the liquid pressure close to the device throat. Further details of the vortex cavitation device can be found in studies devoted to the analysis of the features of this system [17–21].

Cavitation inception in vortex-type cavitation is achieved at lower pressure drops than those required in Venturi or orifice plate configurations [18]. The cavitation bubbles or cavities formed in the throat region are sharply compressed when the liquid pressure is recovered far from the throat, leading to bubble collapse. During this period, extreme temperatures and pressures are locally reached. By single bubble dynamics simulation, Pandit et al. [22] have estimated peak temperatures over 5000 K and local pressures of around 1000 Pa for this device geometry, likely leading to the formation of supercritical water (SCW) domains prior the collapse. Simulations indicated that bubble diameters may reach ~20 μm, which are reduced to around 3 μm before collapsing.

In this work, cavitation inception was verified at pressure drop ~0.9 bar for the 1 g/L PAM solution using acoustic and vibration sensors, following Ranade et al. [20]. A change from around 12 to 60 dB was found at inception, consistent with data reported in [20]. It is worthwhile mentioning that Thacker et al. [23] have studied the influence of liquid viscosity on the pressure drop required for cavitation inception in this system, using glycerol aqueous solutions. The pressure drop range studied in this work is above the pressure drop required for cavitation inception in a similar system for solutions with viscosities of 61 mPa s. The PAM solutions used in this work have significantly lower viscosities. Therefore, under the experimental conditions examined, hydrodynamic cavitation was certainly achieved.

2.2. Materials

High-molar mass cationic polyacrylamide (PAM) FZ-81 from Faisan S.A., León XIII N° 154, CP 1888 Florencio Varela, Argentina, was used for preparing the polymer solutions. The skeletal structure of the PAM used is shown in Figure 2. As additives, analytical-grade potassium chloride (KCl) and calcium chloride ($CaCl_2$) were used in 0.01 M concentrations. All solutions were prepared with distilled water.

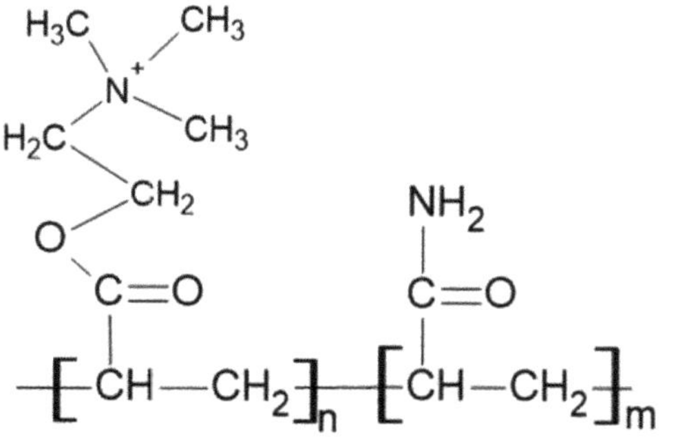

Figure 2. Skeletal structure of the cationic polyacrylamide used.

2.3. Viscosity and Viscoelasticity Measurement

Shear stress versus shear rate curves were measured using a plate-cone (CPE-40) Brockfield LVDV-II+ programmable viscometer and WINGATHER® and DVLoader® programs (AMETEK Brookfield, 11 Commerce Boulevard, Middleboro MA 02346, USA). The influence of PAM concentration on the polymer aqueous solution viscosity was first assessed, as well as the effect of adding 0.01 M of KCl or $CaCl_2$. Then, the solutions were subjected to HC for a given number of passes to analyze the influence on the rheological properties. Viscosity measurements were performed under a controlled temperature of 20 ± 0.1 °C. The rheological curves were analyzed using the Ostwald–de Waele model (i.e., the power law model), as expressed in Equation (2) [24]:

$$\tau = k\,\gamma^n \tag{2}$$

where τ is the shear stress (Pa), γ is the shear rate (Hz), k is the consistency index ($Pa{\cdot}s^n$), and n is the flow behavior index (dimensionless).

Viscoelastic properties were assessed for the solutions before HC treatment and after 500 passes, using a Modular Compact Rheometer (Model MCR 102e, Anton Paar, Graz, Austria) controlled by RheoCompass software (21 CFR Part 11, 1.21.805 version, Anton Paar). Linear viscoelastic properties were determined through small-amplitude oscillatory shear (SAOS) tests [12,25,26]. A strain sweep was imposed at a fixed angular frequency to determine the linear viscoelastic range (LVR). The applied strain of shear amplitude (γ_0) and angular frequency (ω) varies with time (t) according to Equation (3).

$$\gamma = \gamma_0 \sin(\omega t) \tag{3}$$

Within the LVR, the resulting shear stress (τ) changes with the same frequency but may be shifted in phase. The oscillating shear stress function can be expressed in terms of the sine and cosine of the angular velocity, as expressed in Equation (4), where G' is called the storage modulus and G'' is the loss modulus. G' and G'' can also be a function of the angular frequency.

$$\tau = \gamma_0 \left(G' \sin(\omega t) + G'' \cos(\omega t)\right) \tag{4}$$

G' and G'' were determined from the SAOS test with a strain sweep (0.1–100%) at a fixed frequency of 2 Hz. Briefly, the G' modulus is related to the solid-like elastic behavior, the trend to return to the initial configuration after the deformation, while the G'' modulus is related to the viscous component. Thus, if $G' > G''$, the sample behaves more like a rigid solid experiencing reversible deformation (elasticity). If $G' < G''$, the sample behaves like a fluid, experiencing irreversible deformation. A useful magnitude to summarize the analysis is the ratio G''/G', also referred to as $\tan(\delta)$, where δ is the phase shift. If this ratio is < 1, the sample tends to behave as an elastic solid in the LVR. If >1, the sample behaves more like a fluid [25,26].

3. Results and Discussion

3.1. Polymer Solution Viscosity

The PAM solutions' rheological curves can be well represented by Equation (2) with increasing deviations from Newtonian behavior as the PAM concentration increases. Thus, the consistency index increases while the index of flow behavior becomes progressively lower than 1 for higher polymer concentrations (Table 1 and Figure 3a).

Table 1. Effect of PAM concentration and addition of salts on indexes of Equation (2) used to describe the rheological behavior of the aqueous solutions.

PAM Concentration	Salt Concentration	Consistency Index, k (Pa·s^n)	Index of Flow Behavior, n (Dimensionless)
0.02 g/L	–	0.003 ± 0.0002	0.91 ± 0.05
0.10 g/L	–	0.009 ± 0.0005	0.76 ± 0.04
0.50 g/L	–	0.025 ± 0.002	0.67 ± 0.03
0.80 g/L	–	0.064 ± 0.003	0.57 ± 0.03
1.10 g/L	–	0.119 ± 0.008	0.53 ± 0.03
1.30 g/L	–	0.299 ± 0.015	0.40 ± 0.02
1.00 g/L	–	0.092 ± 0.007	0.55 ± 0.03
1.00 g/L	0.01 M KCl	0.043 ± 0.003	0.65 ± 0.03
1.00 g/L	0.01 M $CaCl_2$	0.026 ± 0.002	0.70 ± 0.04

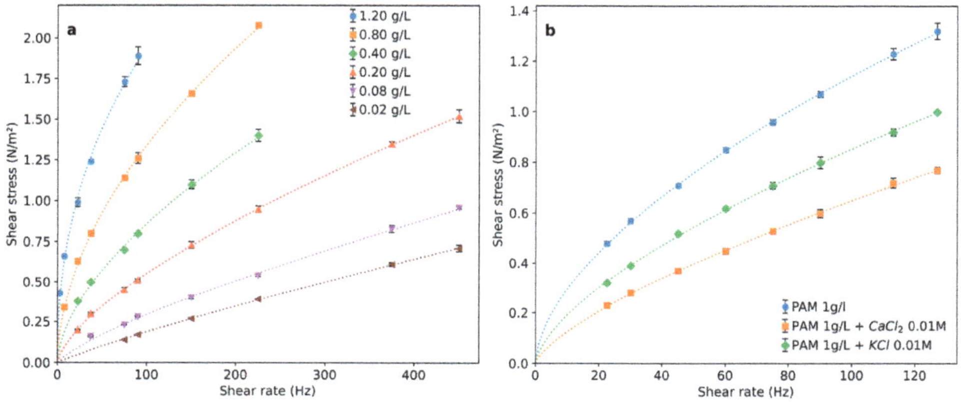

Figure 3. Rheological behavior of PAM solutions for (**a**) different concentrations and (**b**) with addition of salts with different ionic forces.

Salt addition induces a significant modification of the flow behavior (Table 1 and Figure 3b), which depends on the cation charge. The solution viscosity decreases, and the flow index becomes closer to those of Newtonian solutions. This effect is coincident with the one reported for PAM solutions containing high salt concentrations characteristic of oil-well wastewaters [12,27].

Polymers are generally present as individual chains in dilute and semidilute solutions. Their conformations can vary from highly extended to tightly coiled, depending on the interactions among chains (attractive or repulsive) and between the polymer and the solvent. PAM is generally in the form of extended linear chains in semidilute aqueous solutions. The addition of an inert salt, like NaCl or $CaCl_2$, negatively affects the interactions among the polymer chains, reducing the repulsion due to electrostatic screening. For this reason, the addition of an inert salt induces a decrease in PAM solutions' viscosity [12]. Recently, Arok et al. [26] have evidenced that PAM aqueous solutions above a certain concentration tend to form gel-like structures, as evidenced from the viscoelastic characteristic of the solutions (elastic modulus larger than viscous modulus) and from light scattering and electrophoretic experiments. When NaCl was added to the solutions above a certain concentration, electrostatic forces induce the disappearance of the gel-like structure, simultaneously decreasing the solution viscosity.

3.2. Effect of the Number of Passes and the HC Pressure Drop on Polymer Rheology

Figure 4 illustrates the effect of treating a 1 g/L PAM solution with HC during a different number of passes through the vortex-type HC device. Different flow rates through the vortex diode lead to different pressure drops between 2 and 3.5 bars being imposed.

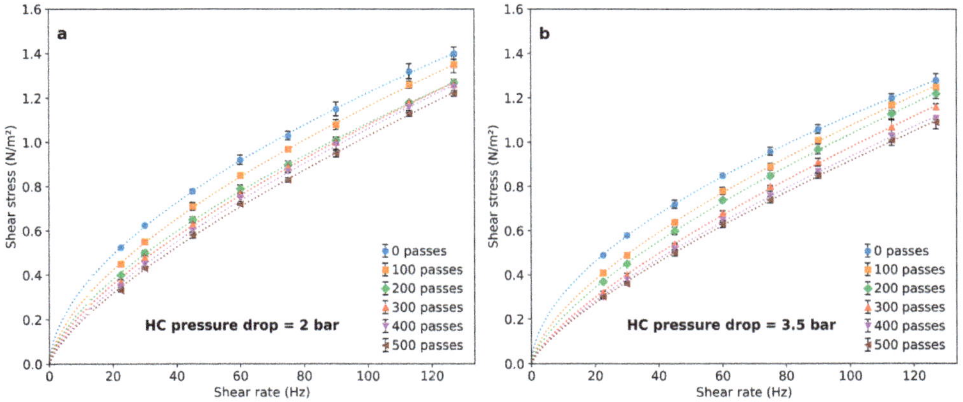

Figure 4. Rheological behavior of PAM solutions after a different number of passes through the cavitation device for (**a**) 2 bar pressure drop and (**b**) 3.5 bar pressure drop.

In all cases, polymer solutions exhibited a decrease in viscosity following HC treatment. This effect was dependent on the number of passes and only slightly associated with the pressure drop in the HC device. The effect was notable after 100 passes and became progressively less important for an increasing number of passes, leading to marginal differences above 400 passes. The flow indexes calculated by fitting results to Equation (2) are shown in Figure 5. The results indicate that the polymer viscosity can be reduced by imposing HC with a vortex diode device, and the number of passes is a significant variable to set for achieving rheological modifications.

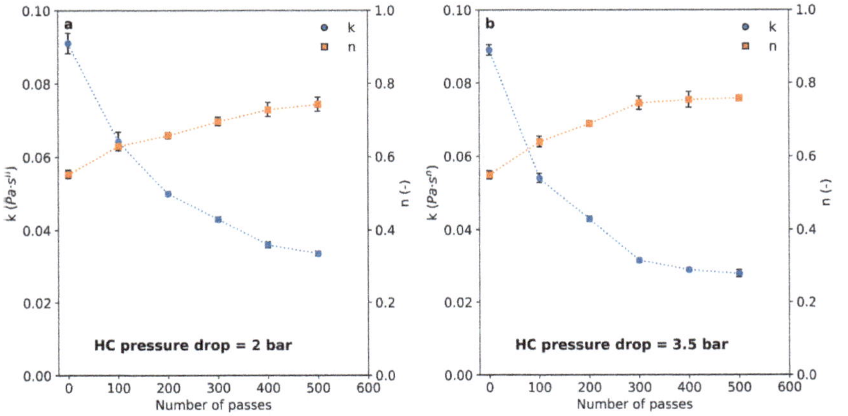

Figure 5. Consistency (k) and flow behavior (n) indexes (Equation (2)) calculated for the PAM solutions after a different number of passes through the cavitation device for (**a**) 2 bar pressure drop and (**b**) 3.5 bar pressure drop through the vortex diode.

The consistency index (k) decreased significantly in all cases, while the index of flow behavior slowly increased, reaching a plateau close to 0.8 for all the examined pressure drops within the range between 2.0 bar and 3.5 bar, remaining almost unchanged for more than 400 passes. The effect of pressure drop through the HC device did not appear to have a significant impact when considering the measurement uncertainties, suggesting a mechanism distinct from those typically involved in HC-enhanced chemical reactions.

Previous studies [28] identified an optimal pressure drop of 2.8 bar for vortex-type hydrodynamic cavitation (HC) devices used in the degradation of organic compounds. A strong dependence on head loss (or pressure drop) is commonly reported in the literature, as it is associated with cavitation collapse intensity, radical formation, pyrolysis, and the generation of supercritical water (SCW) [15,22,28,29]. Polymer molecules are prone to cleavage by the exerted mechanical forces during the collapse or through chemical degradation by pyrolysis, radical attack, and other reactions, such as those observed in SCW reactors [30]. The interaction between bubble size and polymer chain size may influence the predominant mechanism. Mechanical forces can selectively affect molecules above a certain size threshold, which cannot be entrapped within the bubbles. In contrast, chemical mechanisms would be responsible for the degradation of short-chain molecules. Pandit et al. [22] have estimated cavitation bubbles with sizes in the order of a few microns before collapsing for this system. In addition, the extreme conditions that can be attained suggest that SCW domains are likely to appear. These domains within the cavities may contribute to capturing polymer chains that are below a critical length and subject them to chemical degradation. Intermediate length chains, too large to get entrapped in the bubbles but too small to be affected by mechanical forces, may remain stable in the liquid bulk. This phenomenon is sketched in Figure 6. Large polymer chains, which are unlikely to enter cavitation bubbles, experience intense localized mechanical stress that leads to chain cleavage, as predicted by Zhang et al. [14]. Additionally, the formation of SCW may further promote depolymerization through the oxidation of smaller chains [30]. Previous works that used acoustic or hydrodynamic cavitation to treat polymer aqueous solutions have evidenced molecular size reduction by size-exclusion chromatography [1,9]. The trend towards molar weight reduction was evidenced in these works. In addition, Yan et al. [9] found that the molar mass distributions of chitosan became narrower for increasing HC time (i.e., larger number of passes), starting from a mean of circa 10^6 Da and concentrating progressively in the range of 0–50,000 Da, suggesting that there is a molecular size limit below which there is a minimal effect of HC. These authors showed that after 30 min HC, they obtained a broad molar mass distribution, reaching values up to ~2×10^6 Da, with a mean of around 10^6 Da. In contrast, for 120 min HC, the distribution reduced to a range up to ~2.5×10^5 Da, with a mean of ~50,000 Da and a mode one order of magnitude larger.

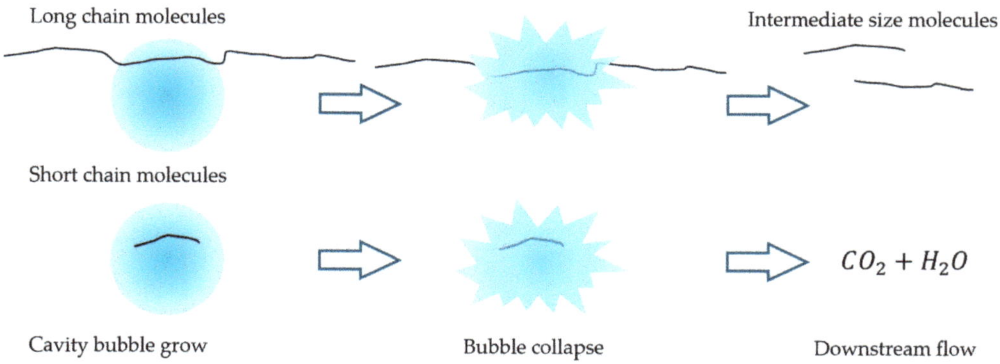

Figure 6. Sketch of the idealized polymer molecules' evolution under HC. Long-chain molecules can be cleaved to relatively stable intermediate-size molecules by mechanical shear or by partially entering the bubble. Short-chain molecules can enter the bubbles and be further oxidized. As a result, polymer molecules of intermediate size are formed, and further depolymerization by HC cannot be achieved.

Hence, while relatively small and large chains are degraded by pyrolysis or cleaved by shear forces, respectively, there may exist an intermediate size range that is too large to enter the bubble and undergo pyrolysis, yet too small to experience sufficient mechanical shear. For this reason, after a certain number of passes, the effect of HC becomes negligible—even if the solution's viscosity remains well above that of water. Therefore, HC can reduce polymer molecular size only up to a certain limit. However, this limit could potentially be shifted by redesigning the HC device to generate mechanical stresses at a different characteristic length scale, consistent with the Kolmogorov scale of turbulence energy dissipation.

3.3. *Effect of the Number of Passes and the Salt Concentration on Polymer Rheology*

The effect of having salts dissolved in the 1 g/L PAM solutions on the rheological modifications occasioned by a different number of passes through the HC device are illustrated in Figure 7. The initial solution viscosity was already reduced when salt was added, particularly for the higher-charged cations (Figure 3b). After subjecting them to HC, a marked reduction in the solution viscosity was observed. Apparently, the presence of charged ions in the solution enhances the effect of HC, which is particularly high during the first 100 passes. The flow behavior index tends to be one after 500 passes, indicating that the solutions approach Newtonian behavior. This is particularly evident for the PAM solution containing $CaCl_2$ (Figure 7b).

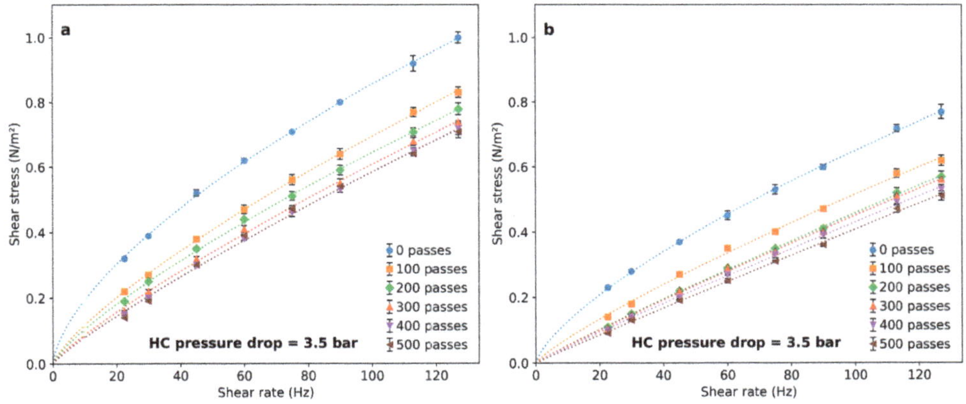

Figure 7. Rheological behavior of PAM solutions with salt addition, after a different number of passes through the cavitation device. Final salt concentration: (**a**) 0.01 M KCl; (**b**) 0.01 M $CaCl_2$.

The addition of salt significantly decreased the apparent viscosity, as it alters the structure and distribution of PAM chains. According to Arok et al. [26], the ions in the solution promote a charge-screening effect, weakening intermolecular interactions and inducing a transition from a gel-like macromolecular network to smaller, cluster-like structures. Regarding HC, salt addition accelerated the depolymerization rate of PAM in all cases and extended the effectiveness of HC over a greater number of passes, as the solutions continued to approach the viscosity of water. It has been proposed that salt addition induces a salting-out effect, enhancing the solubility of polymer short chains within the SCW domain inside the cavity bubbles while they grow [7]. Thus, increased PAM solubility in SCW, combined with enhanced chain mobility, likely increased the probability of partial or complete chain incorporation into cavitation bubbles, as sketched in Figure 8. Another factor to consider is the decrease in PAM hydrodynamic radius observed with the addition of bivalent ions (Ca^{2+} and Mg^{2+}) [31,32]. This would enhance the PAM chains' mobility and facilitate their entry, partially or entirely, into cavitation bubbles.

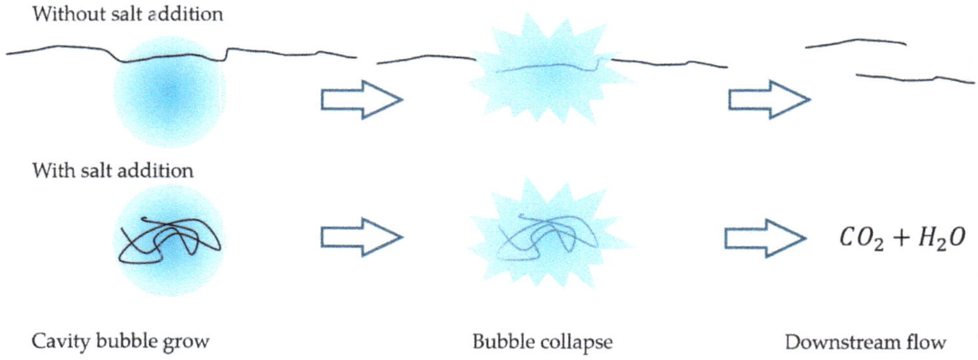

Figure 8. Sketch of the salt effect on HC performance. The addition of salt reduces the hydrodynamic radius of polymer chains, allowing larger molecules to enter cavitation bubbles and undergo degradation. As a result, the critical size limit decreases, enabling further depolymerization and leading to lower final viscosities.

The flow behavior index (n), which tends toward more Newtonian-like values with an increasing number of passes and with salt addition (Figure 9), suggests a reduction in both the average molecular weight and the strength of water–polymer interactions. This effect was clearly more pronounced in salt-containing solutions, with the $CaCl_2$-PAM solution nearly reaching ideal Newtonian behavior after 500 passes (Figure 9b).

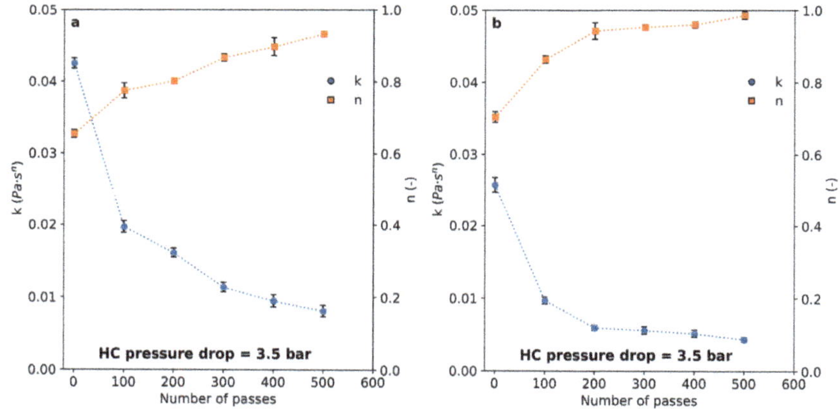

Figure 9. Consistency (k) and flow behavior (n) indexes (Equation (2)) calculated after a different number of passes through the cavitation device for PAM solutions with (**a**) 0.01 M KCl or (**b**) 0.01 M $CaCl_2$ added. Pressure drop through the vortex diode = 3.5 bar.

3.4. Polymer Solution Viscoelasticity

Changes in viscoelasticity were observed as a result of the presence of electrolytes and the application of HC. Table 2 presents the results for the loss and storage modulus ratio under the initial condition and after 500 passes through the HC device. The effects of electrolytes differed markedly between Ca^{2+} and K^+. While Ca^{2+} slightly increased the elasticity of the initial solution, K^+ decreased it, although both ions reduced the average viscosity. This suggests that the reported thinning effect [26] may not be the only mechanism involved and may depend on the salt species. Within the linear viscoelastic zone, the studied concentrations of Ca^{2+} and K^+ exhibited opposite effects: Ca^{2+} enhanced rigidity and structural stability, whereas K^+ reduced both. Thus, Ca^{2+} appears to exert a "cementing effect", while K^+ promotes dispersion. This behavior can be explained by the ability of Ca^{2+} to cross-link polyacrylamide chains [12,32].

Table 2. Loss modulus G″ to storage modulus G′ ratio for the different compositions and operating conditions.

Condition	Composition	Head Loss	G″/G′
Initial	1 g/L PAM	-	0.57 ± 0.07
Initial	1 g/L PAM + 0.01 M $CaCl_2$	-	0.19 ± 0.04
Initial	1 g/L PAM + 0.01 M KCl	-	6.56 ± 1.58
500 passes	1 g/L PAM	2	18,023 ± 3250
500 passes	1 g/L PAM	2.8	>20,000
500 passes	1 g/L PAM	3.5	>20,000
500 passes	1 g/L PAM + 0.01 M $CaCl_2$	3.5	>20,000
500 passes	1 g/L PAM + 0.01 M KCl	3.5	>20,000

Nonetheless, the effects of HC largely overshadowed the differences caused by salt addition. Significant reductions in elasticity were observed in all cases, with some values exceeding the instrument's detection limits. The increase in the G″/G′ ratio for treated PAM solutions without salt addition depended only marginally on the applied pressure drop. This shift reflects a tendency toward purely viscous behavior. Meanwhile, no significant differences were found between PAM solutions with and without added KCl or $CaCl_2$ after HC treatment. HC effectively eliminated all elastic properties, transforming the PAM solution into an almost purely viscous fluid. This transformation cannot be attributed solely to the presence of dissolved salts, particularly because, at the tested concentrations, the gel-like structure of the polymer remains comparable to that in distilled water [26], with only the charge screening effect being present. The changes are likely due to a generalized weakening of polymer chain-to-chain interaction forces.

4. Conclusions

HC effectively reduced the viscosity and viscoelasticity of PAM solutions, with the extent of modification strongly influenced by the number of passes. A critical molecular size limit is suggested, below which further depolymerization becomes minimal. This limitation appears to result from an intermediate size range of polymer chains that are too small to undergo mechanical shear-induced cleavage but too large to enter cavitation bubbles and be degraded by chemical mechanisms.

Salt addition was found to enhance the degradation of PAM significantly. This effect was attributed to a combination of charge screening and salting-out phenomena, which reduce intermolecular interactions, decrease hydrodynamic radius, and increase the solubility and mobility of polymer chains in the supercritical water domains attained within the cavity bubbles before implosion.

HC eliminated the elastic behavior of PAM solutions, transforming them into nearly purely viscous fluids. This transformation is likely related to a decrease in the average molecular weight. The observed plateau in the consistency index supports the idea that the HC treatment narrows the molecular weight distribution, not just causing uniform chain cleavage.

In summary, HC is a powerful tool for tailoring the rheological properties of PAM solutions. However, intrinsic limitations related to chain size and cavitation dynamics must be considered. Future optimization of HC device design may allow for further improvements in depolymerization performance.

Author Contributions: Conceptualization, methodology, investigation, software, validation, data curation, formal analysis, and writing—original draft preparation: S.N.F. and M.d.P.B.; investigation, resources, data curation, writing—review and editing, visualization, supervision, project administration, and funding acquisition: M.A.A. and M.C. All authors have read and agreed to the published version of the manuscript.

Funding: This research was funded by Universidad de Buenos Aires, grant number UBACyT 20020220100154BA.

Data Availability Statement: The original contributions presented in this study are mainly included in the article. Further inquiries can be directed to the corresponding author. The raw data supporting the conclusions of this article will be made available by the authors upon request.

Conflicts of Interest: The authors declare no conflicts of interest. The funders had no role in the design of this study; in the collection, analyses, or interpretation of data; in the writing of the manuscript; or in the decision to publish the results.

References

1. Zupanc, A.; Petkovšek, M.; Zdovc, B.; Žagar, E.; Zupanc, M. Degradation of hydroxypropyl methylcellulose (HPMC) by acoustic and hydrodynamic cavitation. *Ultrason. Sonochemistry* **2024**, *109*, 107020. [CrossRef]
2. Wang, X.; Jin, A.; Zhu, M.; Feng, C.; He, H.; Huang, Z.; Li, K.; Wang, L. Study on low-temperature plasma γ-Al_2O_3 catalytic viscosity reduction of polyacrylamide solution. *Environ. Sci. Pollut. Res.* **2023**, *30*, 36098–36111. [CrossRef] [PubMed]
3. Lim, K.Y.; Foo, K.Y. Innovation in Depolymerization Techniques. In *Depolymerization: Concepts, Progress, and Challenges Volume 2: Advances and Breakthroughs*; ACS Symposium Series; American Chemical Society: Washington, DC, USA, 2025; Chapter 2; pp 17–40. [CrossRef]
4. Gągol, M.; Przyjazny, A.; Boczkaj, G. Wastewater treatment by means of advanced oxidation processes based on cavitation–A review. *Chem. Eng. J.* **2018**, *338*, 599–627. [CrossRef]
5. Galloni, M.G.; Fabbrizio, V.; Giannantonio, R.; Falletta, E.; Bianchi, C.L. Applications and applicability of the cavitation technology *Curr. Opin. Chem. Eng.* **2025**, *48*, 101129. [CrossRef]
6. Kawadkar, A.S.; Gogate, P.R. Intensified depolymerization using ultrasound—A review of mechanisms, reactors, operating conditions and applications. *Chem. Eng. Process.-Process Intensif.* **2023**, *191*, 109446. [CrossRef]
7. Fleite, S.N.; Ayude, M.A.: Ranade, V.V.; Cassanello, M.C. Hydrodynamic cavitation effects on advanced oxidation processes and mass transfer: A conceptual model. *Chem. Eng. J. Adv.* **2024**, *18*, 100603. [CrossRef]
8. Pavlenko, A. Numerical Modeling of the Behavior of Bubble Clusters in Cavitation Processes. *Energies* **2024**, *17*, 1741. [CrossRef]
9. Yan, J.; Ai, S.; Yang, F.; Zhang, K.; Huang, Y. Study on mechanism of chitosan degradation with hydrodynamic cavitation. *Ultrason. Sonochemistry* **2020**, *64*, 105046. [CrossRef]
10. Prajapat, A.L.; Gogate, P.R. Depolymerization of carboxymethyl cellulose using hydrodynamic cavitation combined with ultraviolet irradiation and potassium persulfate. *Ultrason. Sonochemistry* **2019**, *51*, 258–263. [CrossRef]
11. Yin, J.-Y.; Ma, L.-Y.; Siu, K.-C.; Wu, J.-Y. Effects of Ultrasonication on the Conformational, Microstructural, and Antioxidant Properties of Konjac Glucomannan. *Appl. Sci.* **2019**, *9*, 461. [CrossRef]
12. Spildo, K.; Sæ, E.I.Ø. Effect of Charge Distribution on the Viscosity and Viscoelastic Properties of Partially Hydrolyzed Polyacrylamide. *Energy Fuels* **2015**, *29*, 5609–5617. [CrossRef]
13. Prajapat, A.L.; Gogate, P.R. Intensified depolymerization of aqueous polyacrylamide solution using combined processes based on hydrodynamic cavitation, ozone, ultraviolet light and hydrogen peroxide. *Ultrason. Sonochemistry* **2016**, *31*, 371–382. [CrossRef] [PubMed]
14. Zhang, X.; Fan, Z.; Cai, L.; Fu, Y.; Liu, J.; Liu, Q.; Tong, Q.; Qiao, S.; Sun, A. Polyacrylamide structural instability and degradation induced by nanobubbles: A molecular simulation study. *Phys. Fluids* **2025**, *37*, 022024. [CrossRef]
15. Wang, B.; Su, H.; Zhang, B. Hydrodynamic cavitation as a promising route for wastewater treatment—A review. *Chem. Eng. J.* **2021**, *412*, 128685. [CrossRef]
16. Ranade, V.V.; Bhandari, V.M.; Nagarajan, S.; Sarvothaman, V.P.; Simpson, A.T. *Hydrodynamic Cavitation a Systematic Introduction to Critical Technologies and Applications of Hydrodynamic Cavitation*; John Wiley & Sons: Hoboken, NJ, USA, 2022.
17. Simpson, A.; Ranade, V.V. 110th Anniversary: Comparison of Cavitation Devices Based on Linear and Swirling Flows: Hydrodynamic Characteristics. *Ind. Eng. Chem. Res.* **2019**, *58*, 14488–14509. [CrossRef]
18. Simpson, A.; Ranade, V.V. Flow characteristics of vortex based cavitation devices: Computational investigation on influence of operating parameters and scale. *AIChE J.* **2019**, *65*, e16675. [CrossRef]

19. Ranade, V.V.; Kulkarni, A.A.; Bhandari, V.M. Vortex Diodes as Effluent Treatment Devices. U.S. Patent No. 9,422,952B2, 23 August 2016. Available online: https://patentimages.storage.googleapis.com/7f/ae/c2/b1a4a224890f81/US9422952.pdf (accessed on 18 June 2025).
20. Ranade, N.V.; Sarvothaman, V.; Ranade, V.V. Acoustic Analysis of Vortex-based Cavitation Devices: Inception and extent of cavitation. *Ind. Eng. Chem. Res.* **2021**, *60*, 8255–8268. [CrossRef]
21. Gode, A.; Madane, K.; Ranade, V.V. Design of vortex-based cavitation devices/reactors: Influence of aspect ratio, number of inlets and shape. *Ultrason. Sonochemistry* **2023**, *101*, 106695. [CrossRef]
22. Pandit, A.V.; Sarvothaman, V.P.; Ranade, V.V. Estimation of chemical and physical effects of cavitation by analysis of cavitating single bubble dynamics. *Ultrason. Sonochemistry* **2021**, *77*, 105677. [CrossRef]
23. Thaker, A.H.; Madane, K.R.; Ranade, V.V. Influence of viscosity and device scale on pressure drop and cavitation inception: Vortex based cavitation devices. *Chem. Eng. J.* **2023**, *474*, 145943. [CrossRef]
24. Whorlow, R.W. *Rheological Techniques*, 2nd ed.; Ellis Horwood Series in Physics and Its Applications; Ellis Horwood Ltd.: Hertfordshire, UK, 1992.
25. Laun, M.; Auhl, D.; Brummer, R.; Dijkstra, D.J.; Gabriel, C.; Mangnus, M.A.; Rüllmann, M.; Zoetelief, W.; Handge, U.A. Guidelines for checking performance and verifying accuracy of rotational rheometers: Viscosity measurements in steady and oscillatory shear (IUPAC Technical Report). *Pure Appl. Chem.* **2014**, *86*, 1945–1968. [CrossRef]
26. Árok, Z.V.; Sáringer, S.; Takács, D.; Bretz, C.; Juhász, Á.; Szilagyi, I. Effect of salinity on solution properties of a partially hydrolyzed polyacrylamide. *J. Mol. Liq.* **2023**, *384*, 122192. [CrossRef]
27. Hu, J.; Fu, M.; Li, M.; Luo, Y.; Ni, S.; Hou, L. Main Controlling Factors Affecting the Viscosity of Polymer Solution due to the Influence of Polymerized Cations in High-Salt Oilfield Wastewater. *Processes* **2024**, *12*, 791. [CrossRef]
28. Sarvothaman, V.P.; Simpson, A.T.; Ranade, V.V. Modelling of vortex based hydrodynamic cavitation reactors. *Chem. Eng. J.* **2019**, *377*, 119639. [CrossRef]
29. Fleite, S.N.; Torres, R.; Lagorio, M.G.; Ranade, V.V.; Cassanello, M.C. Hydrodynamic cavitation effects over complex organic mixtures. *Chem. Eng. Res. Des.* **2024**, *204*, 371–381. [CrossRef]
30. Xu, X.; Zhang, P.; Luo, X. Synergistic mechanism of polymer and asphaltene in supercritical water oxidation treatment of polymer-containing oily sludge: Degradation pathway and nitrogen transfer. *J. Hazard. Mater.* **2025**, *494*, 138627. [CrossRef] [PubMed]
31. Davlyud, D.N.; Vorobeva, E.V.; Laevskaya, E.V.; Cherednichenko, D.V.; Vorobev, P.D. Flocculating and hydrodynamic properties of aqueous-salt solutions of acrylamide polymers. *Russ. J. Appl. Chem.* **2019**, *92*, 1135–1142. [CrossRef]
32. Bessaies-Bey, H.; Baumann, R.; Schmitz, M.; Radler, M.; Roussel, N. Effect of polyacrylamide on rheology of fresh cement pastes. *Cem. Concr Res.* **2015**, *76*, 98–106. [CrossRef]

fluids

MDPI

Article

Hysteresis-Induced Onset and Progressive Decay of Periodic and Metastable Sheet Cavitation in a Chamfered Circular Orifice

Min Son *,†, Michael Börner, Wolfgang Armbruster and Justin S. Hardi

Institute of Space Propulsion, German Aerospace Center (DLR), D-74239 Lampoldshausen, Germany; michael.boerner@dlr.de (M.B.); wolfgang.armbruster@dlr.de (W.A.); justin.hardi@dlr.de (J.S.H.)
* Correspondence: min.son@unibw.de
† Current address: Institut für Thermodynamik, Fakultät für Luft- und Raumfahrttechnik, Universität der Bundeswehr München, D-85577 Neubiberg, Germany.

Abstract: This study investigates the onset and decay mechanisms of sheet cavitation within a chamfered orifice under turbulent conditions, using high-speed backlight imaging for detailed frame-by-frame analysis. A distinctive metastable sheet cavitation regime was identified, distinguished by its unique hysteresis behavior during onset conditions, with the ability to control periodicity through variations in cavitation numbers. This new sheet cavitation regime appears at high cavitation numbers, contrary to typical expectations of cavitation inception, highlighting a new potential risk within the range of safe operation for hydraulic systems equipped with control valves. Furthermore, linear growth and rapid collapse of the bubble sheet were observed, which differs from the conventional periodic behavior of sheet cavitation on hydrofoils. The new mechanism to intentionally initiate and control this sheet cavitation regime by manipulating the pressure drop across the orifice could potentially be adopted for industrial applications, particularly in the generation of controlled and dispersed bubbles. Future research should include quantifying bubble dynamics within this regime and assessing the effects of fluid properties and orifice geometries on cavitation characteristics. In summary, this study introduces a new perspective on metastable sheet cavitation, emphasizing its potential applications and importance in the design and operation of fluid systems.

Keywords: sheet cavitation; hysteresis behavior; bubble dynamics; high-speed imaging

1. Introduction

Cavitation occurs when the pressure in a liquid falls below its vapor pressure at constant temperature, leading to the formation of vapor bubbles. This process is driven by the liquid reaching its tensile strength and rupturing due to pressure reduction [1]. Cavitation commonly arises in mechanical devices inducing high pressure drops, such as hydraulic turbines [2], pumps [3], venturis [4], and orifices [5], potentially damaging these devices. In diesel injectors, cavitation bubbles formed from high pressure drops can erode the injector nozzle's internal surfaces [6], while the hydraulic flip induced by strong cavitation alters spray characteristics [7]. In liquid rocket engines, cavitation not only can cause mechanical damage but can also trigger combustion instability, potentially leading to complete engine destruction. Cavitation may accidentally occur in the propellant supply system, especially in turbopumps, as the temperatures of cryogenic propellants are very close to their saturation temperature [8]. Despite careful design to prevent cavitation, it may still occur during transient starting conditions when the pressure drop through the supply line to the chamber is significantly higher than the design point, and the propellant temperature is higher than the design point due to not fully chilled supply lines and injector head [9].

Conversely, cavitation can be intentionally generated to harness its energy. Using the pressure wave and temperature variation from collapsing cavitation bubbles can enhance chemical reactions, turning the bubble-generating instrument into a reactor [10,11]. This approach finds applications in water waste purification [12,13] and in the food and beverage industries for chemical and physical processes [14]. Moreover, cavitation bubbles can be downsized to the nanoscale with precise design and control [15].

Son et al. [16] identified cavitation from the liquid oxygen (LOX) orifice as a potential source of rocket combustion instability by visualizing the orifice flow [16], showing periodic fluctuations that could induce oscillations in the injection flow rate, affecting the heat release rate within the combustion chamber [17,18]. Cavitation initiates at a cavitation number (K) of approximately 2.0, with its strength increasing as the cavitation number decreases with increasing pressure drop, a finding consistent with other studies [19]. Interestingly, a new type of cavitation regime appears when the pressure drop decreases after cavitation occurs, characterized by a metastable bubble sheet with cyclic growth and decay periods.

Metastable sheet cavitation under laminar flow conditions in a venturi has been previously studied [20], yet the sheet cavitation in this study occurs under turbulent flow conditions. Additionally, periodic sheet cavitation observed with a hydrofoil [21] differs in decay mechanism from that in the current study. While recent work by Ge et al. [22] identified periodic cavitation intensity modulated by thermal effects in a Venturi reactor, this study reveals a distinct pressure-driven hysteresis mechanism inducing sheet cavitation below the initial inception pressure. While hysteresis characteristics of cavitation have been observed on hydrofoil surfaces [23,24], this new regime in the orifice has not been previously reported or studied.

This study investigates periodic sheet cavitation induced at high cavitation numbers due to hysteresis. Using high-speed imaging with backlight, the onset and progressive decay processes are analyzed as a function of time. Furthermore, the bubble sheet length and periodic characteristics are compared as a function of the cavitation number.

2. Experimental Setup and Methods

2.1 Experimental Setup

The experimental setup principally comprises four parts: the upstream valve, the manifold, the transparent orifice part, and the extended downstream module, as illustrated in Figure 1. Water, serving as the working fluid, was not specially treated to remove dissolved gases. The flow was controlled by a manual ball valve located upstream of the test module at a sufficient distance to avoid disturbances from the valve, which was smoothly adjusted to induce hysteresis. At the inlet section of the module, water flows through four radial holes to mitigate the pressure and mass flow fluctuation of inflow on the orifice flow. The transparent block, fabricated from polymethyl methacrylate (PMMA), is designed to capture the cavitation flow within the orifice; its dimensions are 40 × 40 mm. The orifice has a diameter of 2.0 mm, and a 45-degree chamfer with a depth of 0.5 mm was applied at the orifice entrance, as illustrated in Figure 2. This orifice design was inspired by the LOX injector used in rocket engine combustion chamber experiments [18]. An additional nozzle with a hydraulic diameter of 3.5 mm, closely matching the 3.6 mm tube diameter, is installed at the downstream module's end. Static pressures upstream (P_u) and downstream (P_d) were measured using 'Kistler static pressure sensors (PiezoSmart 4043, Winterthur, Switzerland), and the flow rate was gauged by a turbine flow meter from Badger Meter (Vision 2006 2F66, Milwaukee, WI, USA).

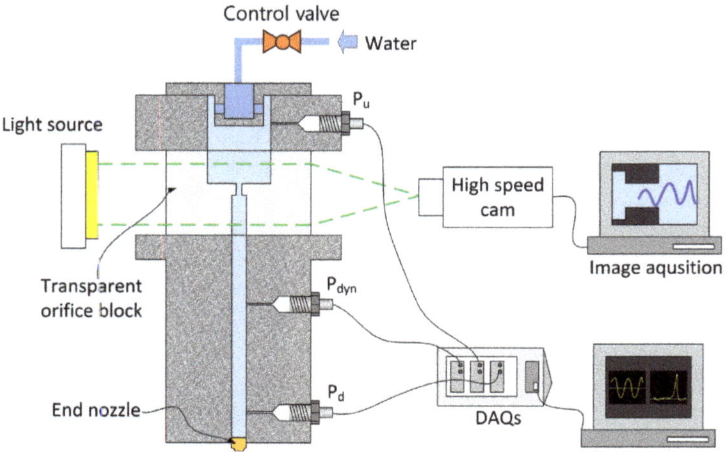

Figure 1. Experimental setup for flow visualization and data acquisition.

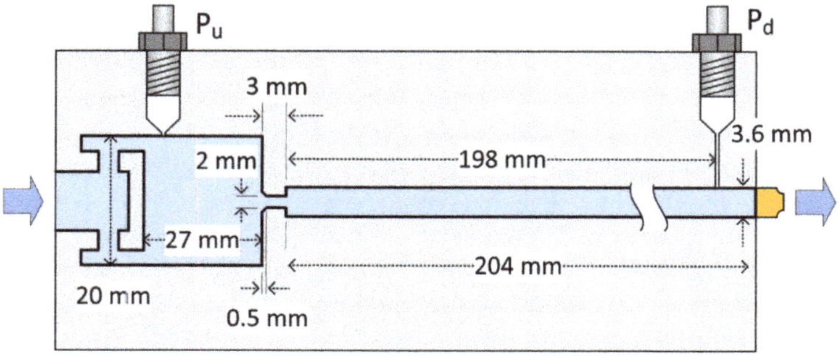

Figure 2. Dimensions of internal geometries of the flow path.

To visualize cavitation behavior, a backlight imaging technique coupled with a high-speed camera was used. A Photron UX100 camera (Tokyo, Japan), capturing at a frame rate of 40,000 fps and an exposure time of 1 μs, was used. The camera was manually triggered once the flow reached a steady state after the flow conditions were changed. A multi-array LED lamp with a diffuser, specified at 37.5 W, 4350 lm, and 6500 K, was used as the backlight source. The LED lamp directly illuminated the cavitation flow through the transparent orifice block.

2.2. Experimental Conditions

Cavitation characteristics were investigated in relation to the cavitation number, which is defined by the difference between the flow pressure and the vapor pressure (P_v) of the working fluid. The definition of the cavitation number varies according to the study's purpose. The definition based on the pressure difference divided by the kinetic energy of the orifice flow, as in Equation (1), has primarily been used for wake cavitation of hydrofoils and venturi-like channels. Here, P_{ref} is chosen as the upstream pressure [20] or free flow pressure [23]. Another definition, as shown in Equation (2), was used for valves [25] and orifices [26]. In this study, the latter definition is adopted because the velocity in the orifice is not clearly defined due to the bubbly flow. $P_v = 2.34$ kPa for water at 293 K was used, with the temperature measurement error impacting the cavitation number by less than $\pm 1\%$ within a ± 5 K range. The detailed experimental conditions for the sheet cavitation regime are listed in Table 1, where dP represents the pressure drop across the orifice, $\dot{m}$ is the measured mass flow rate, Re is the Reynolds number in the orifice, and C_d is the discharge coefficient. *Re* is defined using the bulk velocity, which is calculated from the measured mass flow rate, $\dot{m}$. In this study, the pressure drop was mainly controlled by the upstream valve, and cavitation numbers and Reynolds numbers were calculated based on the measured pressures and mass flow rates.

$$K = \frac{P_{ref} - P_v}{\frac{1}{2}\rho V^2} \tag{1}$$

$$K = \frac{P_u - P_v}{P_u - P_d} \tag{2}$$

Table 1. Experimental conditions for sheet cavitations.

Target *dP* [bar]	Measured *dP* [bar]	$\dot{m}$ [g/s]	*K*	*Re*	C_d
0.1	0.097	15.3	10.88	1.14×10^4	0.74
0.4	0.396	28.6	3.48	2.13×10^4	0.81
0.5	0.496	36.7	3.03	2.74×10^4	0.93
0.6	0.616	40.9	2.66	3.05×10^4	0.95
0.8	0.798	44.6	2.31	3.22×10^4	0.94

The initiation or inception of bubbly cavitation at $K = 2.0$ has been previously investigated by Son et al. [16], where the pressure drop was 1.0 bar, as illustrated in Figure 3. As the pressure drop increases, cavitation bubbles erupt downstream, as shown in Figure 4, potentially causing acoustic resonance along the downstream tube [16]. While cavitation is not expected when the cavitation number exceeds the inception cavitation number, in this case, $K = 2.0$, a unique regime of metastable sheet cavitation was observed at higher cavitation numbers. This regime was induced solely by a hysteresis procedure when the pressure drop decreased after strong cavitation had already been initiated.

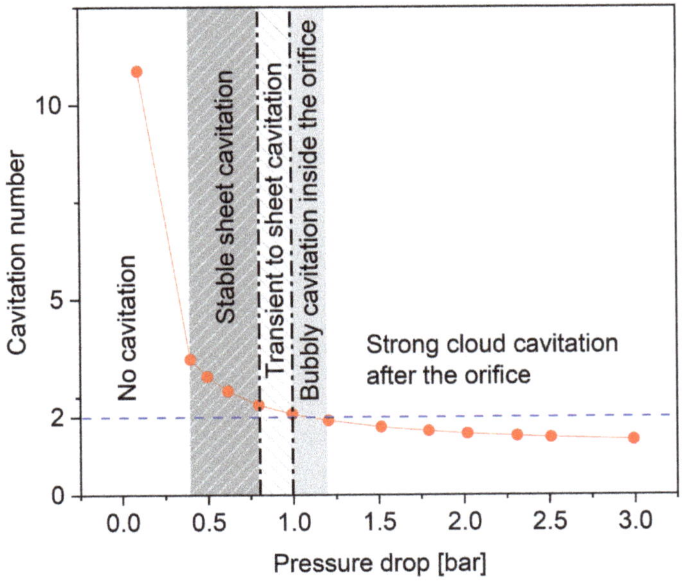

Figure 3. Cavitation regime based on orifice pressure drops.

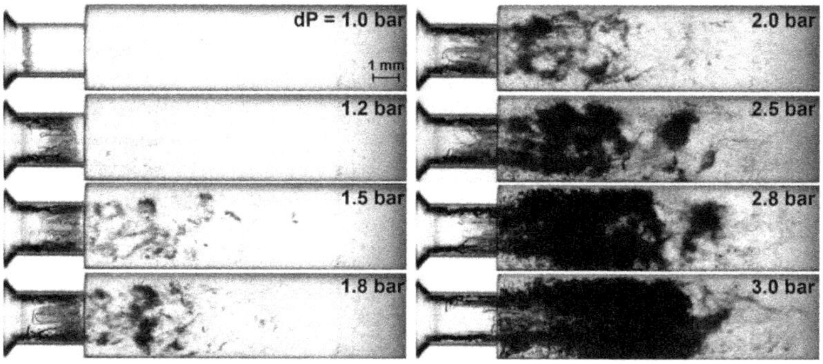

Figure 4. Visualized cloud cavitation as the pressure drop increases [16].

3. Results and Discussion

3.1. Onset and Decay Characteristics of Sheet Cavitation

Metastable sheet cavitation does not appear when the valve is initially opened and the pressure increases. Before the cavitation starts, the cavitation onset is solely controlled by the instantaneous pressure conditions without any memory effect. The onset process is illustrated in Figure 5a, with corresponding snapshots shown in Figure 5b. Initially, bubbly cavitation occurs suddenly at **A** when the pressure drop increases. In this study, the sudden inception point appears at $dP = 1.0$ bar [16]. Once cavitation with small bubbles is initiated, the bubble cloud transitions to **B**, generating strong cloud cavitation. Subsequently, when the pressure drop decreases back to **A**, closing the valve, the cloud cavitation reverts to small bubble cavitation at **A** again. The process between **A** and **B** is reversible and does not exhibit hysteresis. In this regime, the length of the bubble cloud is proportional to the pressure drop or the Reynolds number. Figure 6 clearly shows that the bubble length increases as the flow Reynolds number decreases, and the sheet almost reaches the orifice exit at $dP = 1.0$ bar.

When the valve is carefully and slowly closed, the pressure drop falls below the inception point of **A** ($dP = 1.0$ bar in this study). In contrast to the cloud cavitation regime, this slow decrease in pressure allows the flow to retain the memory of bubble nucleation, thus introducing a hysteresis effect. As pressure drop and Reynolds number decrease further below the inception point, the small nucleated bubbles coalesce and grow into a larger bubble attached at the end of the chamfer, as shown in snapshot **C** in Figure 5b. Figure 5c illustrates this process. As the pressure drop continues to decrease, the integrated bubble expands in the circumferential direction, forming a uniform ring-shaped bubble sheet at approximately $dP = 0.8$ bar and $Re = 3.22 \times 10^4$. Here, the conservative forces, such as surface tension and reduced pressure gradients in the recirculation zone, stabilize the bubble sheet, maintaining its structure even below the original inception pressure. In this sheet cavitation regime, the bubble sheet length increases with decreasing pressure drop at point **C**, in contrast to the cloud cavitation regime. The transition from **A** to **D** does not occur before the initial cavitation inception and is induced only by this hysteresis-driven onset process, where the prior state of bubble nucleation controls the subsequent cavitation regimes. A detailed discussion of these mechanisms is presented in Section 3.2.

Sheet cavitation presents different attachment characteristics compared to cloud cavitation. Observing bubble traces near the orifice wall, as indicated by the red arrows in Figure 7a, shows a clear detachment point in cloud cavitation, while in sheet cavitation, the bubble sheet remains attached on the wall. Due to the re-entrant flow of the downstream jet, cloud cavitation detaches earlier and releases bubble clumps [27]. However, the low-pressure region with sheet cavitation at the chamfer exit sustains vaporization, while the absence of strong re-entrant jets, unlike in cloud cavitation, allows further elongation, as shown in Figure 7b.

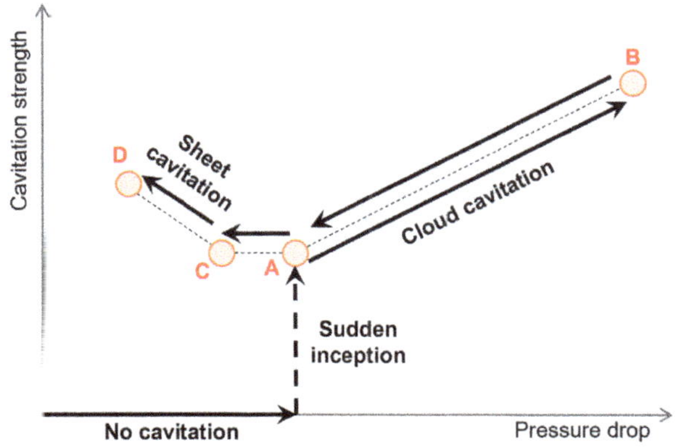

(a)

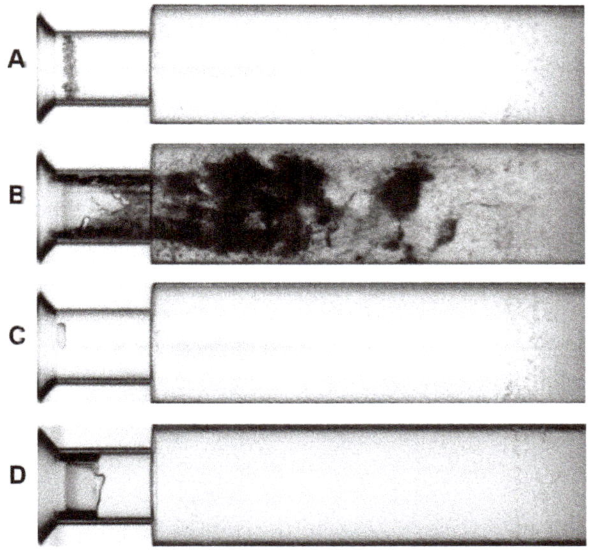

(b)

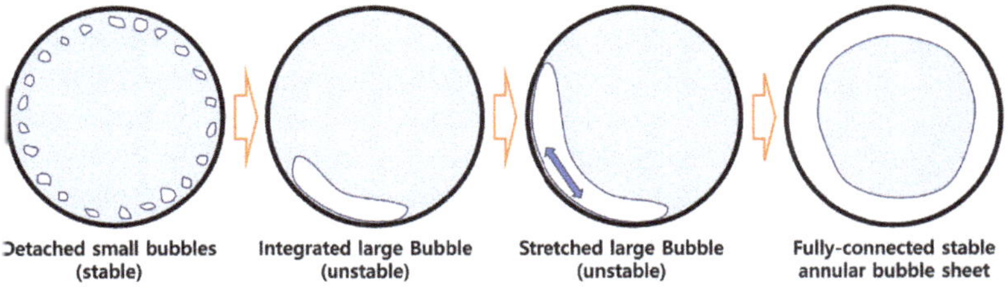

(c)

Figure 5. Onset mechanism of stable sheet cavitation: (**a**) a schematic of the corresponding procedure, (**b**) visualized cavitation onset, and (**c**) sequential onset steps through **A-C-D** in cross-sectional views at the orifice inlet.

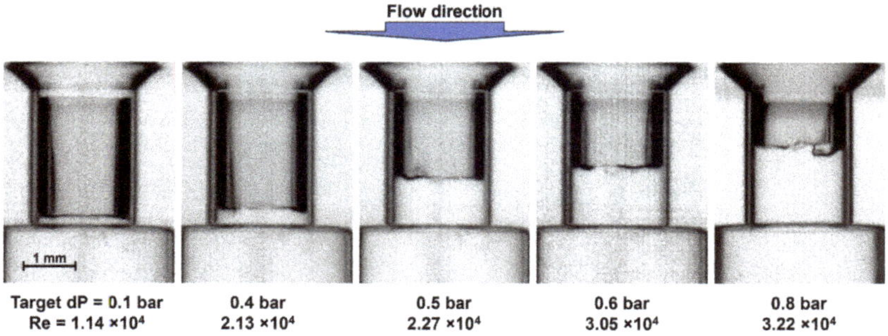

Figure 6. Snapshots of the maximum bubble lengths at various flow conditions.

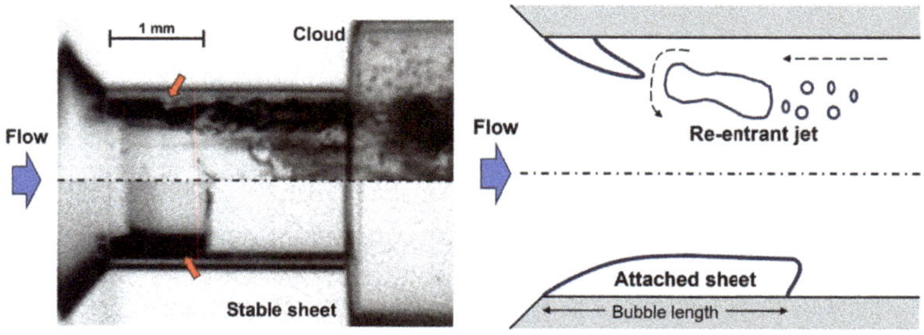

(a) (b)

Figure 7. Comparison of cloud and sheet cavitation: (**a**) captured images and (**b**) schematics of cavitation structures.

Sheet cavitation experiences periodic growth and decay, as shown in Figure 8. Starting from a very short bubble layer, it gradually grows for about 400 ms before collapsing suddenly in just a few milliseconds. The bubble sheet repeats this cycle periodically. This behavior is highlighted when measuring the bubble sheet length over time, as presented in Figure 9. The length of the bubble sheet is defined as the distance from the end of the chamfer to the tip of the sheet, as illustrated in Figure 7b. The bubble sheet extends linearly over time and collapses from the maximum length to the minimum length. Similar periodic behavior in cloud cavitation, where decay results from the re-entrant jet, has been documented [27,28]. However, unlike previous studies in which cloud cavitation cycles had similar durations for growth and decay, decay occurs extremely rapidly in this study. During the decay, very small bubbles form, escaping into the downstream tube. These downstream bubbles are much smaller and more widely dispersed compared to those in strong cloud cavitation, as shown in Figure 5b.

Close investigation of the decay moment, as shown in Figure 10, reveals how the bubble initiates and propagates. When the bubble layer reaches its maximum length at 396.175 ms, the sheet tip develops wrinkles due to Kelvin–Helmholtz instabilities caused by velocity gradients between the bubble sheet and the orifice flow. A pronounced wrinkle, indicated by the red arrow at 396.2 ms, marks the beginning of a rapid collapse at 396.3 ms. The collapse point penetrates the bubble sheet by 396.575 ms, deepening into a V-shape by 396.75 ms. This collapsing momentum can be explained by a micro jet generated by a shock wave from the energy released during bubble collapse [29,30]. The collapse spreads circumferentially due to shock waves from energy release, disrupting the entire sheet. Furthermore, this process may also induce a strong turbulent flow downstream, disturbing the flow equilibrium within the orifice and leading to the circumferential collapse.

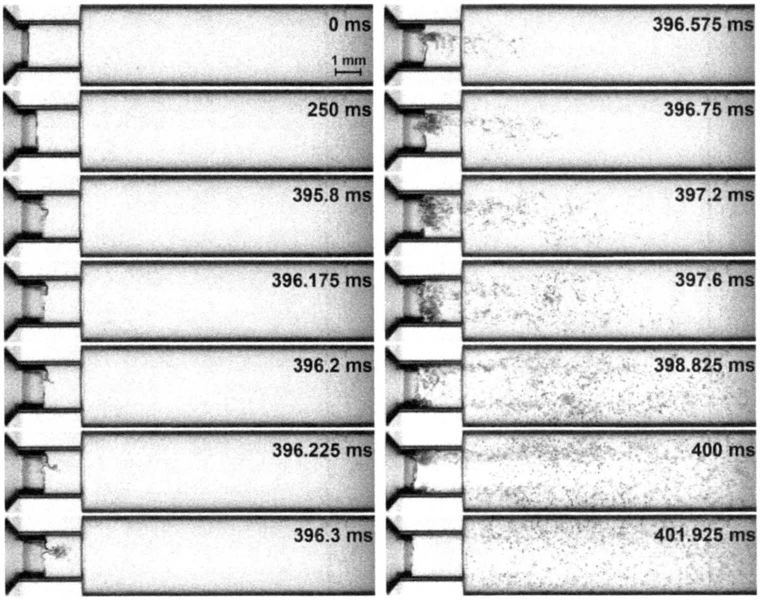

Figure 8. Sequential snapshots of sheet cavitation at $dP = 0.8$ bar and $Re = 3.22 \times 10^4$.

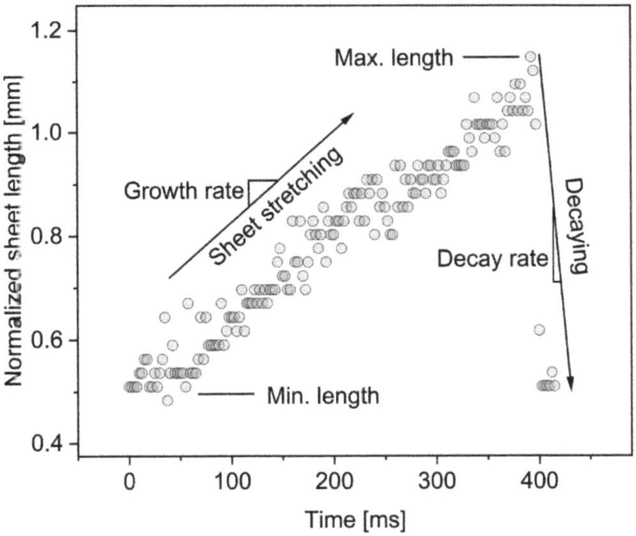

Figure 9. Measured bubble sheet lengths in a cycle at $dP = 0.8$ bar and $Re = 3.22 \times 10^4$.

The top of Figure 11 summarizes the maximum and minimum lengths of the bubble sheets caused by periodic growth and decay under all flow conditions. When the Reynolds number increases, both the maximum and minimum lengths decrease. In addition, a limit for the sheet length is observed. When the Reynolds number is reduced to 2.13×10^4, the maximum and minimum lengths converge to a single value similar to the orifice length. The gap between the maximum and minimum lengths approaches zero, indicating that the decay amount is immeasurably small.

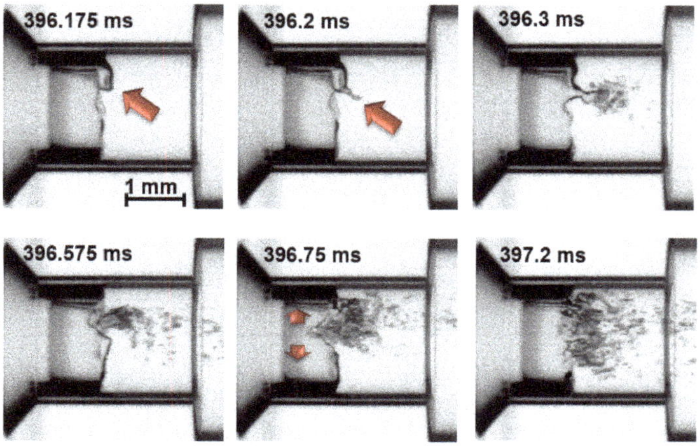

Figure 10. Close-up of the orifice at decaying moment at $dP = 0.8$ bar and $Re = 3.22 \times 10^4$ (Video S1).

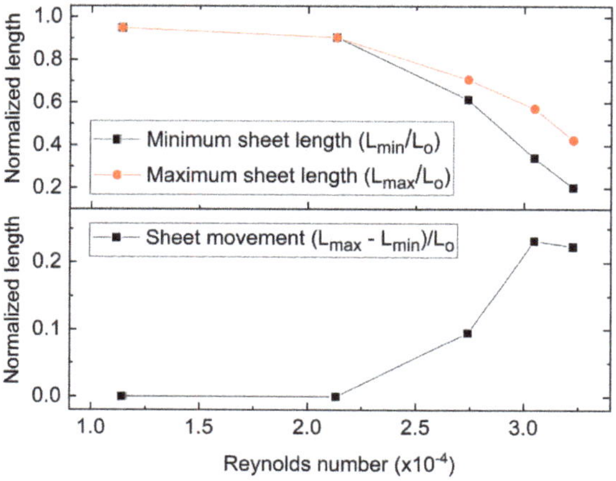

Figure 11. The maximum and minimum sheet lengths during cyclic growth and decay according to Reynolds numbers. The lengths are normalized by the orifice length ($L_o = 2.5$ mm), except for the chamfer length.

Figure 12 illustrates this immeasurable decay with sequential images at $dP = 0.4$ bar and $Re = 2.13 \times 10^4$. Under this condition, the bubble sheet stretches similarly to Figure 9, but only a small bubble separates from the sheet, unlike the destructive decay observed at $Re = 3.22 \times 10^4$. Therefore, the change in bubble sheet length is negligible. The separated bubble disintegrates into smaller bubbles downstream.

The sheet movement distance, defined as the difference between the maximum and minimum sheet lengths, is strongly influenced by the Reynolds number, as shown at the bottom of Figure 11. The results also indicate an upper limit for movement distance, which is reached at $Re = 3.05 \times 10^4$. Beyond this point, further increases in the Reynolds number do not significantly affect movement distance. However, it should be noted that the minimum and maximum sheet lengths continue to decrease as the Reynolds number increases beyond 3.05×10^4.

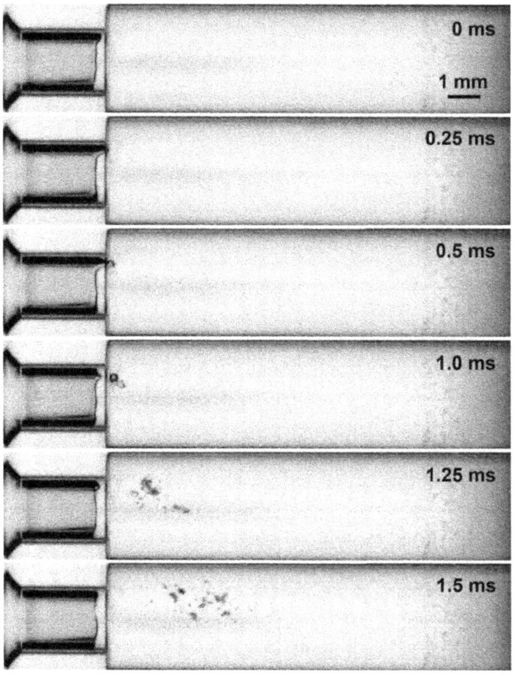

Figure 12. Decay sequence of sheet cavitation at $dP = 0.4$ bar and $Re = 2.13 \times 10^4$.

Based on the definitions in Figure 9 and the decay cycle frequencies, the growth rate and decay rate are calculated, as shown in Figure 13. The decay rate is approximately 60 times faster than the growth rate. Both rates are proportional to the Reynolds number; however, measurements below $Re = 2.13 \times 10^4$ are not feasible. The growth and decay rates follow similar trends, but both reach the maximum values at $Re = 3.05 \times 10^4$.

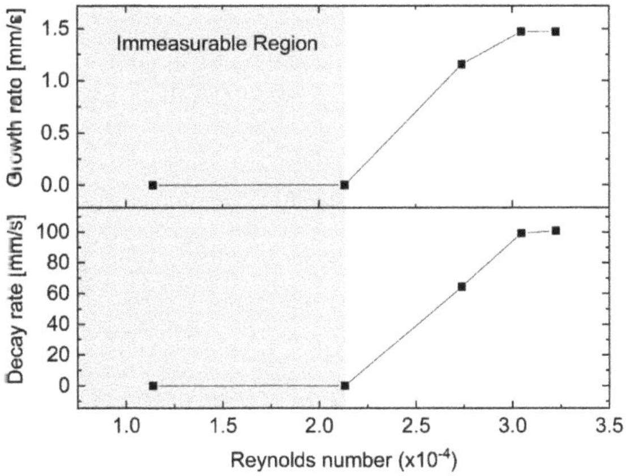

Figure 13. Growth and decay rates of bubble sheets according to Reynolds numbers.

In this study, we do not focus on downstream bubbles; however, Figure 12 provides a rough comparison of the characteristics of the bubbles that escape from the orifice. Note that the snapshots in Figure 14 are not compared at the same periodic moment, and quantitative analysis is not feasible due to differences in the focusing points on the circumferential decaying regions. However, increases in pressure drop and Reynolds number result in an increased amount of downstream bubbles and reduced bubble sizes. Future studies are expected to quantify the downstream bubbles, which could be used for cyclic bubble generation control in terms of size, quantity, and frequency.

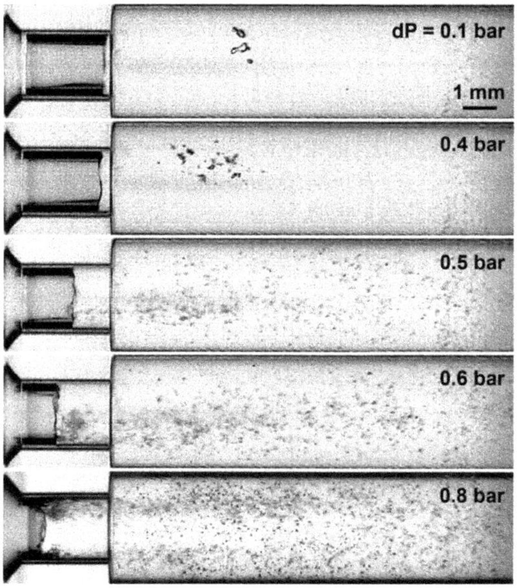

Figure 14. Comparison of downstream bubbles after the start of the decay process.

3.2. Discussion

The sheet cavitation observed in this study exhibits distinctive characteristics. Remarkably, it remains stable even at Reynolds numbers high enough to induce turbulent flow within the orifice, contrasting with the laminar sheet cavitation documented in previous research [20].

This new regime may be explained by bubbles trapped within the recirculation zone; however, the bubble sheet length exhibits a distinct trend compared to the reattachment behavior of a separated boundary layer in forward-facing step flows. Although the reattachment length of the recirculation zone in forward-facing step flows increases with increasing Reynolds numbers [31], the maximum bubble length in this study instead decreases, as shown in Figure 11. It should be noted that the Reynolds numbers in this study exceed 20,000, probably higher than the critical Reynolds number at which the reattachment length stops increasing [31]. Hence, the recirculation zone itself can be assumed to remain relatively similar in size for all cases in this study, and the maximum bubble sheet length is controlled by another mechanism.

The sheet cavitation regime arises from a metastable equilibrium between surface tension, flow-induced shear, and pressure gradients, as shown in Figure 15. Initially, small bubbles nucleated by sudden inception merge into a circumferentially connected sheet. This coalescence is stabilized by surface tension, which minimizes interfacial energy while resisting the shear force of the orifice flow and the turbulence from the recirculation zone downstream. The continuous annular geometry of the sheet enhances stability, as surface tension forces scale with curvature, creating a restoring force that counteracts flow-driven elongation.

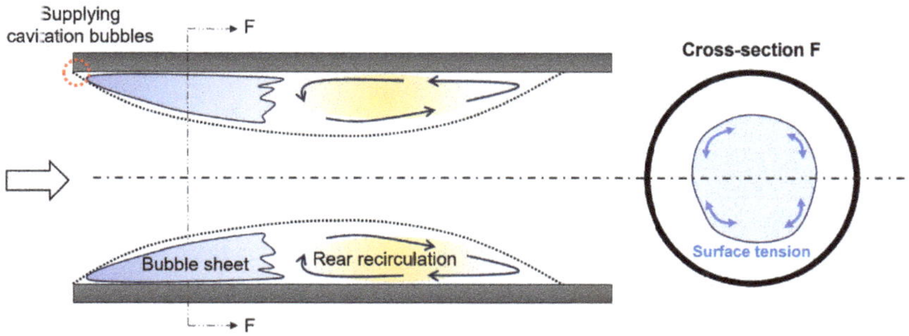

Figure 15. Schematic of bubble sheet generation inside the recirculation zone.

As the sheet elongates, the recirculation region contracts, amplifying shear stresses at the sheet tip. This explains the inverse proportional relationship between the maximum sheet length and Reynolds number. The higher *Re* enhances turbulence strength in the recirculation behind the bubble sheet, reducing the critical length at which shear force overwhelms the resisting force of surface tension. The abrupt decay occurs when Kelvin–Helmholtz instabilities at the vapor–liquid interface at the end of the sheet exceed the attenuating capacity of surface tension, triggering microjet-driven disintegration [30].

In addition, the onset condition of this cavitation regime shows hysteresis. During slow valve closure, the cavitating flow can transition from cloud cavitation to a metastable sheet by reducing the turbulent energy that disturbs bubble coalescence. Once the metastable annular sheet is formed, it remains even below the inception pressure ($dP < 1.0$ bar in this study) due to the balance between disturbing and resisting forces. This metastability can have a critical impact on flow systems. For example, in rocket engines, this cavitation regime could induce instability during throttle adjustments in the start-up transient phase, even when the evaluated cavitation number is higher than the typical limit.

On the other hand, the periodic generation of micro bubbles, as shown in Figure 12, can offer advantages in various fields. For instance, in biomedical applications, such as ultrasound imaging, sonoporation, tumor ablation, and targeted drug delivery, the precise control of bubble size and frequency by the new bubble sheet regime in this study may improve diagnostic accuracy and therapeutic efficiency without external actuators [32]. Additionally, in chemical reactors and industrial environments, this controlled bubble generation can improve mixing efficiency and mass transfer, supporting localized chemical reactions and efficient cleaning processes while reducing energy consumption and preventing bubble coalescence. Moreover, for optical flow diagnostics, controlled sheet cavitation could act as natural tracers in cases where seeding is not possible [33].

A limitation of this study should be noted; the dissolved gas was not controlled. Since dissolved gas can promote the inception of cavitation [34], the inception point of cavitation may change quantitatively. However, because the new regime begins after cavitation has already been initiated, the impact of dissolved gas can be considered minimal. Nevertheless, the effect of dissolved gas should be evaluated in future work.

4. Conclusions

This study has investigated the onset and decay mechanisms of sheet cavitation in a chamfered orifice. Through high-speed imaging and frame-by-frame analysis, a unique metastable sheet cavitation regime was identified, demonstrating notable stability due to equilibrium between surface tension, flow shear, and pressure gradients in recirculation behind the sheet even under turbulent conditions. This regime is characterized by its hysteresis behavior during the onset condition, and the periodicity can be controlled by cavitation numbers.

Key findings of our investigation include the following:

1. The new sheet cavitation regime appears at high cavitation numbers where cavitation inception is typically unexpected, presenting a distinct hysteresis effect governed by path-dependent energy barriers. This suggests a potential risk to hydraulic systems with control valves, particularly in transient operations like rocket engine throttling.
2. The behavior of sheet cavitation, particularly its linear growth and rapid collapse driven by tip instabilities and microjet penetration, provides insights into the dynamic stability of cavitation bubbles and their interactions with flow conditions.
3. Because the metastable sheet cavitation regime can be intentionally initiated and controlled through the manipulation of pressure drop across the orifice, it reveals new possibilities for utilizing the new cavitation regime in various research fields through controlled and dispersed bubble generation.

For future work, exploring the quantitative aspects of bubble dynamics within the metastable sheet cavitation regime, including bubble size distribution and velocity, could provide further insights into optimizing cavitation for specific applications. Additionally, investigating the impact of varying fluid properties (e.g., dissolved gas content) and orifice geometries, such as chamfer angle and L/D ratio, on the stability and characteristics of sheet cavitation could broaden the understanding and applicability of these findings.

In conclusion, this study identifies a new hysteresis-induced metastable sheet cavitation regime in an orifice, which persists below the cavitation inception pressure. This novel regime challenges conventional cavitation number thresholds and presents a dual role as both a potential factor in flow systems and a new bubble generation mechanism for research applications. Further research in this regime could improve our understanding of cavitation and lead to improved design and operation of fluid systems.

Supplementary Materials: The following supporting information can be downloaded at: https://www.mdpi.com/article/10.3390/fluids10030059/s1, Video S1: Close-up video of the bubble sheet decay process at $dP = 0.8$ bar, corresponding to Figure 10.

Author Contributions: Conceptualization, M.S. and J.S.H.; methodology, M.S.; software, M.S.; validation, W.A. and J.S.H.; formal analysis, M.S., M.B. and W.A.; investigation, M.S. and M.B.; resources, M.B.; data curation, M.S.; writing—original draft preparation, M.S.; writing—review and editing, M.S., M.B., W.A. and J.S.H.; visualization, M.S.; supervision, J.S.H.; project administration, J.S.H. All authors have read and agreed to the published version of the manuscript.

Funding: This activity was performed in the frame of the DLR internal project LUMEN.

Data Availability Statement: The data presented in this study are available on request from the corresponding author and with written permission of the DLR e.V.

Acknowledgments: Authors thank Arthur Walz-Steinbach, Jonas Vogel, Markus Dengler, and Michael Zepmeisel for technical support.

Conflicts of Interest: The authors declare no conflicts of interest.

References

1. Brennen, C.E. *Cavitation and Bubble Dynamics*; Cambridge University Press: Cambridge, UK, 2014.
2. Kumar, P.; Saini, R. Study of cavitation in hydro turbines—A review. *Renew. Sustain. Energy Rev.* **2010**, *14*, 374–383. [CrossRef]
3. Binama, M.; Muhirwa, A.; Bisengimana, E. Cavitation effects in centrifugal pumps-A review. *Int. J. Eng. Res. Appl.* **2016**, *6*, 52–63.
4. Garcia, R.; Hammitt, F.G. Cavitation damage and correlations with material and fluid properties. *J. Basic Eng.* **1967**, *89*, 753–763. [CrossRef]

5. Testud, P.; Moussou, P.; Hirschberg, A.; Aurégan, Y. Noise generated by cavitating single-hole and multi-hole orifices in a water pipe. *J. Fluids Struct.* **2007**, *23*, 163–189. [CrossRef]
6. Asi, O. Failure of a diesel engine injector nozzle by cavitation damage. *Eng. Fail. Anal.* **2006**, *13*, 1126–1133. [CrossRef]
7. Abderrezzak, B.; Huang, Y. Investigation of the effect of cavitation in nozzles with different length to diameter ratios on atomization of a liquid jet. *J. Therm. Sci. Eng. Appl.* **2017**, *9*, 031014. [CrossRef]
8. Hashimoto, T.; Yoshida, M.; Watanabe, M.; Kamijo, K.; Tsujimoto, Y. Experimental study on rotating cavitation of rocket propellant pump inducers. *J. Propuls. Power* **1997**, *13*, 488–494. [CrossRef]
9. Hardi, J.; Martin, J.; Son, M.; Armbruster, W.; Deeken, J.C.; Suslov, D.; Oschwald, M. Combustion Stability Characteristics of a sub-scale LOX/LNG Rocket Thrust Chamber. In Proceedings of the Aerospace Europe Conference 2020, Bordeaux, France, 25–28 February 2020.
10. Kumar, K.S.; Moholkar, V.S. Conceptual design of a novel hydrodynamic cavitation reactor. *Chem. Eng. Sci.* **2007**, *62*, 2698–2711. [CrossRef]
11. Zheng, H.; Zheng, Y.; Zhu, J. Recent developments in hydrodynamic cavitation reactors: Cavitation mechanism, reactor design, and applications. *Engineering* **2022**, *19*, 180–198. [CrossRef]
12. Tao, Y.; Cai, J.; Huai, X.; Liu, B.; Guo, Z. Application of hydrodynamic cavitation to wastewater treatment. *Chem. Eng. Technol.* **2016**, *39*, 1363–1376. [CrossRef]
13. Abdrashitov, A.; Gavrilov, A.; Marfin, E.; Panchenko, V.; Kovalev, A.; Bolshev, V.; Karaeva, J. Cavitation Reactor for Pretreatment of Liquid Agricultural Waste. *Agriculture* **2023**, *13*, 1218. [CrossRef]
14. Asaithambi, N.; Singha, P.; Dwivedi, M.; Singh, S.K. Hydrodynamic cavitation and its application in food and beverage industry: A review. *J. Food Process Eng.* **2019**, *42*, e13144. [CrossRef]
15. Xiong, Y.; Peng, F. Optimization of cavitation venturi tube design for pico and nano bubbles generation. *Int. J. Min. Sci. Technol.* **2015**, *25*, 523–529. [CrossRef]
16. Son, M.; Börner, M.; Armbruster, W.; Hardi, J.S. Orifice Flow Dynamics in a Rocket Injector as an Excitation Source of Injector-Driven Combustion Instabilities. *Aerospace* **2023**, *10*, 452. [CrossRef]
17. Gröning, S.; Hardi, J.S.; Suslov, D.; Oschwald, M. Injector-driven combustion instabilities in a hydrogen/oxygen rocket combustor. *J. Propuls. Power* **2016**, *32*, 560–573. [CrossRef]
18. Armbruster, W.; Hardi, J.S.; Suslov, D.; Oschwald, M. Injector-Driven Flame Dynamics in a High-Pressure Multi-Element Oxygen–Hydrogen Rocket Thrust Chamber. *J. Propuls. Power* **2019**, *35*, 632–644. [CrossRef]
19. Esposito, C.; Steelant, J.; Vetrano, M.R. Impact of Cryogenics on Cavitation through an Orifice: A Review. *Energies* **2021**, *14*, 8319. [CrossRef]
20. Croci, K.; Ravelet, F.; Danlos, A.; Robinet, J.C.; Barast, L. Attached cavitation in laminar separations within a transition to unsteadiness. *Phys. Fluids* **2019**, *31*, 063605. [CrossRef]
21. Pelz, P.; Keil, T.; Groß, T. The transition from sheet to cloud cavitation. *J. Fluid Mech.* **2017**, *817*, 439–454. [CrossRef]
22. Ge, M.; Sun, C.; Zhang, G.; Coutier-Delgosha, O.; Fan, D. Combined suppression effects on hydrodynamic cavitation performance in Venturi-type reactor for process intensification. *Ultrason. Sonochem.* **2022**, *86*, 106035. [CrossRef] [PubMed]
23. Amini, A.; Reclari, M.; Sano, T.; Iino, M.; Dreyer, M.; Farhat, M. On the physical mechanism of tip vortex cavitation hysteresis. *Exp. Fluids* **2019**, *60*, 118. [CrossRef]
24. Khoo, M.; Venning, J.; Pearce, B.; Brandner, P. Statistical aspects of tip vortex cavitation inception and desinence in a nuclei deplete flow. *Exp. Fluids* **2020**, *61*, 145. [CrossRef]
25. McCloy, D.; Beck, A. Some cavitation effects in spool valve orifices. *Proc. Inst. Mech. Eng.* **1967**, *182*, 163–174. [CrossRef]
26. Esposito, C.; Mendez, M.; Steelant, J.; Vetrano, M.R. Spectral and modal analysis of a cavitating flow through an orifice. *Exp. Therm. Fluid Sci.* **2021**, *121*, 110251. [CrossRef]
27. Callenaere, M.; Franc, J.P.; Michel, J.M.; Riondet, M. The cavitation instability induced by the development of a re-entrant jet. *J. Fluid Mech.* **2001**, *444*, 223–256. [CrossRef]
28. Trummler, T.; Schmidt, S.J.; Adams, N.A. Investigation of condensation shocks and re-entrant jet dynamics in a cavitating nozzle flow by Large-Eddy Simulation. *Int. J. Multiph. Flow* **2020**, *125*, 103215. [CrossRef]
29. Sarkar, P.; Ghigliotti, G.; Franc, J.P.; Fivel, M. Mechanism of material deformation during cavitation bubble collapse. *J. Fluids Struct.* **2021**, *105*, 103327. [CrossRef]
30. Obreschkow, D.; Tinguely, M.; Dorsaz, N.; Kobel, P.; De Bosset, A.; Farhat, M. Universal scaling law for jets of collapsing bubbles. *Phys. Rev. Lett.* **2011**, *107*, 204501. [CrossRef]
31. Sherry, M.; Jacono, D.L.; Sheridan, J. An experimental investigation of the recirculation zone formed downstream of a forward facing step. *J. Wind Eng. Ind. Aerodyn.* **2010**, *98*, 888–894. [CrossRef]
32. Rodríguez-Rodríguez, J.; Sevilla, A.; Martínez-Bazán, C.; Gordillo, J.M. Generation of microbubbles with applications to industry and medicine. *Annu. Rev. Fluid Mech.* **2015**, *47*, 405–429. [CrossRef]

33. Liu, Z.; Zheng, Y.; Jia, L.; Zhang, Q. Study of bubble induced flow structure using PIV. *Chem. Eng. Sci.* **2005**, *60*, 3537–3552. [CrossRef]
34. Li, B.; Gu, Y.; Chen, M. An experimental study on the cavitation of water with dissolved gases. *Exp. Fluids* **2017**, *58*, 164. [CrossRef]

MDPI AG
Grosspeteranlage 5
4052 Basel
Switzerland
Tel.: +41 61 683 77 34

Fluids Editorial Office
E-mail: fluids@mdpi.com
www.mdpi.com/journal/fluids

www.ingramcontent.com/pod-product-compliance
Lightning Source LLC
LaVergne TN
LVHW070452061126
851858LV00011B/223